Intelligent Systems Modeling and Decision Support in Bioengineering

This book is part of the
Artech House Engineering in Medicine & Biology Series,
Martin L. Yarmush and Christopher J. James, Series Editors.
For a list of recent related Artech House titles, please see the back of the book.

Intelligent Systems Modeling and Decision Support in Bioengineering

Mahdi Mahfouf

ARTECH
HOUSE
BOSTON | LONDON
artechhouse.com

Library of Congress Cataloging-in-Publication Data
A catalog record of this book is available from the U.S. Library of Congress.

British Library Cataloguing in Publication Data
A catalogue record for this book is available from the British Library.

ISBN-10: 1-58053-998-x
ISBN-13: 978-1-58053-998-2

Cover design by Igor Valdman

685 Canton Street
Norwood, MA 02062

10 9 8 7 6 5 4 3 2 1

Contents

Preface

This book, in the form of a research monograph, reflects my research activities and results for more than two decades in two areas, namely, intelligent systems and biomedicine, where it has long been predicted that interesting and effective synergies have had and will continue to have a significant impact on the quality of human life. I believe that such synergies are so natural that they were waiting to be discovered rather than being concocted as a plausible story (or a series of stories) that somewhat made sense or somehow justified the current academic tendencies. Because such areas are potentially broad in their respective remit, it is perhaps useful to emphasize that intelligence here refers to structures or algorithms that can mimic the way the human brain works (i.e., self-learning, self-organizing, hierarchical, flexible, and generalizing). In addition to those paradigms that rely explicitly on heuristics such as fuzzy logic–based systems, biological systems such as neural networks, or the theory of evolution such as evolutionary computation, this book will also include equally intelligent structures that rely on performance measures whose solution(s) may be derived analytically using physical-based model formulations. Biomedicine in this book will cover the themes of anesthesia, general intensive care, and the effect of physical stress (workload) on humans.

The research work covered in this book consists of three major interrelated facets. Modeling and control of anesthesia fall within the first facet where the utilization of model-based predictive control, fuzzy logic–based classification, modeling and identification, neural networks, and genetic algorithms will be highlighted (Chapters 3 to 6). The second facet includes modeling and decision support in the intensive care unit (Chapter 7) using physical-based models as well as neuro-fuzzy architectures, whereas the third facet relates to the identification of human physiological markers for physical stress (workload) via hybrid models (Chapters 8 and 9). It is worth noting that "hybrid" here refers to those models that represent a combination of physical-based structures and data-driven architectures.

Acknowledgments

I wish to express my sincere thanks to a number of people who have helped in the research work behind this book. First, I thank Professor D. A. Linkens for the many years of academic interactions in the domain of biomedicine and intelligent systems. I wish to thank Dr. A. J. Asbury, Dr. G. H. Mills, and Dr. J. E. Peacock for their valuable expertise in interacting with me and my research group and for arranging all the clinical trials that proved to be so vital in validating the models, control, and decision support systems. I also acknowledge the help provided by Professor D. G.

von Keyserlingk, Dr. M. F. Abbod, and Mr. G. Panoutsos in preparing Chapter 2, Dr. H. F. Kwok and Dr. K. Goode during the preparation of Chapter 7, and my former Ph.D. students Dr. C. S. Nunes and Dr. E. El-Samahy for their assistance with Chapters 6, 8, and 9.

Thanks should also go to several organizations and governments that funded a great proportion of the research presented in this book, namely, the U.K. Engineering and Physical Sciences Research Council (EPSRC), the Leverhulme Trust, the Egyptian government, and the Fundacao para a Ciencia e a Tecnologia (Portugal).

Last, but not least, I would like to thank my wife, Amanda, my children, Omar-Tarek and Kareem-Mahdi, and my extended family for their support and patience during the long hours of research, many hours of which were devoted particularly to research in intelligent systems with special applications to biomedicine.

CHAPTER 1

Introduction

This chapter serves as an introduction to the book, discusses the book's motivations, and outlines the way that it is generally organized with respect to general themes and specific chapters.

There is no doubt that clinical medicine is among the last areas where systems engineering tools—such as data-driven modeling, system identification, signal processing, and feedback control—have proliferated. While this positive and, indeed, bold shift towards drawing engineering and clinical medicine even closer together is most welcome by researchers such as myself, it is also important to pause for a moment and be reminded of the *real* motivation behind such a change. Clinical medicine needs to be convinced not only that all the above systems engineering tools can be effective and reliable but, most importantly, that they are safe—although I strongly believe that safety, reliability, and effectiveness go hand in hand.

If one focuses, for example, on feedback control, it is worth noting that the link between the mathematics of control theory and drug strategies was discovered and published in 1969 [1]. It is hardly surprising that the clinical staff I have worked with for over 20 years, including junior and senior doctors as well as nurses, were found to be surprisingly receptive and amenable to designs which are not necessarily transparent (or directly interpretable), provided that the relevant notions of *stability*, *cost function*, and *Bode diagrams* were explained to them before implementing a scheme for which their enthusiastic collaboration was needed.

While I agree with Professor Karl J. Astrom, who argues that control engineering is often perceived as the hidden technology, I will also argue that there is no better and more emphatic way of demonstrating the usefulness of feedback control, modeling, or signal processing than on the most beautiful machine ever created—that is, the human body. This application embodies all the challenges known to system's theory all in one setting: nonlinearities, complex and ill-defined multivariable interactions, long and varying time delays, sudden disturbances, inter- and intradynamics variability, and uncertainties due to external disturbances as well as measurements. On the subject of measurement, clinical medicine is perhaps the one area where Galileo's principle of measuring everything that is measurable and making available all the variables which have not been measured so far will prove most difficult to apply.

Because of the above difficulties facing researchers in clinical medicine, there has been a wide shift by the research community towards paradigms that attempt to alleviate some of the shortfalls of conventional tools for modeling, signal processing, and control, including the curse of dimensionality, the inability to incorporate experts' knowledge systematically, and the difficulty of taking into account uncertainties often present in the environment under investigation.

Such paradigms are embodied in the area of intelligent systems or, as was referred to for the first time by Professor L. A. Zadeh, as *soft computing* (i.e., software computing). Soft computing techniques, as opposed to hard computing (i.e., hardware computing) include three main areas: fuzzy logic (FL), artificial neural networks (ANN), and evolutionary computing (EC). They all include to ability to (logically) understand and reason and to learn or adapt.

This book represents a culmination of more than a decade of research work in the field of biomedicine. It has been written for several reasons, including the author's desire to communicate widely what is believed to be important research findings in the challenging but interesting area of biomedicine, the worldwide interest in the field of intelligent systems in general, and the widespread efforts by the clinical research environment to explore the qualitative-led engineering, especially since the need to transfer human operator's expertise reasoning is growing.

The subject of intelligent systems is wide and has many facets to its credit. The following sections focus on three themes, which will be used as the basis for some of the research described in this book.

1.1 Fuzzy Logic

The idea of fuzzy logic was introduced by Zadeh [2], who argued that real-world problems are too complicated to be represented by a crisp formalism, so in order to obtain a reasonable representation, *fuzziness* must be introduced. Because a practical system makes use of all available information, including mathematical models, sensor measurements, and human experts, a fuzzy system can combine all such available knowledge in a systematic way to form an effective and "complete" decision maker.

In order to model linguistic knowledge successfully, fuzzy logic relies on fuzzy sets. A fuzzy set, A, is a mapping from a set of generic objects, $\{u\}$, so-called universe of discourse, U, which can be discrete or continuous, into the interval [0,1], and this mapping is characterized by a membership function, μ_A, namely, μ_A: U $\rightarrow$ [0,1].

The definitions and plots of common membership functions that are usually used in a large group of applications are given in (1.1) through (1.3), corresponding to Figures 1.1 through 1.3.

$$\text{Triangle}\,(x;a,b,c) = \max\left\{\min\left(\frac{x-a}{b-a},\frac{c-x}{c-b}\right),0\right\} \tag{1.1}$$

$$\text{Trapezoid}\,(x;a,b,c,d) = \max\left\{\min\left(\frac{x-a}{b-a},1,\frac{d-x}{d-c}\right),0\right\} \tag{1.2}$$

$$\text{Gaussian}\,(x;c,\sigma) = \exp\left\{-\left(\frac{x-c}{\sigma}\right)^2\right\} \tag{1.3}$$

The center of a fuzzy inference system is the fuzzy *if-then* rules in which the concept of linguistic variables plays a significant role. A linguistic variable is a variable whose values are words, and these words are characterized by fuzzy sets defined in

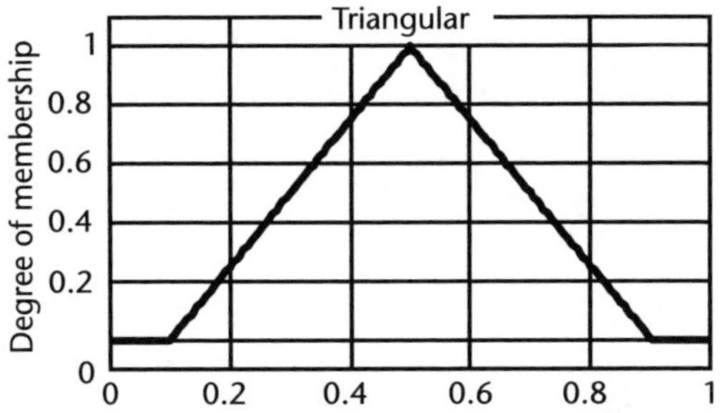

Figure 1.1 A triangular fuzzy set.

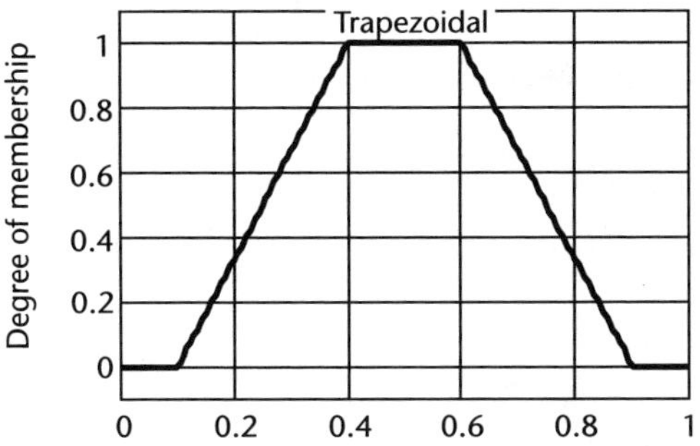

Figure 1.2 A trapezoidal fuzzy set.

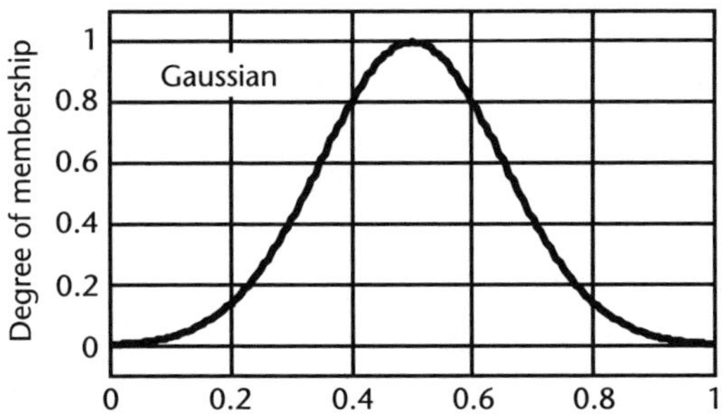

Figure 1.3 A Gaussian fuzzy set.

the universe of discourse in which the variable is defined. For example, "temperature" is a linguistic variable that may take "cold," "warm," and "hot" as its labels, which are characterized by their related membership functions. A fuzzy if-then rule can be expressed symbolically as:

If <antecedent>, then <consequent>

The antecedent is a combination of fuzzy propositions, and the consequent for a Takagi-Sugeno-Kang (TSK) type fuzzy system is a function of inputs [3]:

If *Error is NB and Change of Error is NB,* then $y = f\,(Error,\ Change\ of\ Error)$

where *NB* (which means Negative Big) is one of the selected labels for the fuzzy variables *Error* and *Change of Error*, and $f(\cdot)$ is usually a linear function of the input variables.

However, for a Mamdani type fuzzy system the consequent is a single fuzzy proposition as follows [4]:

If *Error is NB and Change of Error is NB*, then *Output is NB*

Advantages of fuzzy systems include: (1) their ability to combine *quantitative* and *qualitative* information in a systematic fashion, and (2) their ability to reconcile *accuracy* (precision) with *transparency* (interpretability).

1.2 Artificial Neural Networks

The following are the three well-known definitions of a neural network (NN):

1. A new form of computing inspired by biological models;
2. A mathematical model, composed of a large number of processing elements organized into layers;
3. A computing system, made up of a number of simple, highly interconnected processing elements, which processes information by its dynamic state response to external inputs.

The reader is referred to the original seminal work on the neuron model [5].

The structure of the ANN consists of artificial neurons or processing elements (PE) arranged as shown in Figure 1.4.

In relation to Figure 1.4, the output is evaluated as follows:

$$y = w_1 x_1 + w_2 x_2 + \ldots + w_n x_n \tag{1.4}$$

where w_i $(i = 1, 2, \ldots, n)$ are referred to as *weights*, and T is the *threshold* which would or would not allow signals to propagate along the structure. The most celebrated ANN structure is the feedforward ANN, which is represented by a network of neurons such as the one shown in Figure 1.5.

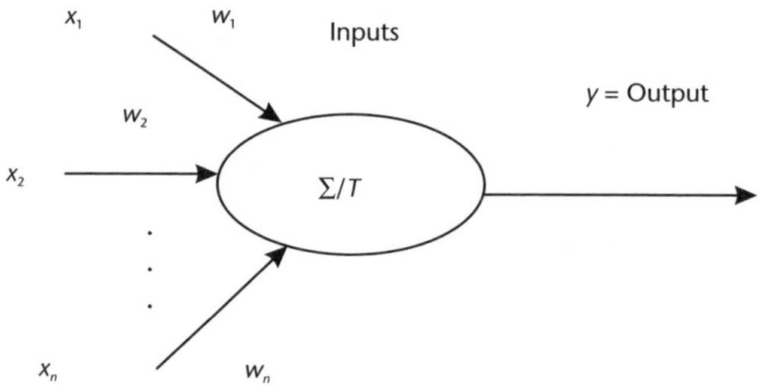

Figure 1.4 The structure of a neural network.

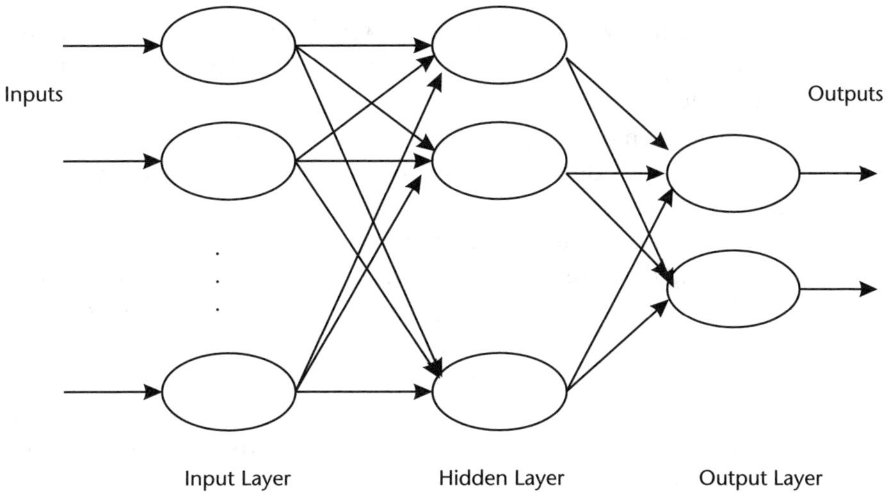

Figure 1.5 A feedforward ANN.

An advantage of neural networks is their ability to represent powerful interpolators by eliciting accurate input/output mappings.

1.3 Evolutionary Computing

Evolutionary computing is the general term used to describe several computational techniques that are based, to a certain extent, on the evolution of biological processes in the natural world. Genetic algorithms (GA) [6] are an EC technique that is based on Darwin's theory of evolution. As its name implies, the technique starts generally with a suboptimal solution and evolves into an optimal (or near optimal) solution as the process of evolution gets under way. GA will be the subject of an appendix in Chapter 6. Here it is sufficient to list the main advantages of EC techniques. Mainly used for problem optimization, advantages of EC techniques are that (1) they do not require function gradient information, and (2) they avoid being stuck in local areas of optimization more often than conventional techniques.

1.4 Book Organization

This book includes nine other chapters that outline research work in three major areas of medicine: anesthesia, general intensive care unit (ICU), and physiology. These are outlined as follows:

- *Chapter 2.* This chapter surveys the utilization of fuzzy logic based control and monitoring techniques in medical sciences with an analysis of its possible future penetration into the discipline. The survey, while not necessarily exhaustive, involved a search of the medical databases MEDLINE and INSPEC, as well as published books through 2004. The main topics cited were classified into 10 fields, including conservative medicine, invasive medicine, regionally defined medical disciplines, neuro-medicine, image and signal processing, laboratory, basic science, nursing, healthcare, and Eastern medicine.

- *Chapter 3*. In order to enable readers of this book to experience the merits of modeling and control designs based on several ideas and paradigms, two studies, which were reported in the author's first book [7], are reproduced in this chapter. The studies describe the development and application of the single-input single-output (SISO) and multi-input multi-output (MIMO) generalized predictive control (GPC) algorithm for online administration of a muscle relaxant as well as anesthetic drugs during surgery on humans. The necessary control systems are presented and explained, and the simulation and the real-time results obtained at two U.K. hospitals (the Sheffield Royal Hallamshire Hospital and the Glasgow Western Infirmary) are analyzed and discussed.
- *Chapter 4*. The identification problem relating to recirculatory physiological models of drug models, particularly analgesics, is explored, and a new *generic* approach is proposed that bridges the gap between the *white box* and the popular *black box* model representations. Parallels are also drawn between this representation and so-called compartmental modeling. The modeling approach is extended to include an architecture deemed more amenable to systematic control design.
- *Chapter 5*. The power of the model architecture introduced in Chapter 4 (recirculatory physiological models) to design monitoring and closed-loop control strategies for drug administration during surgery is reviewed. The subject of anesthesia is approached from the so-called golden standard viewpoint by monitoring arterial blood pressure only. The unconstrained and constrained versions of GPC are evaluated for online administration of an anesthetic drug (isoflurane) during surgery. The clinical trials, which were conducted on humans at the Glasgow Western Infirmary Hospital and which were performed with little involvement of the design engineers, are presented, analyzed, and discussed. In addition, a supervisory layer in the form of a Mamdani-type fuzzy logic based system is included to monitor the normal operation of the control system.
- *Chapter 6*. Anesthesia is viewed here from a different angle as far as quantification of unconsciousness is concerned. The work in this chapter includes three main research topics: (1) classification of depth of anesthesia (DOA); (2) modeling the patient's vital signs; and (3) control of DOA with simultaneous administration of anesthetic and analgesic drugs. First, a fuzzy relational classifier is developed to classify a set of extracted features from the auditory evoked potentials (AEP) into different levels of DOA. Second, a hybrid patient model using fuzzy models is developed. In addition, the surgical stimulus effect is incorporated into the patient model using fuzzy models. Finally, these three parts are incorporated within a closed-loop simulation system environment.
- *Chapter 7*. There is no doubt that the dynamics involved in the interactions between patient and ventilator are complex and not very well understood. This chapter, however, while not promising to provide all the answers to the problems surrounding all the phenomena involved, proposes to shed some

light into some processes with the aim of developing a hybrid knowledge and model-based intelligent advisory system for intensive care ventilators which will be implemented clinically. It reviews the components of a decision support architecture that are formulated using a hybrid approach that combines the themes of fuzzy and neural-fuzzy systems as well mechanistic formulations.

- *Chapter 8.* In 1998 the author established a new research laboratory called the Human Performance Laboratory, whose main objective was to study human-machine interactions and to identify physiological/psychological markers that would lead to the detection and diagnosis of physical as well as psychological exhaustion. The study included in this chapter describes the effect of physical workload on heart rate, blood pressure, and respiration rate in healthy human subject volunteers. The study also describes a hybrid closed-loop physiological model mapping physical stress levels to the previously mentioned physiological variables.
- *Chapter 9.* The study described in the previous chapter is extended here to include more variables in the input/output(s) mapping, such as thermoregulation (body temperature) and brain activity via EEG. To deal with the problem of intersubject parameter variability, an intelligent supervisory structure is superimposed on the basic model architecture to map the subject's features to the closest set of prestored parameters. Finally, the overall model is exploited by determining the levels of physical workload to achieve certain predefined levels of physiological variables.
- *Chapter 10.* This chapter concludes the book by listing the key contributions of the work and by identifying future areas of research in monitoring and modeling.

References

[1] Bluell, J., et al., "Modern Control Theory and Optimal Drug Regimes," *Mathematical Biosciences*, Vol. 5, 1969, pp. 285–296.

[2] Zadeh, L. A., "Fuzzy Sets," *Information and Control*, Vol. 12, 1965, pp. 94–102.

[3] Takagi, T., and M. Sugeno, "Fuzzy Identification of Systems and Its Applications to Modeling and Control," *IEEE Trans. on Systems, Man and Cybernetics*, Vol. SMC-15, 1985, pp. 116–132.

[4] Mamdani, E. H., "Application of Fuzzy Algorithm for Control of Simple Dynamic Process," *Proceedings IEE*, Vol. 121, 1974, pp. 1585–1588.

[5] McCulloch, W. S., and W. Pitts, "A Logical Calculus of the Ideas Immanent in Nervous Activity," *Bulletin of Mathematical Biophysics*, Vol. 5, 1943, pp. 115–133.

[6] Holland, J. H., "Genetic Algorithms and the Optimal Allocation of Trials," *SIAM Journal of Computing*, Vol. 2, 1973, pp. 89–104.

[7] Mahfouf, M., and D. A. Linkens, *Generalised Predictive Control and Bioengineering*, London, U.K., Taylor & Francis Publishers, 1998.

CHAPTER 2

A Survey on the Utilization of Fuzzy Logic–Based Technologies in Medicine and Healthcare

In this chapter, current published knowledge about fuzzy logic in medicine and healthcare is reviewed. The achievements of frontier research in this particular field within medical engineering are described. The justification of such a technology, embedded within the theme of soft computing (terminology first used by Professor L. A. Zadeh), lies in the fact that biological systems are so complex that the development of computerized systems within such environments is not always a straightforward exercise. In practice, a precise model may not exist for such systems, and these may prove very difficult to model. In most cases fuzzy systems are considered to be ideal tools, as human minds work from approximate data, extract meaningful information and produce crisp answers to complex problems. This chapter surveys the utilization of fuzzy logic based control and monitoring techniques in medical sciences with an analysis of its possible future penetration into the discipline.

2.1 Introduction

There is no doubt that the complexity of biomedical systems, including biological systems, can render traditional quantitative approaches of analysis and design inappropriate; for when dealing with such systems, there is almost an unavoidable significant degree of fuzziness (subjectivity) when describing their behavior and analyzing their characteristics. One might ask, legitimately, from where does such fuzziness emanate; it is largely due to the lack of precise mathematical techniques for dealing with systems that comprise a relatively large number of interrelated components or involve a large number of variables in their decision tree [1]. Fuzzy sets are known for their ability to introduce notions of continuity into deductive thinking [2]. In practice, this means that fuzzy sets allow the use of conventional symbolic systems (specified in the form of tabulated rules) in continuous form. This is a particularly important issue since medicine is known to be a continuous domain.[1] Many practical applications of fuzzy logic in medicine are known to use its continuous subset features such as fuzzy scores, continuous version of conventional scoring systems, and fuzzy alarms. Perhaps the most celebrated and successful

1. The steel industry for instance can be a mixture of continuous (micro-nanophenomena evolving) and discrete (stock scheduling) domains.

approach to date relates to fuzzy logic control where a handcrafted and/or automatically optimized rule-based mapping between input and measured (observed) output variables realizes a continuous control law for drug administration among other implementations.

Those of us who have trod in this area of research for many years have come to realize that one does not need to look into micro-phenomena and set micro-objectives to be faced with the more often dreaded problems of incompleteness, uncertainty, and inconsistency of systems. Indeed, more in medicine than in the physical sciences,[2] we have come to accept that the lack of adequate data (information) and its imprecise, and often contradictory, nature is much more a fact of life. This is of particular concern as all these problems have to be taken into account, possibly in a pareto[3] optimal sense, to make a medical decision that may have vital consequences for the object under medical attention. The inherited sources of inaccuracy in the medical sector have long been a subject of debate between various parties, but here I propose to classify them according to my experience and that of the members of my research group who have had the privilege of working in such an interesting synergetic environment as medical engineering, as follows:

- Information about the patient, which given its subjectivity can be channeled towards a number of categories, all of which include uncertainty. Also, as such information passes successively between personnel, it becomes noisy and distorted, hence adding more uncertainty,
- Medical history of patients, which is supplied by the patient, and is usually highly subjective and may include simulated, exaggerated, or understated symptoms; for example, ignorance of previous diseases, failure to mention previous operations, and general recollection often leads to doubts about the patient's medical history.
- Physical examinations, which physicians conduct to obtain objective data. These are subject to mistakes and the overlooking of important indications, and at times may simply be incomplete. Furthermore, physicians may misinterpret other indications because the boundary between normal and pathological status is not always clear.
- Results of laboratory tests are objective data, but they depend on the accuracy of the measurements, and are subject to organizational issues (mislabeling sample, wrong laboratory) and the improper behavior of the patient prior to the examination.
- Results of histological, X-ray, ultrasonic, and other clinical investigations, which depend on correct interpretation by medical staff (different medical personnel from different clinical establishments can give different interpretations to the same results).

This chapter surveys the use of fuzzy logic in medicine, and it is based on searches in such medical databases as MEDLINE and INSPEC, and of published

2. Researchers from other disciplines have to forgive me for this statement as this is not intended to mean that the systems they deal with are less challenging!
3. This refers to the set of best solutions lying in the feasibility region (where all the objectives are satisfied).

books. The purpose of the study was to establish a roadmap that may help forecast the future developments of fuzzy technology in medicine and healthcare. A simple search of the word "fuzzy" obtained many cited papers; half of which used the term to describe an unsharp border of a structure or situation. For the rest, most of them more recent, the word "fuzzy" was used as part of fuzzy sets or fuzzy logic [3]. The main topics that were cited can be classified into ten fields (each of which can be further classified into subheadings) as follows: conservative medicine, invasive medicine, regionally defined medical disciplines, neuro-medicine, image and signal processing, laboratory, basic science, nursing, healthcare, and Eastern medicine.

2.2 Fuzzy Technology in the Identified Fields

The following sections give a description of the key contributions that fuzzy technology has made in each of the subtopics which have been identified in the medical literature. While it is recognized that this is by no means an exhaustive study, it should be seen as an indication of the salient features of each subdescription, which have inherently been amenable to uncertainty reasoning.

2.2.1 Conservative Disciplines

2.2.1.1 Internal Medicine

Internal medicine is a classic field of research in computer-aided diagnosis, which began in the 1960s with high hopes that difficult clinical problems might yield to mathematical formalism. The main areas in internal medicine can be classified into rheumatology, gastroenterology, hepatology, and pulmonology. Developments in knowledge-based expert systems in medicine began with the MYCIN system [4] about a decade after Zadeh's introduction of fuzzy logic. It was based on Boolean algebra, pattern matching, and decision analysis to the diagnostic process. Although MYCIN did not use fuzzy logic, one of its major components was the use of certainty factors—an early recognition of the major role that uncertainty plays in medical decision-making. Since MYCIN was introduced, knowledge-based medical expert systems have abounded. Most of these systems have tried to deal with uncertainty, and some of then have used fuzzy logic directly [5–12]. Designing and tuning fuzzy rule–based systems for medical diagnosis was discussed by Rotshtein [13].

The highest volume of publications under this subheading occurred between 2000 and 2002 (a total of 33). A fuzzy inference system was developed to aid in the diagnosis of pulmonary embolism using ventilation-perfusion scans and correlated chest X-rays. The Mamdani fuzzy model was successfully employed to implement the inference system [14]. An advanced computer-aided diagnosis (CAD) scheme was developed for the automated detection of polyps in computed tomographic (CT) colonography [15]. Other work included (1) the application of the fuzzy "k-nearest neighbor" (k-NN) classifier of pattern recognition theory to explain the abnormal way of breathing that resulted from diaphragm paralysis [16] and (2) the development of an early diagnostic system for postoperative infection [17].

Additionally, a tutorial style paper was written for physicians and biologists who were not necessarily familiar with fuzzy set theory and biomedical applications. The field was introduced in the framework of medical diagnosis problems and illustrated with an application to inflammatory protein variations. Relationships between signs and diagnoses were interpreted as labels of fuzzy sets, and it was shown how diagnoses could be derived from soft matching processes [18]. An early successful application of fuzzy logic in expert systems was in a system called CADIAG, where the medical information was derived from medical records taken from a hospital information system. The information was "fuzzified" and coded in terms of rules. The system used fuzzy logical inference mechanisms to generate diagnostic information. The latest version of the system (CADIAG-2) has applications in the fields of rheumatology, gastroenterology, and hepatology [19, 20]. Other expert systems that utilize fuzzy logic include SPHINX [21], RENOIR [22, 23], and CLINAID [24].

2.2.1.2 Cardiology and Vascular Surgery

The early papers on fuzzy logic and cardiology were concerned with cardiovascular investigation and fuzzy concepts [25], the use of fuzzy set theory in evaluation of cardiac functions [26] and electrocardiographic (ECG) analysis [27], and analysis of cerebrovascular disease [28]. In the mid-1990s several workers pointed to the concept of fuzzy sets in cardiovascular medicine. Implementation of fuzzy control of a total artificial heart was one application [29]. The main task of the artificial heart control system is to maintain sufficient organ perfusion by controlling the pumping rate. Another application that incorporates fuzzy logic is a system called TOTOMES, which was designed to assess cardiovascular dynamics during ventricular assistance [30]. This involves multi-interpretation tasks and dynamic system identification, as well as fuzzy reasoning for realizing state estimation, and detection and diagnosis of malfunctioning. A heart condition classification system using tomography images based on fuzzy logic has achieved 94% accuracy [31]. Other applications include coronary artery disease fuzzy classifier [32–34], ECG classification and diagnosis [35–39], and diagnosis and treatment of heart disease [40]. In another study, two different classifiers for the identification of premature ventricular complexes (PVCs) in surface ECGs were described [41]. A decision-tree algorithm based on inductive learning from a training set and a fuzzy rule–based classifier has been included. The gross sensitivity of the fuzzy rule–based system was evaluated at 81.3%, and the positive predictivity at 80.6%. The successful application of the fuzzy neural network for ECG beat recognition and classification was achieved by Osowski and Linh [42]. Such application confirmed the recognition of different types of beats on the basis of the ECG waveforms. Finally, the particular focus of the work by Pfaff et al. [43] was on the prediction of cardiovascular risk factors in hemodialysis patients, specifically the interventricular septum (IVS) thickness of the hearts of individual patients as an important quantitative indicator to diagnose left ventricular hypertrophy. Data-based clustering, cluster-based rule extraction, rule-based construction, and cluster- and rule-based prediction were used in the developed algorithms. The methods employed included crisp and fuzzy algorithms.

2.2.1.3 Intensive Care

Intensive care applications are close to anesthesia in their medical function; nevertheless, applications can be divided into blood pressure and respiration regulation, EEG monitoring, and pain relief. An expert system based on fuzzy logic was designed as a warning system in the pediatric intensive care unit (PICU). It was able to make assessments at regular intervals concerning the level of abnormality in the EEG [44]. Keeping the oxygenation status of newborn infants within physiologic limits is a crucial task in intensive care. For this purpose, several vital parameters are supervised routinely by monitors such as electrocardiographs, transcutaneous partial oxygen pressure monitors, and pulse oximeters. Many automated systems based on fuzzy logic have been developed, which are capable of distinguishing between critical situations and artifacts [45–47]. In an effort to ease staffing burdens and potentially improve patient outcome in an intensive care unit (ICU) environment, a real-time system was developed to accurately and efficiently diagnose cardiopulmonary emergencies [48]. As for expert systems, an emergency-oriented system was developed to respond to inquiries concerning dangerous conditions [49]. FLORIDA is another expert system, which determines the physiological condition of patients in ICU using fuzzy logic and knowledge bases [50, 51]. For online monitoring of patients in ICU, a system for breath detection was developed based on fuzzy sets and noninvasive sensor fusion [52]. It must be said, perhaps, that relative to "adult" intensive care research, research into neonates and infants is not so flourishing, partly because of the difficulty in accessing data, as well as the absence of multidisciplinary expertise collaboration. However, among other research initiatives, the work by Olliver [53] describes the use of a fuzzy logic-based weaning platform and further develops parametrizable components for weaning newborns of differing body size and disease-state in the PICU.

2.2.1.4 Pediatrics

Although pediatrics is similar to internal medicine in methods and thinking, the application of fuzzy sets is scanty. There is a very limited amount of application using expert systems, fuzzy modeling, and fuzzy control [54]. For unborn babies, fuzzy logic has been used to analyze the pattern and interpolation of changes in the cardiotocogram (fetal heart rate pattern plus uterine concentration) during labor, which can lead to unnecessary medical intervention or fatal injury [55]. A knowledge-based system using fuzzy logic for classifying/detecting distorted plethysmogram pulses in neonates and pediatric patients has also been developed [56], and it has been shown to classify 679 (82%) valid segments and 543 (93%) distorted segments correctly.

2.2.1.5 Endocrinology

Endocrinology is an important part of the broad medical branch belonging to internal medicine. Although fuzzy logic has not been applied directly in this field, it has been considered for thyroid diseases, but with little success. In 1978 a version of an inference engine for dealing with imprecision and uncertainties was applied to the treatment of diabetic patients [57]. An expert system (PROTIS) used for deduction of fuzzy rules was developed for treatment of diabetes [58, 59]. Another decision support system for treatment of diabetic outpatients using fuzzy classification was described by Stadelmann [60]. Fuzzy inference was utilized for a diagnostic system of diabetic patients by quantitative analysis of the dynamical responses of glucose tolerance tests [61]. A knowledge-based system was also developed for monitoring diabetics, which consisted of fuzzy rules and hierarchical neural networks [62]. Subsequent work included the application of fuzzy logic–based ideas and neural network techniques to modify intravenous insulin administration rates during glucose infusion [63]. Forty critically ill, fasted diabetic subjects entered the study and submitted to glucose and potassium infusion.

2.2.1.6 Oncology

The use of fuzzy logic for oncology has been concerned with classification (for discriminating cancer from normal tissue) and therapy advice. Therapy was mainly based on advanced image processing [64]. Magnetic resonance image analysis for tumor treatment planning [65] and ovarian cancer [66] has been attempted. Generally, image-clustering analysis is mostly applied for determining brain tumor segmentation [67]. Also, a fuzzy reasoning algorithm has been used to diagnose breast tumors using three-dimensional ultrasonic echographic images [68]. The advantages of using fuzzy logic based structures has also been highlighted by Hiltner [69] and Fletcher-Heath [70] when they presented research looking into the use of fuzzy descriptions for the segmentation and analysis of medical image data (brain structures and tumors in MRI-data). This has provided better results than the exclusive use of standard methods.

2.2.1.7 Gerontology

Originally, gerontology applications were mainly concerned with the use of fuzzy logic for clustering. The same research group used different approaches for different subjects [71]. This is a good example of how fuzzy logic could be applied in a flexible way. A fuzzy relational system was used to implement the databases for building a veterinary expert system [72]. Later on, a few broader applications started to emerge, such as the work by Lim [73] who analyzed the difficulties of designing an inference system for the diagnosis of arthritic diseases, including variations of disease manifestations under various situations and conditions.

2.2.1.8 General Practice

In general practice, fuzzy logic can be used to assist physicians to find relationships between patients by applying fuzzy clustering [74]. A computerized system for Clinical Practice Guidelines (CPGs) designed to improve quality of care by assisting physicians in their decision-making has been developed using a fuzzy classification procedure [75]. However, the literature search did not find many papers related to this subject.

2.2.2 Invasive Medicine

2.2.2.1 Surgery

Advanced diagnostic tools have contributed immense benefits in surgery. Such tools include advanced image processing, pathophysiological reasoning, as well as improved control mechanisms and simulation systems in anesthesia. In anesthesia, many applications have been reported in the use of fuzzy logic to control drug infusion for maintaining adequate levels of anesthesia, muscle relaxation, and patient monitoring and alarms. This subject will be discussed further in a separate section. In an indirect application of fuzzy logic, it has been used within virtual reality (VR) simulation [76]. If VR surgical simulators are to play an important role in the future, quantitative measurement of competence would have to be part of the system. Because surgical competence is vague and is characterized by such terms as "too long," "too short," "too close," or "too far," it is possible that the principles of fuzzy logic could be used to measure competence in a VR surgical simulator. Plastic surgery is another field where fuzzy logic is utilized as a manipulation tool for creating a new image of the patient using an unsupervised segmentation method [77].

It was Radezky et al. [78, 79] who recognized that traditional models of tissue deformation have difficulties in simulating the appearance of deformation because of the unknown physical parameters of the tissue's elasticity. The authors described a method for elastodynamic shape modeling with neuro-fuzzy systems, which are able to adapt the necessary parameters from real tissues.

The progress of the use of fuzzy logic in the surgery field is limited. There is, however, a wide selection of applications where fuzzy logic could be explored, such as invasive, endoscopic, preoperative planning surgery.

2.2.2.2 Orthopedics

The field of orthopedics is similar to surgery, but there is no obvious fuzzy solution for a typical orthopedic problem. There are two reports that appear to have used fuzzy logic, but both were concerned with medical records and waiting lists for orthopedic patients [80]. However, in the detection of orthopedic symptoms, fuzzy and nonfuzzy rule bases were compared for gait event detection in electrically stimulated walking of paraplegic subjects using fuzzy identification methods [81]. In a different study, von-Altrock [82] considered fuzzy logic methods to study the knee-surgery recovery problem. Patients have to limit the strain on the knee during recovery. The author discusses the use of a pressure sensor that enables one to monitor the pressure one may be putting on a bad knee. The author claims that fuzzy logic in combination with an appropriate tension sensor made it all possible.

2.2.2.3 Anesthesia

The application of fuzzy logic in the field of anesthesia is expanding rapidly. Biomedical engineering improves the patient-monitoring task by measuring a large number of the patient's vital parameters. These measurements improve the safety of the patient during surgical procedures. One feature of fuzzy logic control is its ability to adapt to the pharmacokinetic and pharmacodynamic parameter changes of the patient, which has been proved a system for the delivery of the muscle relaxant pancuronium [83]. As further pathological states can be recognized, it may lead to an increase in cognitive load on the anesthetist. In order to reduce this strain on the anesthetist, and to support intraoperative monitoring, intelligent monitoring and alarm systems have been proposed [84]. The first real-time expert system for advice and control (RESAC) in anesthesia was developed to advise on the concentration of inhaled volatile anesthetics [85]. It merges clinical information and online measurements using Bayesian inference and fuzzy logic. Anesthetists were confident enough to follow the dosage advice given by RESAC in most of the patients. On the other hand, direct application of fuzzy logic rule-based controllers has been implemented for controlling drug infusion to maintain an adequate level of anesthesia by monitoring blood pressure and muscle relaxation [86–90]. Evoked potentials were also used as a measure of depth of anesthesia for controlling drug infusion using fuzzy logic controllers [91, 92], and EEG monitoring [93].

Supervisory control, or a *safety shell*, has been used to oversee the performance of a controller and to direct the controller to take corrective actions in the case of special situations, such as disturbances [94]. More recently, fuzzy logic has been used to monitor different indicators of depth of anesthesia, followed by fusing the indications together to obtain a final depth of anesthesia. Based on this indication, fuzzy logic is used to decide the amount of drug infused to maintain a constant depth of anesthesia [95].

2.2.2.4 Artificial Organs

One of the early applications of fuzzy logic to artificial organs was reported by Saridis who used fuzzy decision-making in prosthetic devices [96, 97]. Feng and Andrews [98] developed an adaptive fuzzy logic controller that has the ability to incorporate expert knowledge in terms of fuzzy rules, which can be online reinforced by a learning algorithm. The system was applied to a swinging leg. Also, the system was used in a later development for predicting the forces required for standing up and developing a closed-loop controller [99, 100]. The rapid developments in computer systems and software architecture such as fuzzy logic and neural network controllers have encouraged the suggestion that artificial organs could be developed rapidly. Nevertheless, more extensive research is needed to reach this stage [101].

Finally, a study to classify myoelectric signals using new fuzzy clustering neural network (NN) architectures to control multifunction prostheses has been described by Karlick [102]. This paper presents a comparative study of the classification accuracy of myoelectric signals using multilayered perceptron NN using back-propagation, conic section function NN, and new fuzzy clustering NNs (FCNNs). The results suggest that FCNN can generalize better than other NN algorithms and help the user learn better and faster.

2.2.3 Regionally Defined Medical Disciplines

The regionally defined medical disciplines are mainly concerned with certain organs of the human body. Special diagnostic methods are often used for this branch, in addition to those used in internal medicine. This field also requires special therapeutic procedures, including surgery, which is conducted by specialized physicians with particular experience.

2.2.3.1 Gynecology

Tumor diagnosis, treatment of intra-uterine fetal distress, monitoring of preterm infants, and differentiating bursts of ultrasonic data are applications that have all used fuzzy mathematics [45]. All of these systems are based on decision support systems, which are very complex and require extensive computing. In turn, this will restrict their dissemination and limit their use to specialized people. A simplified system that any practitioner, even one unaccustomed to computing, would be able to use has been developed and transferred to widely available and inexpensive microcomputers. SELF (System in Fuzzy Set) has been devised as an extensive interactive system for medical decision help [103]. It is generically parameterized and allows many kinds of application, and it works according to such rules as "if there is premise, then there is conclusion," tempered by a coefficient. The knowledge bases are represented on a correspondence chart where columns materialize the premises (clinical signs, laboratory results) and horizontal lines represent all the diagnoses and therapeutic conclusions. The system includes procedures that create, modify, and update basic knowledge. It uses fuzzy set rules to draw conclusions. SELF was first applied to the prescription of contraceptive methods, but it has now been tested in other specialties such as gynecology, penology, and hematology. Another developed system based on fuzzy logic aids the diagnosis of breast cancer by analyzing the lobulation [64]. A computer system called ToxoNet [104] has also been described as being able to process the results of serological antibody tests having been performed during pregnancy by means of a fuzzy knowledge base containing medical knowledge on the interpretation of toxoplasmosis serology tests. The time intervals between two subsequent tests have been modeled as fuzzy sets, since they allow the formal description of the temporal uncertainties. Finally, a manuscript by Sarmento [105] includes interesting bibliographic notes on image denoising methods based on wavelet and fuzzy theories, and examples of applications to ultrasound-mammography imaging modalities are given.

2.2.3.2 Dermatology

A decision support system that processes lexical fuzzy knowledge from standard pathology textbooks has been developed by Kolles and Hubschen [106]. The user can choose among four methods: full text and hypertext approaches (which are classic lexical retrieval methods) and two possibilistic, knowledge-based approaches (fuzzy logic computation and an adapted evidence combination scheme based upon the theory of belief functions). The effectiveness of these approaches is demonstrated by applying them to dermatopathology cases. At the top of the lists of differential diagnoses generated by the system, the degree of conformity between the approaches is high. In terms of classifying the structure and tissues in echocardiogram images, a multiple-feature, hierarchical, fuzzy and neural network fusion solution system has been developed by Brotherton [107].

In a paper by Acha [108], a new system for burn diagnosis is proposed. The aim of the system is to separate burn wounds from healthy skin, and the different types of burns (burn depths) from each other, identifying each one. After the burn is segmented, some color and texture descriptors are calculated, and they are the inputs to a Fuzzy-ARTMAP neural network. Clinical effectiveness of the method was demonstrated on 62 clinical burn wound images obtained from digital color photographs, yielding an average classification success rate of 82% compared to expert classified images.

2.2.3.3 Dental Medicine

Fuzzy inference for personal identification in medical jurisprudence has been demonstrated. A sex determination program was constructed for teeth using a fuzzy inference technique and was compared with known discriminate analysis. The materials examined were mandibular and maxillary dental plaster models of adult males and females, and infant males and females. Teeth characteristics were used as inputs to a fuzzy inference program, and the probability of maleness, obtained from the discriminate analysis, was set as the output object. Finally, a sex determination program for both permanent and deciduous teeth was constructed using a fuzzy inference software development tool. Each measured value was input to the program, and the output results were compared with those of discriminate analysis. The percentages of correct determinations for permanent teeth were 93.0% for males and 89.0% for females using the fuzzy inference program [109]. Another application of fuzzy logic is for diagnostic processes for tooth conservation [110]. In the field of dentistry skills, an expert system was developed for capturing the knowledge of dentists, and then using fuzzy sets to exploit it as a commercial expert system [111]. Another different, more involved study concentrated on developing a computer-assisted inference model for selecting appropriate types of headgear appliance for orthodontic patients and to investigate its clinical versatility as a decision-making aid for inexperienced clinicians [112]. Fuzzy rule bases were created for degrees of overjet, overbite, and mandibular plane angle variables, respectively, according to subjective criteria based on the clinical experience and knowledge of the authors.

2.2.3.4 Ophthalmology

In ophthalmology, fuzzy reasoning has been applied to eye movement. Prochazka [113] used fuzzy logic control to describe in a "biologically compatible" way the sensorimotor behavior. Also, it was utilized for diagnosis of glaucoma, which is one of the leading causes of blindness worldwide [114]. In a different study, a classification system for the effects of diabetes mellitus (DM) on blood flow hemodynamics of the ophthalmic arteries by using a neuro-fuzzy system was described [115]. Test results suggested that 85% success rates were reached from the data of right ophthalmic arteries, and 87.5% success rates were reached from the data of left ophthalmic arteries.

2.2.3.5 Otology, Rhinology, Laryngology

In the field of otology, and laryngology, a fuzzy expert system has been developed for classification of pharyngeal dysphagia. Also, fuzzy analysis has been used to improve hearing aids, fuzzy logic based discrimination of spoken words has been investigated to improve cochlea implementation, and neuro-fuzzy systems have been implemented to localize sounds in space [116]. Because data/knowledge fusion is not a trivial exercise, there is generally little literature on the subject, but the study by Debon [117] is an interesting endeavor as it proposes a three-dimensional segmentation method of oesophagus inner and outer wall from endosonographic sequences (composed of separate slices uniformly distributed), using the cooperation of different models. All these components have been integrated in a coherent architecture hierarchically organized that allows belief updating.

2.2.3.6 Urology

The application of fuzzy set theory in urology is mainly concerned with diagnosis. An electromyogram of the corpora cavernous (CC-EMG) imparts information on the autonomic cavernous innovation and/or the cavernous smooth muscles. The CC-EMG is interpreted mainly by evaluation of signal patterns of higher activity. An evaluation of the patterns is derived from these features using fuzzy logic. A dramatic improvement in the diagnosis of erectile dysfunction has been achieved [118]. A urologic postsurgical monitoring system has been developed to monitor catheterized, postsurgery patients specifically to assist care providers in the reduction of clot retention. Fuzzy logic was also used to correlate with the linguistic descriptors used by medical experts [119]. Another fuzzy expert system has been developed mainly for diagnosis and treatment of a typical adenomatous hyperplasia (AAH) of the prostate and its distinction from well-differentiated prosthetic adenocarcinoma with small acinar pattern [120]. A new system for prostate diagnostics based on multifeature tissue characterization has also been proposed [121]. Two adaptive neuro-fuzzy inference systems (FIS) working in parallel have bee utilized. The system has been evaluated on 100 patients undergoing radical prostatectomy.

2.2.4 Neuromedicine

2.2.4.1 Neurology

Soft computing methods, namely, fuzzy logic and neural networks, have been applied in different ways to neurology. The techniques used in these applications are: measurement and instrumentation [122], signal and image processing including classification [123], clustering analysis, linguistic modeling, pharmacotherapy prediction, control of body movements by electrical stimulation [124], and self-optimizing and neuro-fuzzy networks for decision support [125–127].

2.2.4.2 Psychology

The paradigm of the fuzzy logical model of perception (FLMP) has been extended to the domain of perception and recognition of facial affect. The FLMP fits the judgments from two features experiments significantly better than an additive model

[128]. A neural network based cognitive architecture termed Concept Hierarchy Memory Model (CHMM) for conceptual knowledge representation and commonsense reasoning with fuzzy relations between a concept was presented in terms of weights and on the links connecting them [129]. Using a unified inference mechanism based on code firing, CHMM performs an important class of commonsense reasoning, including concept recognition and property inheritance. Fuzzy sets have also been used to conduct a comparative study on traditional crisp set to fuzzy set representations of quantitative estimates of drug use [130]. This is to investigate what survey respondents recall about their drug use. All numeric estimates of drug use showed fuzzy set properties. Compared to traditional self-reports of drug use, fuzzy set representations provide a more complete and detailed description of what participants recall about past drug use.

2.2.4.3 Psychiatry

A complex psychiatric computer expert system, including functions that help the physicians and the hospital staff in administrative, diagnostic, therapeutic, statistical, and scientific work has been developed. The diagnostic decision support system is based on fuzzy logic and backwards chaining [131]. A multivariate classification technique based on fuzzy set mathematics was applied to the demographic, historical, and mental-state data on dementia praecox cases, and manic-depressive insanity cases [132].

Logic and analogy pose a fundamental problem in psychiatry. The latter has been explored again with the use of fuzzy logic. Marchais and Grize [133] put forward the hypothesis of a common root using the systemal method, an original strategy and a particular test. Several questions are tackled: definitions, hierarchy, linking, linearity or circularity, closed or open aspects of thought process, integration, separation or continuity, ways of appearance, discrete organization or organization in networks, different or common origin, and the importance of modal and nonmonotonic logic.

In addition to the above "sparse" contributions, this category also includes a clinical prediction study relating to suicide [134]. This study focused on evaluating the accuracy of Fuzzy Adaptive Learning Control Network (FALCON) neural networks, a nonlinear algorithm, in identifying of a group of patients. The trained FALCON may therefore assist in identification of a subgroup of individuals who remain unrecognized by clinicians and contribute to prevention of suicide.

2.2.5 Image and Signal Processing

2.2.5.1 Signal Processing

Dynamic state recognition and event prediction are fundamental tasks in biomedical signal processing. Fuzzy logic was early applied in medical systems for signal processing. Most of the work reported in such papers utilize fuzzy logic for pattern recognition and fuzzy clustering of EEG, ECG, and evoked potentials in conjunction with or without neural networks [135].

Some of the fields of application for signal processing are listed here: monitoring the electrical responses of nerve fibers [136], R-spike detection for rhythm monitoring [137], clustering of gradient-echo functional MRI in the human visual cortex [138], magnetic resonance imaging [139], clustering approach to evoked potentials [140], expert system [141] or evaluation tools for EEG interpretation [142], matrix-assisted laser desorption/ionization mass spectrometry [143], feedback control [144], improved monitoring of preterm infants [45], detection of rapid eye movement [145] or visual evoked potential [146], discrimination of vowels in the cochlear implant signal [147], pattern recognition methods to classify esophageal motility records [148], rule-based labeling of computed tomography images [149–151], segmentation of ventricular angiographics images [152], and medical image restoration [153].

2.2.5.2 Radiation Medicine

Diagnostic imaging and treatment planning for radiation therapy has been the most commonly applied topic in this field using fuzzy logic. It is mainly concerned with image processing using fuzzy techniques (i.e., matching two image volumes in terms of their surface elements—tiles). Fuzzy logic has been utilized to register the surfaces of two volumes acquired by different medical imaging modalities [154]. Apart from image processing, diagnosis of chronic liver disease from liver scintiscans based on scintigraphic results by fuzzy reasoning gave a better accuracy in diagnosis than a conventional scoring system [155]. In terms of treatment, a fuzzy system for modeling and optimization of radiotherapy treatment planning was developed by Sadati and Mortazavi [156]. A similar system was also developed which was based on three-dimensional radiotherapy of cancer patients [157]. Further, an expert system was developed for prediction of radiation damage in organs [158]. Leszczynski [159] developed a portal image classifier based on the fuzzy k-NN algorithm. The fuzzy k-NN portal classifier was capable of identifying almost all the truly unacceptable portals with an acceptably low false alarm rate. Yan [160] developed a fuzzy logic technique to optimize the weighting factors in the objective function of an inverse treatment planning system for intensity-modulated radiation therapy (IMRT). This system was tested using one simulated (but clinically relevant) case and one clinical case. The results indicate that the optimal balance between the target dose and the critical organ dose can be achieved.

2.2.5.3 Radiology

Image processing, in particular magnetic resonance images, is one of the most popular applications of fuzzy logic in medicine [161–166]. Such popularity is due to the pioneering work of pattern recognition using fuzzy logic by Bezdek [167]. Another survey on the use of fuzzy models for segmentation and edge detection in medical image data is reported by Bezdek [168].

Fuzzy c-means clustering segmentation techniques for tissue differentiation in conjunction with a fuzzy model for segmentation and edge detection have been applied very successfully in many applications. The brain and the heart are most often investigated [169], as well as breast cancer diagnosis [64], diagnosis of rheumatic diseases [170], search for lung nodules [171], and interpretation of X-ray fluorescence spectra [172]. Other diagnosis concerns are malformations and degenerative disturbances.

2.2.6 Laboratory

2.2.6.1 Biomedical Laboratory Tests

Classification of constitution of blood, interpretation of pathophysiological background, controlling the bioprocess, analysis of tests, assistance in diagnosis, and solving clinical problems with expert systems are areas where fuzzy logic has been utilized. Fuzzy logic has also been used in conjunction with neural networks in clinical laboratory computing with application to integrated monitoring [173], and also in clinical test database systems [174]. The features that fuzzy logic provides make approximate reasoning and interpolation of laboratory tests suitable for its incorporation into expert systems [175, 176]. This framework makes it possible for precise laboratory tests to be interpreted in terms of fuzzy propositions. Furthermore, with the recent emergence of genomics in the medical scene, fuzzy logic has also been successfully exploited in this are as well [177–179].

2.2.7 Basic Science

2.2.7.1 Medical Reasoning and Decision Support

The use of fuzzy logic in medicine started in the early 1970s. Several papers on fuzzy logic are concerned with fuzzy rule extraction and data mining [180–183], stability, self-organizing, model-based and adaptive systems, supervisory control, hierarchical systems, and synergies with other soft computing techniques such as neural networks and genetic algorithms [110] and expert systems [184–186]. Even the topic of facilitating the production and spread of technological information is discussed [187].

Basic rule-based fuzzy logic controllers such as PI, PD, and PID are designed either from the expert (anesthetists) or crafted by hand depending on the experience of the programmer. This includes tuning the membership functions in terms of the shape, width, and position. This type of controller is widely used and is the most applicable control type in anesthesia [188].

Self-learning systems are concerned with the control of systems with unknown or time-varying structure or parameters. The self-organizing fuzzy logic controller has the ability to realize adaptation by building its fuzzy rules online as it controls the process, altering and adding as many rules as it judges necessary from off-line criteria. This approach has many successful applications in the control of muscle relaxation [86], and simultaneous control of blood pressure and muscle relaxation [189].

Model-based and adaptive systems are most successful when a physician plays a part in the closed-loop. The adaptive scheme plays an important role in adapting the controller to changes in the process (patient) and its disturbances. Fuzzy modeling and control are often based on qualitative assessment of the patient's condition using fuzzy inductive reasoning [190, 191]. A self-adaptive fuzzy controller with reinforcement learning is yet another technique applied to simulation of a paraplegic standing up [99]. Even for patient monitoring, adaptive controllers are being utilized for intelligent monitoring of diabetic patients [192].

Reasoning with fuzzy logic is possible without the need for much data because the backbone of the logic is expressed as if-then rules. However, the rules cannot be expressed when there are unknown logical relationships unless the logic is defined. Thus, attempts are being made to combine different techniques such as neural networks and genetic algorithms, with fuzzy logic organizing the mapping relationship by learning. Neuro-fuzzy networks were developed by fusing the ideas that originated in the fields of neural and fuzzy systems [193]. A neuro-fuzzy network attempts to combine the transparent, linguistic, and symbolic representation associated with fuzzy logic with the architecture and learning rules commonly used in neural networks. These hybrid structures have both a qualitative and a quantitative interpretation, and they can overcome some of the difficulties associated with solely neural algorithms, which can usually be regarded as black box mappings, and with fuzzy systems, where few modeling and learning theories exist. Many applications are being reported using fuzzy-neural control [194] and modeling [195].

Although writing fuzzy rules is easy, specific forms of membership functions are much harder to derive. In this case a genetic algorithm is used to adjust the membership function towards convergence. GAs are exploratory search and optimization methods that were devised on the principles of natural evolution and population genetics. Modeling clinical data can be achieved using genetic-fuzzy logic techniques [196].

Cascading two techniques is another approach to hybrid fuzzy control; for example, by using the discrete wavelet transform analysis to extract features from the clinical data, then feeding the features to a fuzzy logic systems (clustering or neuro-fuzzy) to extract the final output [197]. This methodology was also applied for forecasting generalized epileptic seizures from the EEG signal by wavelet analysis and dynamic unsupervised fuzzy clustering [198].

A totally fuzzy logic based hierarchical architecture for manipulating procedures on a complex process (i.e., the patient) has been developed [199]. The novel hierarchical architecture for fuzzy logic monitoring and control of intravenous anesthesia has two main objectives: the primary task is to utilize auditory evoked response signals for augmenting cardiovascular and body function signs into a multisensor fuzzy model-based strategy for anesthesia monitoring and control. The secondary task is to extend an existing fuzzy patient model for use as a training simulator.

As for supervisory control, a multiple drug hemodynamic control system by means of a fuzzy rule-based adaptive control system has been developed for controlling mean arterial pressure and cardiac output. Supervisory capabilities are added to ensure adequate drug delivery [200].

2.2.7.2 Anatomy, Pathology, Forensic Medicine, and Genetics

There are many applications of fuzzy logic to these fields of medicine, mostly based on image analysis and fuzzy clustering. A typical application is a vessel tracking algorithm for retinal images based on fuzzy clustering [201]. The application of a fuzzy c-means segmentation technique for tissue differentiation in magnetic resonance images of a hemorrhagic glioblastoma multiform was also reported [202].

In histopathology, an expert system was developed which considers certainty factors and possibility theory [203]. Fuzzy logic was utilized for reasoning with uncertainties, and an analysis was presented of five strategies and their suitability in pathology [204].

In forensic medicine, the application of fuzzy inference to personal identification (sex determination) was reported by Takeuchi [109]. Also, Schafer and Lemke [205] reported the use of fuzzy logic and fractal geometry in forensic medicine.

Fuzzy logic has not been applied directly, but instead has been utilized to develop control strategies for the application of pH-state to fed-batch cultivation of genetically engineered Escherichia coli [206]. Fuzzy mathematics was also used to classify South American Indian tribes according to genetic characteristics [207].

2.2.7.3 Physiology

The use of fuzzy logic in physiology has been cited in rehabilitation, including functional electrical stimulation (FES) [98, 208–212].

2.2.7.4 Pharmacology/Biochemistry

Pharmacology and biochemistry have also benefited from the application of fuzzy set theory. One application is the prediction of the rodent carcinogenicity of organic compounds using a fuzzy adaptive least-squares method [213]. Other applications are pharmacokinetic modeling, classification of nucleotide sequence, assignment of peptides with nonstandard amino acids, and segmentation of protein surfaces [214–218].

2.2.7.5 Education

The use of fuzzy logic for education has different facets. Fuzzy mathematics was utilized for training and evaluating the teaching of students in a clinical setting [219–221]. Evaluation of the self-taught ability of nursing administrators with fuzzy medicine was also reported [222]. Virtual reality is an emerging technology that can teach surgeons new procedures and can determine their level of competence before they operate on real patients [76].

2.2.8 Nursing

Psychological analysis is increasingly applied to nursing issues. Im and Chee [223] made a case for more fuzzy logic based reasoning in nursing in their paper. Whereas preanalytical discussions might be unclear, analysis should bring rigor. Rolfe [224] examines some recent innovations in computer logic, and argues that nursing can learn from a new breed of fuzzy computer programs which appear to be able not only to perform better than experts, but to verbalize their decision-making processes.

Some of the mistakes and misconceptions concerning nurse-related reasoning are discussed by Cave [225]. As a running theme, the paper in the *Journal of Advanced Nursing* by Rolfe was chosen as an example. The mistakes and misleading remarks found in the paper are not uncommon; so the paper's correction aim is targeted more widely than a single author critique. The mistakes and misconceptions could initially appear abstract and without practical outcome. In order to avoid this, a nursing case study is presented, displaying some undesirable consequences for a nurse who is misled by her reading.

Assessment of nursing quality using fuzzy mathematics is a different subject, which is more related to education and was discussed further by Fei [226].

2.2.9 Public Health and Health Policy and Management

Healthcare studies have been increasing recently due to a greater interest among people. There are many uncertainties in healthcare that can be detected. Some topics that can be reviewed include evaluation of drinking water quality, driving fatigue, health risks in work environment in terms of injury [227], illnesses [228], safety [229], evaluation for health policy [230], security issues among hospitals in Japan and its feasible solutions [231], and health services management in the tropics [232].

The first application of fuzzy logic in healthcare was the development of an index to assess the health of the patient. The index was developed using a fuzzy approach since the boundaries of status are not sharply defined [233].

In terms of hospital healthcare, waiting list management is another topic to which fuzzy mathematical programming has been applied [234]. Another expert system was designed to help physicians and hospital staff in administrative, diagnostic, therapeutic, statistical, and scientific work. In this system, there are separate data-storing, health insurance supporting, and simple advisory programs [131].

Other applications relate to designing control strategies for administering measles vaccines [235], optimizing food diets [236], diagnosing bacterial infection in hospitalized patients [237], medical record validation [238], and prioritization of organ transplant patient waiting lists [239].

2.2.10 Eastern Medicine

The use of fuzzy logic in Chinese medical diagnostic systems was based on utilizing the theory as a mathematical tool for diagnosis inside an expert system [240, 241]. Another expert system for traditional Chinese medicine was developed using the PROLOG language [242]. One important organ in Chinese medicine diagnosis is the tongue. The chromatic distribution helps in the diagnosis procedure. Fuzzy sets were used to fit a model for the distribution [243]. The preparation of the drug was also another area where fuzzy logic was used. A machine was designed based on fuzzy logic for decocting the medicine, and it achieved almost the same drug produced by ancient process [244]. There also exist many expert systems based on fuzzy logic for Chinese medicine, diagnosis, and treatment [245, 246].

2.2.11 Bibliographic Papers and Books

The first comprehensive bibliographic paper covering all fields of fuzzy logic was compiled by Kandal and Yager [247]. Another survey that covers 20 years of fuzzy set theory application in medicine was written by Maiers [248]. It summarizes the medical applications of fuzzy set theory, assesses the value of fuzzy sets in medical applications, and suggests their future potential. An introduction to fuzzy sets, approximate reasoning and fuzzy rules as a tool for modeling sets with ill-defined or flexible boundaries with uncertainties has been presented by Dubois and Prade [249–255]. Other papers focus on surveying special fields like virtual reality and robotics in medicine [256], general application of fuzzy logic in medicine [257], and a case study of different medical application domains [258]. A good source of references and an excellent book in the field of fuzzy logic is by Zimmermann [259]. A book edited by Cohen and Hudson [260] contains a collection of papers that deal with medical decision-making relating to heart disease. Different approaches were reported: knowledge-based systems, statistical approaches, modeling, and hybrid systems.

Another book that has an excellent survey of the fuzzy logic in medicine was edited by Toedorescu, Kandel, and Jain [261]. It also covers recent applications in modeling of the brain, tumor segmentation, dental developments, myocardial ischaemia diagnosis, heart disease diagnosis, and usage in the ICU. Neuro-fuzzy system applications were also reported in the book. In 1989, a group of Japanese scientists established a learned society for fuzzy systems and biomedical applications: the Japanese Bio-Medical Fuzzy Systems Association (BMFSA), which has an international journal *Biomedical Soft Computing and Human Sciences* published in Japanese with a yearly issue in English.

2.3 Discussion

There is no doubt that fuzzy logic based technology has had the success in medical diagnosis that was predicted by its founder Professor L. A. Zadeh back in the early days of introducing the theory itself [1]. One might even argue, and for good reason, that such success and impact were indeed far more than expected! I can safely say that no other technology has had such a far-reaching impact on a particular discipline than fuzzy technology has had on medicine and healthcare of the life sciences. Why is this? The answer lies in the fact that all the subheadings that were considered share some common denominators:

1. These are systems that are characterized by complexities and uncertainties.
2. Not all processes involved can be described by physical equations only, but expert's knowledge can often be made available.
3. Data can be sparse and input/output mappings are often of a high dimensional nature.

Although the literature search in this chapter was not exhaustive, it has been organized hierarchically according to carefully identified medical categories, which are correlated. Such a correlation should encourage successful applications to be transferred from one domain to another. The two-dimensional representation of the data (time and sector) not only shows the current state of the art but also the dynamic process of the information spread in each category from which future trends can be extrapolated. Tables 2.1 and 2.2 represent a summary of a number of

Table 2.1 Number of Applications of Fuzzy Logic in Each Year for the Specified Medical Subheadings (Up to 1989 and Until 1998)

Publication Year	*Up to and 1989*	*1990*	*1991*	*1992*	*1993*	*1994*	*1995*	*1996*	*1997*	*1998*
Conservative disciplines										
Internal medicine	35	3	2	—	2	9	7	5	2	3
Cardiology	9	—	3	2	1	7	2	3	1	11
Intensive care	3	—	—	1	2	2	—	2	2	2
Pediatrics	1	—	—	—	—	4	—	3	—	—
Endocrinology	3	1	—	—	1	—	1	—	1	2
Oncology	1	—	—	—	—	1	—	—	7	10
Gerontology	1	—	7	4	3	3	6	4	4	—
General practice	2	—	—	—	1	—	—	—	—	—
Invasive medicine										
Surgery	—	—	—	—	—	1	—	—	1	2
Orthopedics	—	—	—	—	—	2	—	—	—	2
Anesthesia	6	1	1	3	1	8	3	6	9	6
Artificial organs	—	—	—	—	—	1	—	2	3	1
Regionally defined medical disciplines										
Gynecology	—	2	—	—	1	—	—	2	3	3
Dermatology	—	—	—	—	—	2	—	—	—	1
Dental medicine	—	—	—	—	1	1	—	—	1	1
Ophthalmology	1	—	1	—	—	1	—	1	1	1
Otology, rhinology, etc.	—	—	—	1	—	—	2	1	—	—
Urology	1	1	—	—	1	2	1	1	3	—
Neuromedicine										
Neurology	5	—	—	—	—	1	2	2	7	—
Psychology	5	—	1	3	1	3	4	6	4	—
Psychiatry	4	—	1	—	—	1	3	1	1	—
Image and signal processing										
Signal processing	4	3	3	2	—	3	1	3	5	7
Radiation medicine	4	1	—	—	—	—	2	2	2	2
Radiology	3	3	—	—	2	4	5	7	10	4
Laboratory										
Biochemical and tests	1	1	3	2	1	1	2	4	2	1
Basic science										
Medical information	29	2	3	5	7	6	16	11	9	—
Anatomy, pathology, etc.	2	2	1	4	2	1	—	2	2	3
Physiology	—	1	—	—	—	1	2	1	—	3
Pharmacology	2	—	—	1	2	6	7	4	5	1
Education	1	1	—	—	1	1	1	2	2	1
Nursing	1	—	—	—	—	—	1	—	1	—
Healthcare	2	—	—	—	2	1	3	1	6	1
Year totals	125	20	26	28	32	73	71	76	94	68

applications of fuzzy logic for each criterion classified on a yearly basis using MEDLINE and INSPEC. Figure 2.1 shows the total number of publications each year in each database. It is worth noting that each database is different, since

Table 2.2 Number of Applications of Fuzzy Logic in Each Year for the Specified Medical Subheadings (1999 Until 2004)

Publication Year	*1999*	*2000*	*2001*	*2002*	*2003*	*2004*	*Total*
Conservative disciplines							
Internal medicine	3	10	11	12	3	5	112
Cardiology	9	4	12	7	4	6	81
Intensive care	3	2	4	3	3	5	34
Pediatrics	—	—	3	1	1	4	17
Endocrinology	1	2	1	—	1	—	14
Oncology	4	3	5	11	16	9	67
Gerontology	—	—	1	1	—	—	34
General practice	—	—	1	—	—	—	4
Invasive medicine							
Surgery	2	2	2	1	—	—	11
Orthopedics	4	2	4		—	3	17
Anesthesia	9	3	6	4	—	2	68
Artificial organs	—	1	—	—	1	1	10
Regionally defined medical disciplines							
Gynecology	3	4	3	3	4	3	31
Dermatology	—	—	—	2	4	—	9
Dental medicine	1	1	—	1	1	—	8
Ophthalmology	—	3	—	3	3	1	16
Otology, rhinology, etc.	2	1	1	—	—	—	8
Urology	—	1	—	—	2	—	13
Neuromedicine							
Neurology	1	3	4	1	3	1	30
Psychology	—	—	1	3	2	1	34
Psychiatry	1	—	1	—	—	2	15
Image and signal processing							
Signal processing	6	5	6	11	3	5	67
Radiation medicine	2	1	—	—	4	—	19
Radiology	14	18	14	16	21	18	139
Laboratory							
Biochemical and tests	5	7	1	9	4	1	45
Basic science							
Medical information	2	2	3	2	6	7	110
Anatomy, pathology, etc.	2	5	6	2	—	4	37
Physiology	3		3	3	3	2	21
Pharmacology	2	1	5	2	4	7	50
Education	—	2	—	—	1	1	13
Nursing	—	—	—	—	1	—	4
Healthcare	3	2	3	2	—	1	27
Year totals	82	85	101	100	95	89	1165

MEDLINE mainly reviews medical journal papers, whereas INSPEC reviews computer science journals and conference papers. The trace in Figure 2.1 referred to as "common" refers to those publications that were cited in both databases and as can be seen the intersection between the two databases, which remains limited despite the fact that the number of publications increases (see Table 2.3). The only exception lies in a few researchers having the privilege of wearing the two hats, corresponding to the two databases. The decrease in the number of publications as per INSPEC is not necessarily a negative aspect and it might be due to the fact that those who are from a computer science background are taking stock of developments in the medical sector to reassess their research priorities. The positive aspect reflected in Figure 2.1 is the upward trend of the MEDLINE-based publications.

In the light of the above, future developments of fuzzy technology in medicine and healthcare can be tentatively forecast:

1. Internal medicine, anesthesia, radiology, electrophysiology, pharmacokinetics, and neuromedicine use fuzzy logic methods to a considerable degree. Such technology evolves around fuzzy control, fuzzy signal and image processing, fuzzy expert systems, fuzzy modeling, and fuzzy neural simulation.
2. In the field of surgical disciplines, dental medicine, general practice, education, and nursing, applications are scarce and nonspecific.
3. Papers in the field of medical reasoning and decision support sciences are growing rapidly. This will be used as a landmark to other sectors for growing themselves.

Table 2.3 Total Number of Paper Publications Per Year as Cited by INSPEC and MEDLINE Databases, Including Those Papers Which Intersect Between the Two Databases

Year	*1990*	*1991*	*1992*	*1993*	*1994*	*1995*	*1996*	*1997*	*1998*	*1999*	*2000*	*2001*	*2002*	*2003*	*2004*
MEDLINE	18	24	27	27	47	68	66	67	36	61	53	78	66	66	72
INSPEC	3	2	4	4	28	6	17	27	34	26	38	34	41	35	22
Common	1	0	0	0	2	2	4	4	2	5	6	11	6	6	5
Total	20	26	31	31	73	72	79	90	68	82	85	101	101	95	89

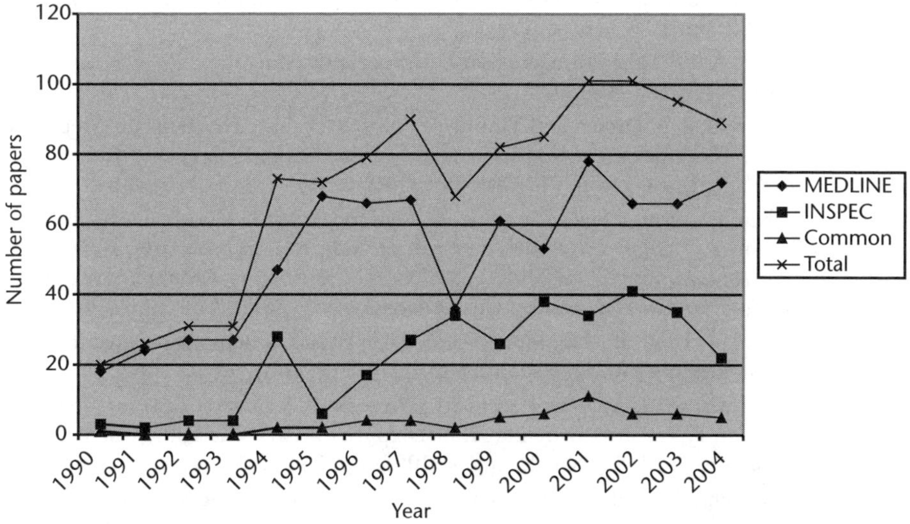

Figure 2.1 Number of paper publications of fuzzy logic in medicine and healthcare as cited by MEDLINE and INSPEC.

As already stated, this literature search cannot in any way represent a complete review, but it is sufficient to point at the vectors that govern the penetration of the fuzzy logic–based technology in medicine and healthcare. It is hoped that the categories that were identified here will lead to other matching techniques and subject areas that will lend themselves naturally to this technology.

References

[1] Zadeh, L., "Biological Application of the Theory of Fuzzy Sets Systems," *Proc. Int. Symp. Biocybernetics of the Central Nervous System*, Boston, MA: Little, Brown, 1969, pp. 199–212.

[2] Martin, J. F., "Fuzzy Control in Anesthesia," *J. Clin. Monit.*, Vol. 10, 1994, pp. 77–80.

[3] von Keyserlingk, D. G., "Roadmap for ERUDIT," *Verlsge Mainz,* Aachen, Germany, February 1998.

[4] Shortliffe, E. H., *Computer-Based Medical Consultations, MYCIN*, New York: Elsevier, 1976.

[5] Sanchez, E., "Inverses of Fuzzy Relations: Application to Possibility Distributions and Medical Diagnosis," *Proceedings of the 1977 IEEE Conference on Decision and Control*, IEEE, New York, 1977, pp. 1384–1389.

[6] Sanchez, E., "Resolution of Eigen Fuzzy Sets Equations," *Fuzzy Sets and Systems*, Vol. 1, No. 1, 1978, pp. 69–74.

[7] Sanchez, E., "Inverses of Fuzzy Relations: Application to Possibility Distributions and Medical Diagnosis," *Fuzzy Sets and Systems*, Vol. 2, No. 1, 1979, pp. 75–86.

[8] Bartolin, R., et al., "The Fuzzy Set Theory as a Biomedical Diagnostic Aid," *Sem. Hop.*, Vol. 58, No. 22, 1982, pp. 1361–1365.

[9] Adlassnig, K. P., "A Survey of Medical Diagnosis and Fuzzy Subsets," in M. M. Gupta and E. Sanchez, (eds.), *Approximate Reasoning in Decision Analysis*, New York: North Holland, 1982, pp. 203–217.

[10] Esogbue, A. O., "Measurements and Validation of Fuzzy Mathematical Model for Medical Diagnosis," *Fuzzy Sets and Systems*, Vol. 10, 1983, pp. 223–242.

[11] Lesmo, L., L. Saitta, and P. Torasso, "Dealing with Uncertain Knowledge in Medical Decision-Making: A Case Study in Hepatology," *Artificial Intelligence in Medicine,* Vol. 1, No. 3, 1989, pp. 105–116.

[12] Rocha, A. F., "The Brain as a Symbol-Processing Machine," *Prog. Neurobiol.*, Vol. 53, No. 2, 1997, pp. 121–198.

[13] Rotshtein, A., "Design and Tuning of Fuzzy Rule-Based Systems for Medical Diagnosis," in H. N. Teodorescu, A. Kandel, and L. C. Jain, (eds.), *Fuzzy and Neuro-Fuzzy Systems in Medicine*, Boca Raton, FL: CRC Press, 1999.

[14] Serpen, G., et al., "Fuzzy Inference System for PIOPED-Compliant Diagnosis of Pulmonary Embolism," *Smart Engineering System Design: Neural Networks, Fuzzy Logic, Evolutionary Programming, Data-Mining, and Complex Systems, Proceedings of the Artificial Neural Networks in Engineering Conference ANNIE 2000*, Vol. 10, 2000, pp. 753–758.

[15] Yoshida, H., et al., "Computer-Aided Diagnosis Scheme for Detection of Polyps at CT Colonography," *Radiographics*, Vol. 22, No. 4, July/August 2002, pp. 963–79.

[16] Sokolowska, B., A. Jozwik, and M. Pokorski, "A Fuzzy-Classifier System to Distinguish Respiratory Patterns Evolving After Diaphragm Paralysis in the Cat," *Japanese Journal of Physiology*, Vol. 53, No. 4, August 2003, pp. 301–307.

[17] Sakaguchi, S., et al., "Development of an Early Diagnostic System Using Fuzzy Theory for Postoperative Infections in Patients with Gastric Cancer," *Digestive Surgery*, Vol. 21, No. 3, 2004, pp. 210–214.

[18] Sanchez, E., "Fuzzy Logic and Inflammatory Protein Variations," *Clinica. Chimica. Acta.*, Vol. 270, 1998, pp. 31–42.

[19] Adlassnig, K. P., "A Fuzzy Logical Model of Computer-Assisted Medical Diagnosis," *Methods of Information in Medicine*, Vol. 19, No. 3, 1980, pp. 141–148.

[20] Adlassnig, K. P., "Fuzzy Set Theory in Medical Diagnosis," *IEEE Trans. on Systems, Man and Cybernetics*, SMC16, No. 2, 1986, pp. 260–265.

[21] Fieschi, M., et al., "SPHINX: A System for Computer-Aided Diagnosis," *Methods of Information in Medicine*, Vol. 21, No. 3, 1982, pp. 143–148.

[22] Belmonte-Serrano, M., C. Sierra, and R. Lopezde-Mantaras, "RENOIR: An Expert System Using Fuzzy Logic for Rheumatology Diagnosis," *International Journal of Intelligent Systems*, Vol. 9, No. 11, 1994, pp. 985–1000.

[23] Godo, L., et al., "Pneumon-IA and Terap-IA: Three Medical Applications Based on Fuzzy Logic," *Artificial Intelligence in Medicine*, Vol. 21, No. 1–3, January/March 2001, pp. 153–162.

[24] Kohout, L. J., et al., "CLINAID: A Knowledge-Based Decision Support System for Use in Medicine, Computer Assisted Decision Making: Expert Systems, Decision Analysis, Mathematical Programming," *Proceedings of UNICOM Seminar*, Amsterdam, the Netherlands: North Holland, 1986, pp. 133–146.

[25] Kalmanson, D., and F. Stegall, "Cardiovascular Investigation and Fuzzy Concepts," *Amer. J. Cardio.*, Vol. 35, 1975, pp. 30–34.

[26] Joly, H., et al., "Application of Fuzzy Set Theory to the Evaluation of Cardiac Functions," *MEDINFO '80, Proc. 3rd World Conference on Medical Information*, Tokyo: North Holland Publications, Vol. 1, 1980, pp. 91–95.

[27] Jagannathan, V., et al., "Artificial Intelligence Methods in Quantitative Electroencephalogram Analysis," *Computer Programs in Biomedicine*, Vol. 15, No. 3, 1982, pp. 249–257.

[28] Dong, W., S. Zhou, and L. Wang, "A Fuzzy Relation Model for Multifactorial Analysis of Cerebrovascular Diseases," *Proceedings of the Ninth Annual Conference of the IEEE Engineering in Medicine and Biology Society*, IEEE, New York, Vol. 1, 1987, pp. 307–308.

[29] Kauffmann, R., H. Reul, and G. Rau, "The Helholtz Total Artificial Heart Labtype," *Artificial Organs*, Vol. 18, No. 7, 1994, pp. 537–542.

[30] Yoshizawa, M., et al., "Assessing Cardiovascular Dynamics During Ventricular Assistance," *IEEE Engineering in Medicine and Biology*, Vol. 13, No. 5, 1994, pp. 687–692.

[31] Cios, K. J., et al., "A Novel Algorithm for Classification of SPECT Images of a Human Heart," *Proceedings of the Ninth IEEE Symposium on Computer-Based Medical Systems*, IEEE Comput. Soc. Press, Los Alamitos, CA, 1996, pp. 1–5.

[32] Jain, R., J. Mazumdar, and W. Moran, "Application of Fuzzy-Classifier System to Coronary Artery Disease and Breast Cancer," *Australasian Physical & Engineering Sciences in Medicine*, Vol. 21, No. 3, 1998, pp. 141–147.

[33] Lovelace, J. J., et al., "Internet-Based System for Diagnosis of Coronary Artery Disease," *Computers in Cardiology 1998*, IEEE, New York, Vol. 2, 1998, pp. 45–48.

[34] Akay, Y. M., et al., "Non-Invasive Detection of Coronary Artery Disease," *IEEE Engineering in Medicine and Biology Magazine*, Vol. 13, No. 5, 1994, pp. 761–764.

[35] Pedrycz, W., G. Bortolan, and R. Degani, "Classification of Electrocardiographic Signals: A Fuzzy Pattern Matching Approach," *Artificial Intelligence in Medicine*, Vol. 3, No. 4, 1991, pp. 211–226.

[36] Kerre, E. E., "Outline of an Expert System for ECG Diagnosis Using Fuzzy Sets," *Artificial Intelligence in Medicine*, Vol. 1, No. 3, 1989, pp. 139–144.

[37] Degani, R., and G. Bortolan, "Fuzzy Decision-Making in Electrocardiography," *Artificial Intelligence in Medicine*, Vol. 1, No. 2, 1989, pp. 87–91.

[38] Degani, R., and G. Bortolan, "Methodology of ECG Interpretation in the Padova Program," *Methods of Information in Medicine*, Vol. 29, No. 4, 1990, pp. 386–392.

[39] Degani, R., "Computerized Electrocardiogram Diagnosis: Fuzzy Approach," *Methods of Information in Medicine*, Vol. 31, No. 4, 1992, pp. 225–233.

[40] Hudson, D. L., M. E. Cohen, and P. C. Deedwania, "A Hybrid System for Diagnosis and Treatment of Heart Disease," *Proceedings of the 16th Annual International Conference of the IEEE Engineering in Medicine and Biology Society, Engineering Advances: New Opportunities for Biomedical*, IEEE, New York, Vol. 2, 1994, pp. 1368–1369.

[41] Wieben, O., V. X. Afonso, and W. J. Tompkins, "Classification of Premature Ventricular Complexes Using Filter Bank Features, Induction of Decision Trees and a Fuzzy Rule-Based System," *Medical & Biological Engineering & Computing*, Vol. 37, No. 5, Spetember 1999, pp. 560–565.

[42] Osowski, S., and T. H. Linh, "ECG Beat Recognition Using Fuzzy Hybrid Neural Network," *IEEE Trans. on Biomedical Engineering*, Vol. 48, No. 11, November 2001, pp. 1265–1271.

[43] Pfaff, M., et al., "Prediction of Cardiovascular Risk in Haemodialysis Patients by Data Mining," *Methods of Information in Medicine*, Vol. 43, No. 1, 2004, pp. 106–113.

[44] Si, Y., et al., "An Expert System for EEG Monitoring in the Pediatric Intensive Care Unit," *Medical Trials, Electroencephalogr. Clin. Neurophysiol.*, Vol. 106, No. 6, 1998, pp. 488–500.

[45] Wolf, M., et al., "Improved Monitoring of Preterm Infants by Fuzzy Logic," *Technol. Health Care*, Vol. 4, No. 2, 1996, pp. 193–201.

[46] Henkind, S. J., et al., "A Clinical Alarm System Using Techniques from Artificial Intelligence and Fuzzy Set Theory," in E. Sanchez and L. A. Zadeh, (eds.), *Approximate Reasoning in Intelligent Systems, Decision and Control*, Oxford, U.K.: Pergamon, 1987, pp. 91–104.

[47] Becker, K., et al., "Fuzzy Logic Approaches to Intelligent Alarms," *IEEE Engineering in Medicine and Biology Magazine*, Vol. 13, No. 5, 1994, pp. 710–716.

[48] Ying, H., et al., "Fuzzy Control of Mean Arterial Pressure in Postsurgical Patients with Sodium Nitroprusside Infusion," *IEEE Trans. on Biomedical Engineering*, Vol. 39, No. 10, 1992, pp. 1060–1070.

[49] Kacprzyk, J., and R. R. Yager, "Emergency-Oriented Expert Systems: A Fuzzy Approach," *Information Sciences*, Vol. 37, No. 1–3, 1985, pp. 143–155.

[50] Pilz, U., "Structure and Basic Functional Mode of the FLORIDA Expert System for Intensive Medicine," *Elektrie.*, Vol. 49, No. 10–12, 1995, pp. 399–406.

[51] Pilz, U., and L. Engelmann, "Integration of Medical Knowledge in an Expert System for Use in Intensive Care Medicine," in H. N. Teodorescu, A. Kandel, and L. C. Jain, (eds.), *Fuzzy and Neuro-Fuzzy Systems in Medicine*, Boca Raton, FL: CRC Press, 1999.

[52] Cohen, K. P., et al., "Breath Detection Using Fuzzy Sets and Sensor Fusion," *Proceedings of the 16th Annual International Conference of the IEEE Engineering in Medicine and Biology Society, Engineering Advances: New Opportunities for Biomedical.*, IEEE, New York, Vol. 2, 1994, pp. 1067–1068.

[53] Olliver, S., G. M. Davis, and G. E. Hatzakis, "Weaning Infants with Respiratory Syncytial Virus from Mechanical Ventilation Through a Fuzzy-Logic Controller," *Annual Symposium Proceedings/AMIA Symposium*, 2003, pp. 499–503.

[54] Koprinska, I., G. Pfurtscheller, and D. Flotzinger, "Sleep Classification in Infants by Decision Tree-Based Neural Networks," *Artificial Intelligence in Medicine*, Vol. 8, 1996, pp. 387–401.

[55] Ifeachor, E. C., and N. J. Outram, "Fuzzy Logic Concepts for Pattern Analysis and Interpretation of Changes in the Fatal Electrocardiogram," *Proceedings of the International Conference on Neural Networks and Expert Systems in Medicine and Healthcare*, Plymouth, U.K., 1994, pp. 353–362.

[56] Belal, S. Y., et al., "A Fuzzy System for Detecting Distorted Plethysmogram Pulses in Neonates and Paediatric Patients," *Physiological Measurement*, Vol. 22, No. 2, May 2001, pp. 397–412.

[57] Buisson, J. C., et al., "TOULMED: An Inference Engine Which Deals with Imprecise and Uncertain Aspects of Medical Knowledge," *Proceedings AIME '87, European Conference on Artificial Intelligence in Medicine*, Springer-Verlag, Berlin, 1987, pp. 123–140.

[58] Soula, G., B. Vialettes, and J. L. San Marco, "PROTIS, A Fuzzy Deduction Rule System: Application to the Treatment of Diabetes," *Proc. MEDINFO '83*, Amsterdam, 1983, pp. 233–255.

[59] Turnin, M. C., et al., "Telematic Expert System Diabeto: A New Tool for Diet Self Monitoring for Diabetic Patients, *Diabetes Care*, Vol. 15, No. 2, 1992, pp. 204–212.

[60] Stadelmann, A., et al., "DIABETEX Decision Module 2-Calculation of Insulin Dose Proposals and Situation Recognition by Means of Classifiers," *Comput. Methods Programs Biomed.*, Vol. 32, 1990, pp. 333–337.

[61] Arita, S., M. Yoneda, and Y. Hori, "Knowledge Acquisition Study and Accuracy Rate Evaluation for CADIAG-2/REHUMA with 308 Clinical Cases," in R. Lowen and M. Roubens, (eds.), *Fuzzy Logic*, Boston, MA: Kluwer, 1993, pp. 301–310.

[62] Zahlmann, G., et al., "Concept of a Knowledge Based Monitoring System for Glaucoma and Diabetic Retinopathy Using a Telemedicine Approach," *Proceedings of the 18th Annual International Conference of the IEEE Engineering in Medicine and Biology Society*, IEEE, New York, Vol. 3, 1997, pp. 1230–1231.

[63] Dazzi, D., et al., "The Control of Blood Glucose in the Critical Diabetic Patient: A Neuro-Fuzzy Method," *Journal of Diabetes & Its Complications*, Vol. 15, No. 2, March/April 2001, pp. 80–87.

[64] Kovarlerchuk, B., et al., "Fuzzy Logic in Computer-Aided Breast Cancer Diagnosis: Analysis of Lobulation," *Artificial Intelligence in Medicine*, Vol. 11, No. 1, 1997, pp. 75–85.

[65] Vaidyanathan, M., et al., "Quantitation of Brain Tumor in MRI for Treatment Planning," *Proceedings of the 16th Annual International Conference of the IEEE Engineering in Medicine and Biology Society, Engineering Advances: New Opportunities for Biomedical Engineers*, IEEE, New York, Vol. 1, 1994, pp. 555–556.

[66] Wallace, J. C., et al., "Classification of MR Spectra of Biopsies from Untreated and Recurrent Ovarian Cancer Using Linear Discriminant Analysis," *Magnetic Resonance in Medicine*, Vol. 38, No. 4, 1997, pp. 569–576.

[67] Velthuizen, R. P., "Validity Guided Clustering for Brain Tumor Segmentation [Treatment Planning]," *1995 IEEE Engineering in Medicine and Biology 17th Annual Conference* and *21st Canadian Medical and Biological Engineering Conference*, IEEE, New York, Vol. 1, 1997, pp. 413–414.

[68] Cheng, X. Y., et al., "Breast Tumor Diagnosis System Using Three Dimensional Ultrasonic Echography," *Proceedings of the 19th Annual International Conference of the IEEE Engineering in Medicine and Biology Society, Magnificent Milestones and Emerging Opportunities in Medical Engineering*, IEEE, Piscataway, NJ, Vol. 2, 1997, pp. 517–520.

[69] Hiltner, J., M. Fathi, and B. Reusch, "A New Model for Medical Image Analysis Based on Computational Intelligence," *Proceedings of the SPIE, The International Society for Optical Engineering*, Vol. 3979, Pt. 1–2, 2000, pp. 1295–1302.

[70] Fletcher-Heath, L. M., et al., "Automatic Segmentation of Non-Enhancing Brain Tumors in Magnetic Resonance Images," *Artificial Intelligence in Medicine*, Vol. 21, No. 1–3, January/March 2001, pp. 43–63.

[71] Manton, K. G., and M. A. Woodbury, "Grade of Membership Generalizations and Ageing Research," *J. Gerontol, A. Biol. Sci. Mec. Sci.*, Vol. 50, 1995, pp. M242–M251.

[72] McLeish, M., M. Cecile, and S. A. Lopez, "Database Issues for a Veterinary Medical Expert System," *Statistical and Scientific Database Management, Fourth International Working Conference SSDBM Proceedings*, Springer-Verlag, Berlin, 1989, pp. 177–192.

[73] Lim, C. K., et al., "A Proposed Hierarchical Fuzzy Inference System for the Diagnosis of Arthritic Diseases," *Australasian Physical & Engineering Sciences in Medicine*, Vol. 25, No. 3, September 2002, pp. 144–150.

[74] Sedbrook, T. A., H. Wright, and R. Wright, "A Visual Fuzzy Cluster System for Patient Analysis," *Med. Inf. (Lond.)*, Vol. 18, No. 4, 1993, pp. 321–329.

[75] Jaulent, M. C., et al., "Modeling Uncertainty in Computerized Guidelines Using Fuzzy Logic," *Proceedings of the AMIA, Annual Symposium,* 2001, pp. 284–288.

[76] Ota, D., et al., "Virtual Reality in Surgical Education," *Int. J. Biomed. Comput.*, Vol. 36, No. 4, 1995, pp. 281–291.

[77] Raposio, E., et al., *An "Augmented-Reality" Aid for Plastic and Reconstructive Surgeons, Medicine Meets Virtual Reality*, Global Healthcare Grid, Amsterdam, the Netherlands: IOS Press, 1997, pp. 232–236.

[78] Radetzky, A., et al., "Elastodynamic Shape Modeling in Virtual Medicine," *Proceedings Shape Modeling International '99, International Conference on Shape Modeling and Applications*, 1999, pp. 172–178, 276.

[79] Radetzky, A., A. Nurnberger, and D. P. Pretschner, "Elastodynamic Shape Modeler: A Tool for Defining the Deformation Behavior of Virtual Tissues," *Radiographics*, Vol. 20, No. 3, May/June 2000, pp. 865–881.

[80] Loslever, P., "Error and Data Coding in the Multi-Dimensional Analysis of Human Movement Signals," *Proc. Inst. Mech. Eng.*, Vol. 207, No. 2, 1993, pp. 103–110.

[81] Sau, K. N., and H. J. Chizeck, "Fuzzy vs. Non-Fuzzy Rule Base for Gait Event Detection," *Proceedings of the 16th Annual International Conference of the IEEE Engineering in Medicine and Biology Society, Engineering Advances: New Opportunities for Biomedical Engineers*, IEEE, New York, Vol. 2, 1994, pp. 1069–1070.

[82] von-Altrock, C., "Pressure Sensing in a Medical Shoe," *Circuit Cellar Ink*, No. 103, February 1999, pp. 20–2, 24–5.

[83] Kern, S. E., J. O. Johnson, and D. R. Westenskow, "Fuzzy Logic for Model Adaptation of a Pharmacokinetic-Based Closed Loop Delivery System for Pancuronium," *Artificial Intelligence in Medicine*, Vol. 11, No. 1, 1997, pp. 9–31.

[84] Becker, K., et al., "Design and Validation of an Intelligent Patient Monitoring and Alarm System Based on a Fuzzy Logic Process Model," *Artificial Intelligence in Medicine*, Vol. 11, 1997, pp. 33–53.

[85] Greenhow, S. G., D. A. Linkens, and A. J. Asbury, "Pilot Study of an Expert System Adviser for Controlling General Anaesthesia," *Br. J. Anaesth.*, Vol. 71, No. 3, 1993, pp. 359–365.

[86] Linkens, D. A., (ed.), *Intelligent Control in Biomedicine*, London, U.K.: Taylor & Francis, 1994.

[87] Isaka, S. and A. V. Sebald, "An Adaptive Fuzzy Controller for Blood Pressure Regulation," *Images of the Twenty-First Century, Proceedings of the Annual International Conference of the IEEE Engineering in Medicine and Biology Society*, IEEE, New York, Vol. 6, 1989, pp. 1763–1764.

[88] Isaka, S., et al., "A Fuzzy Blood Pressure Controller," *Proceedings of the Annual International Conference of the IEEE Engineering in Medicine and Biology Society*, IEEE, New York, Vol. 3, 1988, pp. 1410–1411.

[89] Karimi, A., and A. V. Sebald, "Computer Aided Design of Closed-Loop Controllers for Biomedical Applications," *Proceedings of the Annual International Conference of the IEEE Engineering in Medicine and Biology Society*, IEEE, New York, Vol. 3, 1988, pp. 1404–1405.

[90] Oshita, S., K. Nakakimura, and T. Sakabe, "Hypertension Control During Anaesthesia, Fuzzy Logic Regulation of Nicardipine Infusion," *IEEE Engineering in Medicine and Biology*, Vol. 13, No. 5, 1994, pp. 667–670.

[91] Boston, J. R., "Rule Based System for Interpretation of Evoked Potential Waveforms," *Images of the Twenty-First Century, Proceedings of the Annual International Conference of the IEEE Engineering in Medicine and Biology*, IEEE, New York, Vol. 6, 1989, pp. 1808–1809.

[92] Nayak, A., and R. J. Roy, "Anesthesia Control Using Midlatency Auditory Evoked Potentials," *IEEE Trans. on Biomedical Engineering*, Vol. 45, No. 4, 1998, pp. 409–421.

[93] Nahm, W., et al., "Concept for an Intelligent Anesthesia EEG Monitor," *Medical Informatics and the Internet in Medicine*, Vol. 24, No. 1, 1999, pp. 1–9.

[94] Linkens, D. A., and M. F. Abbod, "Supervisory Intelligent Control Using a Fuzzy Logic Hierarchy," *Transactions of the Institute of Measurement and Control*, Vol. 15, No. 3, 1993, pp. 112–132.

[95] Abbod, M. F., and D.A. Linkens, "Anesthesia Monitoring and Control Using Fuzzy Logic Fusion," *Journal of Biomedical Engineering—Applications, Basis & Communications*, Special Issue on Control Methods in Anesthesia, Taiwan, Vol. 10 No. 4, August 1998, pp. 225–235.

[96] Saridis, G. N., "Fuzzy Decision Making in Prosthetic Devices and Other Applications," *6th IFAC Congress*, Boston, MA, August 1975.

[97] Saridis, G. N., and H. E. Stephanou, "Fuzzy Decision Making in Prosthetic Devices," in M. M. Gupta, G. N. Saridis, and B. R. Gaines, (eds.), *Fuzzy Automata and Decision Processes*, New York: North Holland Publishers, 1997, pp. 387–402.

[98] Feng, W., and B. J. Andrews, "Adaptive Fuzzy Logic Controller for FES-Computer Simulation Study," *Proceedings of the 16th Annual International Conference of the IEEE Engineering in Medicine and Biology Society, Engineering Advances: New Opportunities for Biomedical Engineers*, IEEE, New York, Vol. 1, 1994, pp. 406–407.

[99] Davoodi, R., and B. J. Andrews, "FES Standing Up in Paraplegia: A Comparative Study of Fixed Parameter Controllers," *Proceedings of the 18th Annual International Conference of the IEEE Engineering in Medicine and Biology Society, Bridging Disciplines for Biomedicine*, IEEE, New York, Vol. 1, 1997, pp. 447–448.

[100] Davoodi, R., and B. J. Andrews, "A Model of FES Standing Up in Paraplegics Including Upper Limb Interaction and Closed-Form Dynamics," *Proceedings of the 18th Annual International Conference of the IEEE Engineering in Medicine and Biology Society, Bridging Disciplines for Biomedicine*, IEEE, New York, Vol. 2, 1997, pp. 579–580.

[101] Dorgan, S., "Modelling, Simulation and Control of Human Musculoskeletal Dynamics," *IEE Colloquium Simulation in Medicine*, London, December 1998, pp. 9/1–9/4.

[102] Karlik, B., M. O. Tokhi, and M. Alci, "A Fuzzy Clustering Neural Network Architecture for Multifunction Upper-Limb Prosthesis," *IEEE Trans. on Biomedical Engineering*, Vol. 50, No. 11, November 2003, pp. 1255–61.

[103] Kohler, F., et al., "Applications of a Simple System of Aid to Medical Decision-Making with Various Patients in Internal Medicine: SELF," *Rev. Med. Interne.*, Vol. 9, No. 2, 1988, pp. 196–207.

[104] Kopecky, D., et al., "Knowledge-Based Interpretation of Toxoplasmosis Serology Test Results Including Fuzzy Temporal Concepts: The ToxoNet System," *Medinfo*, Vol. 10, Pt. 1, 2001, pp. 484–488.

[105] Sarmento, A. D., "Revisiting Wavelet and Fuzzy-Based Denoising of Medical Images from Ultrasound Mammography," *Proceedings of the IEEE 30th Annual Northeast Bioengineering Conference*, IEEE, Cat. No. 04CH37524, 2004, pp. 51–52.

[106] Kolles, H., and U. A. Hubschen, "A Dermatologic Decision Support System that Processes Lexical Knowledge from Standard Textbooks," *Anal. Quant. Cytol. Histol.*, Vol. 16, 1994, pp. 219–225.

[107] Brotherton, T., et al., "Classifying Tissue and Structure in Echocardiograms," *IEEE Engineering in Medicine and Biology Magazine*, Vol. 13, No. 5, 1994, pp. 754–760.

[108] Acha, B., et al., "CAD Tool for Burn Diagnosis," *Information Processing in Medical Imaging*, Vol. 18, July 2003, pp. 294–305.

[109] Takeuchi, T., "Application of Fuzzy Inference to Personal Identification: Sex Determination from Teeth," *Nippon Hoigaku Zasshi*, Vol. 47, No. 3, 1993, pp. 239–249.

[110] Ito, T., "Application of Fuzzy Logic in Diagnostic Processes for the Propriety of Tooth Conservation," *Biosystems*, Vol. 41, No. 3, 1997, pp. 179–190.

[111] Sims-Williams, J. H., N. Mackin, and C. D. Stephens, "Lessons Learned from the Development of an Orthodontic Expert System," *Proceedings of the International Conference on*

Neural Networks and Expert Systems in Medicine and Healthcare, Plymouth, U.K., 1994, pp. 410–417.

[112] Akcam, M. O., and K. Takada, "Fuzzy Modeling for Selecting Headgear Types," *European Journal of Orthodontics*, Vol. 24, No. 1, February 2002, pp. 99–106.

[113] Prochazka, A., "The Fuzzy Logic of Visuomotor Control," *Can. J. Physiol. Pharmacol.*, Vol. 74, 1996, pp. 456–462.

[114] Losch, B., "Application of Fuzzy Sets to the Diagnosis of Glaucoma," *Proceedings of the 18th Annual International Conference of the IEEE Engineering in Medicine and Biology Society, Bridging Disciplines for Biomedicine*, IEEE, New York, Vol. 4, 1997, pp. 1550–1552.

[115] Serhatlioglu, S., et al., "A Neurofuzzy Classification System for the Effects of Diabetes Mellitus on Ophtalmic Artery," *Journal of Medical Systems*, Vol. 28, No. 2, April 2004, pp. 167–76.

[116] Nandy, D., "An Auditory Localization Model Based on High-Frequency Spectral Cues," *Ann. Biomed. Eng.*, Vol. 24, 1996, pp. 621–638.

[117] Debon, R., et al., "Fuzzy Fusion and Belief Updating: Application to Esophagus Wall Detection on Ultrasound Images," *Proceedings of the Third International Conference on Information Fusion*, Cat. No. 00EX438, Vol. 1, 2000, TUC5/17-23.

[118] Gorek, M., C. Hartung, and C. G. Stief, "Electromyograms in Erectile Dysfunction and Computer-Assisted Interpretation," *Biomed. Tech. (Berlin)*, Vol. 42, No. 3, 1997, pp. 48–54.

[119] Johnson, J. F., J. L. Schmalzel, and H. Arnhold, "A Device for Urologic Post-Surgical Monitoring," *Biomed. Sci. Instrum.*, Vol. 30, 1994, pp. 233–238.

[120] Montironi, R., et al., "A Typical Adenomatous Hyperplasia (Adenosis) of the Prostate: Development of a Bayesian Belief Network for Its Distinction from Well-Differentiated Adenocarcinoma," *Hum. Pathol.*, Vol. 27, No. 4, 1996, pp. 396–407.

[121] Scheipers, U., et al., "Ultrasonic Multifeature Tissue Characterization for Prostate Diagnostics," *Ultrasound in Medicine & Biology*, Vol. 29, No. 8, August 2003, pp. 1137–1149.

[122] Gitter, A., and V. Lin, "Automated Sensory Nerve Conduction Testing Using Fuzzy Logic," *American Journal of Physical Medicine & Rehabilitation*, Vol. 78, No. 5, September/October 1999, pp. 425–34.

[123] Bowman, F. D., R. Patel, and C. Lu, "Methods for Detecting Functional Classifications in Neuroimaging Data," *Human Brain Mapping*, Vol. 23, No. 2, October 2004, pp. 109–119.

[124] Tuor, U. I., et al., "Functional Magnetic Resonance Imaging in Rats Subjected to Intense Electrical and Noxious Chemical Stimulation of the Forepaw," *Pain*, Vol. 87, No. 3, Spetember 2000, pp. 315–324.

[125] Rocha, A. F., "Proposed Methodology for Knowledge Acquisition: A Study on Congenital Heart Disease Diagnosis," *Methods of Information in Medicine*, Vol. 29, 1990, pp. 30–40.

[126] Naranjo, C. A., et al., "Using Fuzzy Logic to Predict Response to Citalopram in Alcohol Dependence," *Clin. Pharmacol. Ther.*, Vol. 62, No. 2, 1997, pp. 209–224.

[127] Brasil, L. M., F. M. de Azevedo, and J. M. Barreto, "A Hybrid Expert System for the Diagnosis of Epileptic Crisis," *Artificial Intelligence in Medicine*, Vol. 21, No. 1–3, January/March 2001, pp. 227–233.

[128] Ellison, J. W., and D.W. Massaro, "Featural Evaluation, Integration, and Judgment of Facial Affect," *J. Exp. Psychol. Hum. Percept. Perform.*, Vol. 23, No. 1, 1997, pp. 213–226.

[129] Tan, A. H., and H. S. Soon, "Concept Hierarchy Memory Model: A Neural Architecture for Conceptual Knowledge Representation, Learning and Common Sense Reasoning," *Int. J. Neural Syst.*, Vol. 7, No. 3, 1996, pp. 305–319.

[130] Matt, G. E., et al., "Improving Self-Reports of Drug Use: Numeric Estimates as Fuzzy Sets," *Addiction*, Vol. 98, No. 9, September 2003, pp. 1239–1247.

[131] Kovacs, M., and J. Juranovics, "'Auctoritas' Psychiatric Expert System Shell," *Medinfo*, Vol. 8, No. 2, 1995, p. 997.

[132] Jablensky, A., and M. A. Woodbury, "Dementia Praecox and Manic-Depressive Insanity in 1908: A Grade of Membership Analysis of the Kraepelinian Dichotomy," *Eur. Arch. Psychiatry Clin. Neurosci.*, Vol. 245, No. 4–5, 1995, pp. 202–209.

[133] Marchais, P. and J. B. Grize, "Logic and Analogy in Psychiatry, Their 'Common Root'," *Ann. Med. Psychol. (Paris)*, Vol. 152, No. 2, 1994, pp. 85–94.

[134] Modai, I., et al., "Fuzzy Logic Detection of Medically Serious Suicide Attempt Records in Major Psychiatric Disorders," *Journal of Nervous & Mental Disease*, Vol. 192, No. 10, October 2004, pp. 708–710.

[135] Guiheneuc, P., et al., "Signal Processing in Electromyography: A Review," *Neurophysiol. Clin.*, Vol. 27, No. 6, 1997, pp. 445–470.

[136] al-Holou, N., and D. S. Joo, "Development of a Fuzzy Logic Based System to Monitor the Electrical Responses of Nerve Fiber," *Biomed. Sci. Instrum.*, Vol. 33, 1997, pp. 376–381.

[137] Molina, A., A. Urbaszek, and M. Schaldach, "R-Spike Detection for Rhythm Monitoring in Real Time with a Fuzzy Logic Detector," *Biomed. Tech.*, Vol. 42, 1997, pp. 57–58.

[138] Moser, E., M. Diemling, and R. Baumgartner, "Fuzzy Clustering of Gradient-Echo Functional MRI in the Human Visual Cortex, Part II: Quantification," *J. Magn. Reson. Imaging.*, Vol. 7, No. 6, 1997, pp. 1102–1108.

[139] Mutch, W. A., et al., "Cerebral Hypoxia During Cardiopulmonary Bypass: A Magnetic Resonance Imaging Study," *Ann. Thorac. Surg.*, Vol. 64, No. 3, 1997, pp. 695–701.

[140] Zouridakis, G., B. H. Jansen, and N. N. Boutros, "A Fuzzy Clustering Approach to EP Estimation," *IEEE Trans. on Biomedical Engineering*, Vol. 44, No. 8, 1997, pp. 673–680.

[141] Riddington, E. P., et al., "A Fuzzy Expert System for EEG Interpretation," *Proceedings of the International Conference on Neural Networks and Expert Systems in Medicine and Healthcare*, Plymouth, U.K., 1994, pp. 291–302.

[142] Moreno, L., et al., "Tools for Acquisition, Processing and Knowledge-Based Diagnostic of the Electroencephalogram and Visual Evoked Potentials, *Journal of Medical Systems*, Vol. 25, No. 3, June 2001, pp. 177–194.

[143] Jensen, O. N., et al., "Automation of Matrix-Assisted Laser Desorption/Ionization Mass Spectrometry Using Fuzzy Logic Feedback Control," *Anal. Chem.*, Vol. 69, No. 9, 1997, pp. 1706–1714.

[144] Lin, C. J., "A Fuzzy Adaptive Learning Control Network with On-Line Structure and Parameter Learning," *Int. J. Neural Syst.*, Vol. 7, No. 5, 1996, pp. 569–590.

[145] Wallner, F., "Detection of Rapid Eye Movement with Rapidly Adapting Neuronal Fuzzy Systems in Imprecise REM Syntax," *Biomed. Tech.*, Vol. 41, No. 4, 1996, pp. 84–90.

[146] Yin, H., et al., "Application of Adaptive Noise Cancellation with Neural-Network-Based Fuzzy Inference System for Visual Evoked Potentials Estimation," *Medical Engineering & Physics*, Vol. 26, No. 1, January 2004, pp. 87–92.

[147] Berger-Vachon, C., et al., "Analytic Importance of the Coding Features for the Discrimination of Vowels in the Cochlear Implant Signal," *Ann. Otol. Rhinol. Laryngol. Suppl.*, Vol. 166, 1995, pp. 351–353.

[148] Abou-Chadi, F. E., F. A. Ezzat, and A.A. Sif el-Din, "A Fuzzy Pattern Recognition Method to Classify Esophageal Motility Records," *Ann. Biomed. Eng.*, Vol. 22, No. 1, 1994, pp. 112–119.

[149] Cosic, D., and S. Loncaric, "Rule-Based Labeling of CT Head Images," *Artificial Intelligence in Medicine, Proceedings of the 6th Conference on Artificial Intelligence in Medicine Europe, AIME '97*, Springer-Verlag, Berlin, 1997, pp. 453–456.

[150] Schmitt, T., et al., "Quantitative Evaluation of Emission Computed Tomography Images Based on Fuzzy Segmentation," *Proceedings of the SPIE, The International Society for Optical Engineering*, Vol. 2847, 1996, pp. 438–444.

[151] Lo, T. F., et al., "Signal Analysis of Electromyography of Fasciculation by a Two-Pass System," *1995 IEEE Engineering in Medicine and Biology, 17th Annual Conference, and 21 Canadian Medical and Biological Engineering Conference*, IEEE, New York, Vol. 2, 1997, pp. 969–970.

[152] Ruben, M., et al., "Segmentation of Ventricular Angiographic Images Using Fuzzy Clustering," *1995 IEEE Engineering in Medicine and Biology 17th Annual Conference and 21 Canadian Medical and Biological Engineering Conference*, IEEE, New York, Vol. 1, 1997, pp. 405–406.

[153] Hui, H. W., C. L. Wei, and T. C. Chin, "Knowledge-Based Medical Image Registration," *Proceedings of the 18th Annual International Conference of the IEEE Engineering in Medicine and Biology Society, Bridging Disciplines for Biomedicine*, IEEE, New York, Vol. 3, 1997, pp. 1200–1201.

[154] Butt, A., et al., "Surface Matching of Multi-Modality Image Volumes by a Fuzzy Elastic Registration Technique," *Comput. Med. Imaging Graph.*, Vol. 22, No. 1, 1998, pp. 13–23.

[155] Shiomi, S., et al., "Diagnosis of Chronic Liver Disease from Liver Scintiscans by Fuzzy Reasoning," *J. Nucl. Med.*, Vol. 36, No. 4, 1995, pp. 593–598.

[156] Sadati, N., and S. Z. Mortazavi, "Fuzzy Modeling and Optimization of Planning in Radiotherapy," *Proceedings of the First Regional Conference, IEEE Engineering in Medicine and Biology Society and 14th Conference of the Biomedical Engineering Society of India*, New York, 1995, pp. 2/11–2/14.

[157] Waschek, T., et al., "Target Volume Definition for Three-Dimensional Radiotherapy of Cancer Patients with a Fuzzy Rule-Based System," *Proc. IEEE Int. Conf. on Fuzzy Systems*, 1996, pp. 1719–1725.

[158] Busch, M., and E. Duhmke, "Rule Based Expert System for Computer Aided Prediction of Radiation Damage of Organs," *CAR '97, Computer Assisted Radiology and Surgery, Proceedings of the 11th International Symposium and Exhibition*, Elsevier, Amsterdam, the Netherlands, 1998, pp. 580–585.

[159] Leszczynski, K., et al., "Application of a Fuzzy Pattern Classifier to Decision Making in Portal Verification of Radiotherapy," *Physics in Medicine & Biology*, Vol. 44, No. 1, January 1999, pp. 253–269.

[160] Yan, H., et al., "Fuzzy Logic Guided Inverse Treatment Planning," *Medical Physics*, Vol. 30, No. 10, October 2003, pp. 2675–2685.

[161] Menhardt, W., et al., "Knowledge Based Interpretation of Cranial MR Images," *Proceedings of the IASTED International Symposium, Computers and Advanced Technology in Medicine, Healthcare and Bioengineering*, ACTA Press, Anaheim, CA, 1990, pp. 72–79.

[162] Serpico, S. B., et al., "TI: Fuzzy-Reasoning Approach to Understanding of 2D NMR Images," *Proceedings of the Ninth Annual Conference of the IEEE Engineering in Medicine and Biology Society*, IEEE, New York, Cat. No. 87CH2513-0, Vol. 4 1987, pp. 1980–1981.

[163] Clark, M. C., et al., "MRI Segmentation Using Fuzzy Clustering Techniques," *IEEE Engineering in Medicine and Biology Magazine*, Vol. 13, No. 5, 1994, pp. 730–742.

[164] Colin, A., and J. Y. Boire, "MRI-SPECT Image Fusion for the Synthesis of High Resolution Functional Images: A Prospective Study [Using Possibility Theory]," *Proceedings of the 19th Annual International Conference of the IEEE Engineering in Medicine and Biology Society, Magnificent Milestones and Emerging Opportunities in Medical Engineering*, IEEE, Piscataway, NJ, Vol. 2, 1997, pp. 499–501.

[165] Hillman, G. R., et al., "A Fuzzy Logic Approach to Identifying Brain Structures in MRI Using Expert Anatomic Knowledge," *Computers & Biomedical Research*, Vol. 32, No. 6, December 1999, pp. 503–516.

[166] Algorri, M. E., and F. Flores-Mangas, "Classification of Anatomical Structures in MR Brain Images Using Fuzzy Parameters," *IEEE Trans. on Biomedical Engineering*, Vol. 51, No. 9, September 2004, pp. 1599–608.

[167] Bezdek, J. C., L. O. Hall, and L. P. Clarke, "Review of MR Image Segmentation Techniques Using Pattern Recognition," *Med. Phys.*, Vol. 20, No. 4, 1993, pp. 1033–1048.

[168] Bezdek, J. C., et al., "Medical Image Analysis with Fuzzy Models," *Stat. Methods Med. Res.*, Vol. 6, No. 3, 1997, pp. 191–214.

[169] Axer, H., G. Berks, and D. G. Keyserlingk, "Visualization of Nerve Fiber Orientation in Gross Histological Sections of the Human Brain," *Microscopy Research & Technique*, Vol. 51, No. 5, December 1, 2000, pp. 481–492.

[170] Boegl, K., et al., "New Approaches to Computer-Assisted Diagnosis of Rheumatologic Diseases," *Radiologe.*, Vol. 35, No. 9, 1995, pp. 604–610.

[171] Carmody, D. P., C. F. Nodine, and H. L. Kundel, "Global and Segmented Search for Lung Nodules of Different Edge Gradients," *Invest. Radiol.*, Vol. 15, No. 3, 1980, pp. 224–33.

[172] Abbott, P. H., and M. J. Adams, "An Expert System for Interpretation of X-Ray Fluorescence Spectra," *Laboratory Automation and Information Management*, Vol. 31, No. 3, 1996, pp. 211–220.

[173] Guez, A., and I. Nevo, "Neural Networks and Fuzzy Logic in Clinical Laboratory Computing with Application to Integrated Monitoring," *Clin. Chim. Acta.*, Vol. 248, No. 1, 1996, pp. 73–90.

[174] Nishimura, H., et al., "Fuzzy Realization in Clinical Test Database System," *Int. J. Biomed. Comput.*, Vol. 28, No. 4, 1991, pp. 289–296.

[175] Amaya Cruz, G. P., and G. Beliakov, "Approximate Reasoning and Interpretation of Laboratory Tests in Medical Diagnostics," *Intelligent Engineering Systems Through Artificial Neural Networks, ASME*, New York, Vol. 4, 1994, pp. 773–778.

[176] Amaya Cruz, G. P., and G. Beliakov, "Approximate Reasoning and Interpretation of Laboratory Tests in Medical Diagnostics," *Cybernetics and Systems*, Vol. 26, No. 6, 1995, pp. 713–729.

[177] Sjahputera, O., and J. M. Keller, "Evolution of a Fuzzy Rule-Based System for Automatic Chromosome Recognition," *FUZZ IEEE '99, 1999 IEEE International Fuzzy Systems Conference Proceedings*, Cat. No. 99CH36315, Vol. 1, 1999, pp. 129–134.

[178] Sadegh-Zadeh, K., "Fuzzy Genomes," *Artificial Intelligence in Medicine*, Vol. 18, No. 1, January 2000, pp. 1–28.

[179] Woolf, P. J., and Y. A. Wang, "A Fuzzy Logic Approach to Analyzing Gene Expression Data," *Physiological Genomics*, Vol. 3, No. 1, June 29, 2000, pp. 9–15.

[180] Nauck, D., and R. Kruse, "Obtaining Interpretable Fuzzy Classification Rules from Medical Data," *Artificial Intelligence in Medicine*, Vol. 16, No. 2, June 1999, pp. 149–169.

[181] Delgado, M., et al., "Mining Association Rules with Improved Semantics in Medical Databases," *Artificial Intelligence in Medicine*, Vol. 21, No. 1–3, January/March 2001, pp. 241–245.

[182] Meesad, P., and G. G. Yen, "Combined Numerical and Linguistic Knowledge Representation and Its Application to Medical Diagnosis," *Proceedings of the SPIE, The International Society for Optical Engineering*, Vol. 4733, 2002, pp. 98–109.

[183] Straszecka, E., "Medical Knowledge Representation in Terms of IF-THEN Rules and the Dempster-Shafer Theory," *Artificial Intelligence and Soft Computing ICAISC 2004, Proceedings Lecture Notes in Artificial Intelligence*, Vol. 3070, 2004, pp. 1056–1061.

[184] Sharkov, G. I., and D. S. Dimitrov, "Knowledge Representation and Reasoning in Expert Systems for Biophysics," *Methodologies for Intelligent Systems, Proceedings of the Third International Symposium*, New York: North Holland, 1988, pp. 150–159.

[185] Tseng, H. C., and D. W. Teo, "Medical Expert System with Elastic Fuzzy Logic," *Proceedings of the Third IEEE Conference on Fuzzy Systems, IEEE World Congress on Computational Intelligence*, IEEE, New York, Vol. 3, 1994, pp. 2067–2071.

[186] Hudson, D. L., and M. E. Cohen, "Fuzzy Logic in Medical Expert Systems," *IEEE Engineering in Medicine and Biology Magazine*, Vol. 13, No. 5, 1994, pp. 693–698.

[187] Hurdle, J. F., "Lightweight Fuzzy Processes in Clinical Computing," *Artificial Intelligence in Medicine*, Vol. 11, No. 1, 1997, pp. 55–73.

[188] Derrick, J. L., C. L. Thompson, and T. G. Short, "The Application of a Modified Proportional-Derivative Control Algorithm to Arterial Pressure Alarms in Anesthesiology," *J. Clin. Monit. Comput.*, Vol. 14, No. 1, 1998, pp. 41–47.

[189] Linkens, D. A., and M. F. Abbod, "Medical Applications of Fuzzy Logic Control," in H. B. Verbruggen, and R. Babuska, (eds.), *Fuzzy Logic Control: Advances and Applications*, New York: World Scientific Series in Robotics and Intelligent Systems, Vol. 23, 1999.

[190] Nebot, A., F. E. Cellier, and D. A. Linkens, "Synthesis of an Anesthetic Agent Administration System Using Fuzzy Inductive Reasoning," *Artificial Intelligence in Medicine*, Vol. 8, No. 2, 1996, pp. 147–166.

[191] Nebot, A., F. E. Cellier, and M. Vallverdu, "Mixed Quantitative/Qualitative Modeling and Simulation of the Cardiovascular System," *Comput. Methods Programs Biomed.*, Vol. 55, No. 2, 1998, pp. 127–155.

[192] Bellazzi, R., et al., "Adaptive Controllers for Intelligent Monitoring," *Artificial Intelligence in Medicine*, Vol. 7, No. 6, 1995, pp. 515–540.

[193] Brown, M., and C. J. Harris, "A Perspective and Critique of Adaptive Neurofuzzy Systems Used for Modeling and Control Applications," *Int. J. Neural Syst.*, Vol. 6, No. 2, 1995, pp. 197–220.

[194] Lee, M., et al., "Total Artificial Heart Using Neural and Fuzzy Controller," *Artificial Organs*, Vol. 20, No. 11, 1996, pp. 1220–1226.

[195] Linkens, D. A., and S. B. Hasnain, "Self-Organizing Fuzzy Logic Control and Its Application to Muscle Relaxation Anesthesia," *IEE Proc. PtD, Control Theory and Applications*, Vol. 138, 1991, pp. 274–284.

[196] Linkens, D. A., et al., "Hierarchical Fuzzy-Based Support System for Anaesthesiology Monitoring and Control," *IEEE International Conference on Fuzzy Systems*, New Orleans, LA, September 8–11, 1996.

[197] Linkens, D. A., et al., "Closed-Loop Control of Anesthesia Using Fuzzy Logic," in P. S. Szczepaniak, et al., (eds.), *Fuzzy Systems in Medicine*, New York: Springer-Verlag, 1999.

[198] Geva, A. B., and D. H. Kerem, "Forecasting Generalized Epileptic Seizures from the EEG Signal by Wavelet Analysis and Dynamic Unsupervised Fuzzy Clustering," *IEEE Trans. on Biomedical Engineering*, Vol. 45, No. 10, 1998, pp. 1205–1216.

[199] Linkens, D. A., M. F. Abbod, and J. K. Backory, "Closed-Loop Control of Anaesthesia Using a Wavelet-Based Fuzzy Logic System," *EUFIT'97*, Aachen, Germany, September 8–12, 1997.

[200] Held, C. M., and R. J. Roy, "Multiple Drug Hemodynamic Control by Means of a Supervisory-Fuzzy Rule-Based Adaptive Control System: Validation on a Model," *IEEE Trans. on Biomedical Engineering*, Vol. 42, No. 4, 1995, pp. 371–385.

[201] Tolias, Y. A., and S. M. Panas, "A Fuzzy Vessel Tracking Algorithm for Retinal Images Based on Fuzzy Clustering," *IEEE Trans. on Medical Imaging*, Vol. 17, No. 2, 1998, pp. 263–273.

[202] Phillips, W. E., R. P. Velthuizen, and S. Phuphanich, "Application of Fuzzy C-Means Segmentation Technique for Tissue Differentiation in MR Images of a Hemorrhagic Glioblastoma Multiforme," *Magn. Reson. Imaging.*, Vol. 13, No. 2, 1995, pp. 277–290.

[203] Bartels, P. H., D. Thompson, and J. E. Weber, "Expert Systems in Histopathology. V. DS Theory, Certainty Factors and Possibility Theory," *Anal. Quant. Cytol. Histol.*, Vol. 14, No. 3, 1992, pp. 165–174.

[204] van Ginneken, A. M., and A.W. Smeulders, "Reasoning in Uncertainties: An Analysis of Five Strategies and Their Suitability in Pathology," *Anal. Quant. Cytol. Histol.*, Vol. 13, No. 2, 1991, pp. 93–109.

[205] Schafer, A. T., and R. Lemke, "Use of Fuzzy Logic and Fractal Geometry in Forensic Medicine," *Beitr. Gerichtl. Med.*, Vol. 50, 1992, pp. 31–36.

[206] Jin, S., K. Ye, and K. Shimizu, "Efficient Fuzzy Control Strategies for the Application of pH-Stat to Fed-Batch Cultivation of Genetically Engineered Escherichia Coli," *J. Chem. Technol. Biotechnol.*, Vol. 61, No. 3, 1994, pp. 273–281.

[207] Black, F. L., "Reasons for Failure of Genetic Classifications of South Amerind Populations," *Hum. Biol.*, Vol. 63, No. 6, 1991, pp. 763–774.

[208] Qi, H., D. J. Tyler, and D. M. Durand, "Neurofuzzy Adaptive Controlling of Selective Stimulation for FES: A Case Study," *IEEE Trans. on Rehabilitation Engineering*, Vol. 7, No. 2, June 1999, pp. 183–192.

[209] Kiryu, T., et al., "Providing Appropriate Exercise Levels for the Elderly," *IEEE Engineering in Medicine & Biology Magazine*, Vol. 20, No. 6, November/December 2001, pp. 116–124.

[210] Marzani, F., E. Calais, and L. A. Legrand, "3-D Marker-Free System for the Analysis of Movement Disabilities: An Application to the Legs," *IEEE Trans. on Information Technology in Biomedicine*, Vol. 5, No. 1, March 2001, pp. 18–26.

[211] Manton, K. G., et al., "Fuzzy Set Analyses of Genetic Determinants of Health and Disability Status," *Statistical Methods in Medical Research*, Vol. 13, No. 5, October 2004, pp. 395–408.

[212] Hardalac, F., et al., "The Examination of the Effects of Obesity on a Number of Arteries and Body Mass Index by Using Expert Systems," *Journal of Medical Systems*, Vol. 28, No. 2, April 2004, pp. 129–142.

[213] Moriguchi, I., H. Hirano, and S. Hirono, "Prediction of the Rodent Carcinogenicity of Organic Compounds from Their Chemical Structures Using the FALS Method," *Environ. Health Perspect.*, Vol. 104, No. 5, 1996, pp. 1051–1058.

[214] Lee, B., et al., "Incorporating Qualitative Knowledge in Enzyme Kinetic Models Using Fuzzy Logic," *Biotechnology & Bioengineering*, Vol. 62, No. 6, March 20, 1999, pp. 722–729.

[215] Bellazzi, R., et al., "A Hybrid Input-Output Approach to Model Metabolic Systems: An Application to Intracellular Thiamine Kinetics," *Journal of Biomedical Informatics*, Vol. 34, No. 4, August 2001, pp. 221–248.

[216] Lee, Y. W., et al., "Dose-Response Assessment by a Fuzzy Linear-Regression Method," *Water Science & Technology*, Vol. 43, No. 2, 2001, pp. 133–140.

[217] Belic, A., et al., "Pathways of Paracetamol Absorption from Layered Excipient Suppositories: Artificial Intelligence Approach," *European Journal of Drug Metabolism & Pharmacokinetics*, Vol. 28, No. 1, January/March 2003, pp. 31–40.

[218] Gueorguieva, I. I., I. A. Nestorov, and M. Rowland, "Fuzzy Simulation of Pharmacokinetic Models: Case Study of Whole Body Physiologically Based Model of Diazepam," *Journal of Pharmacokinetics & Pharmacodynamics*, Vol. 31, No. 3, June 2004, pp. 185–213.

[219] Chen, W. L., et al., "Fuzzy Match and Floating Threshold Strategy for Expert System in Traditional Chinese Medicine," *Fuzzy Sets and Systems*, Vol. 17, No. 2, 1985, pp. 143–151.

[220] Kitamura, K., N. Aikawa, and H. Usui, "Development of Recognizing Mental Status System Using Vital Sign," *IEEE EMBS Asian Pacific Conference on Biomedical Engineering 2003*, IEEE Cat. No. 03EX711, 2003, pp. 88–89.

[221] Weidenbach, M., et al., "Intelligent Training System Integrated in an Echocardiography Simulator," *Computers in Biology & Medicine*, Vol. 34, No. 5, July 2004, pp. 407–425.

[222] Wang, R. H., and Z. Q. Sun, "Evaluation of Self-Taught Ability of Nursing Administrators with Fuzzy Medicine," *Chung Hua Hu Li Tsa Chih*, Vol. 31, No. 7, 1996, pp. 395–396.

[223] Im, E. O., and W. Chee, "Fuzzy Logic and Nursing," *Nursing Philosophy*, Vol. 4, No. 1, April 2003, pp. 53–60.

[224] Rolfe, G., "Science, Abduction and the Fuzzy Nurse: An Exploration of Expertise," *J. Adv. Nurs.*, Vol. 25, No. 5, 1997, pp. 1070–1075.

[225] Cave, P., "Fuzzy Thinking," *J. Adv. Nurs.*, Vol. 28, No. 2, 1998, pp. 274–279.

[226] Fei, L. P., "Using Fuzzy Mathematics to Assess Nursing Quality," *Chung Hua Hu Li Tsa Chih*, Vol. 28, No. 3, 1993, pp. 139–141.

[227] Genaidy, A. M., et al., "What Is 'Heavy'?" *Ergonomics*, Vol. 41, No. 4, 1998, pp. 420–432.

[228] Falconer, L., and H. Hoel, "Occupational Safety and Health: A Method to Test the Collection of 'Grey Data' by Line Managers," *Occup. Med. (Oxf.)*, Vol. 47, No. 2, 1997, pp. 81–89.

[229] Lave, L. B., "Health and Safety Risk Analyses: Information for Better Decisions," *Science*, Vol. 236, No. 4799, 1987, pp. 291–295.

[230] Zweifel, P., "Fuzzy Measurement of Output. A Problem of Evaluation for Health Policy," *Soc. Sci. Med.*, Vol. 16, No. 13, 1982, pp. 1291–1300.

[231] Yamamoto, K., et al., "Necessity to Improve Common Understanding About the Security Issues Among Hospitals in Japan and Some Feasible Approaches," *Int. J. Biomed. Comput*, Vol. 35, 1994, pp. 205–212.

[232] Kwankam, S. Y., D. A. Asoh, and S. F. Boyom, "Fault Diagnosis in an Expert System for Health Services Management in the Tropics," *Methods Inf. Med*, Vol. 36, No. 2, 1997, pp. 134–138.

[233] Sekita, Y., and Y. Tabata, "A Health Status Index Model Using a Fuzzy Approach," *European Journal of Operational Research*, Vol. 3, No. 1, 1979, pp. 40–49.

[234] Huang, X. M., and A. Thompson, "Setting Up Waiting-Time Targets to Satisfy Patients: An Application of Mathematical Programming," *IMA J. Math Appl. Med. Biol.*, Vol. 12, No. 304, 1995, pp. 235–247.

[235] Massad, E., M. N. Burattini, and N. R. Ortega, "Fuzzy Logic and Measles Vaccination: Designing a Control Strategy," *International Journal of Epidemiology*, Vol. 28, No. 3, June 1999, pp. 550–557.

[236] Gedrich, K., et al., "How Optimal Are Computer-Calculated Optimal Diets?" *European Journal of Clinical Nutrition*, Vol. 53, No. 4, April 1999, pp. 309–318.

[237] Cundell, D. R., et al., "Generation of an Intelligent Medical System, Using a Real Database to Diagnose Bacterial Infection in Hospitalized Patients," *International Journal of Medical Informatics*, Vol. 63, No. 1–2, September 2001, pp. 31–40.

[238] Supekar, K., et al., "Fuzzy Rule-Based Framework for Medical Record Validation," *Intelligent Data Engineering and Automated Learning IDEAL 2002, Third International Conference Lecture Notes in Computer Science*, Vol. 2412, 2002, pp. 447–453.

[239] Perris, T., and A. W. Labib, "An Intelligent System for Prioritization of Organ Transplant Patient Waiting Lists Using Fuzzy Logic," *Journal of the Operational Research Society*, Vol. 55, No. 2, February 2004, pp. 103–115.

[240] Guo, R. J., et al., "Fuzzy Set Model and Computerized Diagnosis System in Traditional Chinese Medicine," *Chinese Journal of Computers*, Vol. 4, No. 4, 1981, pp. 260–266.

[241] Chen, G. H., "Use of Fuzzy Mathematics to Evaluate the Teaching of Students in the Clinical Setting," *Chung Hua Hu Li Tsa Chih*, Vol. 31, No. 11, 1996, pp. 658–660.

[242] Cai, J., and H. Guo, "The Structure and Design of a Fuzzy Expert System," *Mini Micro Systems*, Vol. 10, No. 2, 1989, pp. 7–11.

[243] Zhu, Z., et al., "Transformation from Tristimulus RGB to Triattribute HVC in Image Systems," *Journal of Tsinghua University*, Vol. 33, No. 4, 1993, pp. 27–34.

[244] Roan, Y. T., J. R. Horng, and T. C. Hsiao, "The Design of Fuzzy Chinese Drugs Decocting Machine," *Soft Computing in Intelligent Systems and Information Processing, Proceedings of the 1996 Asian Fuzzy Systems Symposium*, IEEE, New York, 1996, pp. 176–181.

[245] Hoang, P. N., "Approximate Reasoning for Oriental Traditional Medical Expert Systems," *IEEE International Conference on Systems, Man, and Cybernetics, Computational Cybernetics and Simulation*, IEEE, New York, Vol. 4, 1997, pp. 3084–3089.

[246] Hsing, L. L., et al., "Fuzzy Theory in Traditional Chinese Pulse Diagnosis," *Proceedings of 1993 International Joint Conference on Neural Networks*, IEEE, New York, Vol. 1, 1993, pp. 774–777.

[247] Kandal, A., and R. R. Yager, "A 1979 Bibliography on Fuzzy Sets, Their Applications, and Related Topics," in M. M. Gupta, R. K. Ragade, and R. R. Yager, (eds.), *Advances in Fuzzy Set Theory and Applications*, Amsterdam, the Netherlands: North Holland, 1979.

[248] Maiers, J. E., "Fuzzy Set Theory and Medicine: The First Twenty Years and Beyond," *Proceedings of the Ninth Annual Symposium on Computer Applications in Medical Care*, IEEE, Washington, D.C., 1985, pp. 325–329.

[249] Dubois, D., J. Lang, and H. Prade, "Fuzzy Sets in Approximate Reasoning, Part 2: Logical Approaches," *Fuzzy Sets and Systems*, Vol. 40, 1991, pp. 203–244.

[250] Dubois, D., and H. Prade, *Possibility Theory: An Approach to Computerized Processing of Information*, New York: Plenum Press, 1988.

[251] Dubois, D., and H. Prade, "Representation and Combination of Uncertainty with Belief Function and Possibility Measures," *Comput. Intell.*, Vol. 4, 1988, pp. 244–264.

[252] Dubois, D., and H. Prade, "Weighted Fuzzy Pattern Matching," *Fuzzy Sets and Systems*, Vol. 28, 1988c, pp. 313–331.

[253] Dubois, D., and H. Prade, "What Are Fuzzy Rules and How to Use Them," *Fuzzy Sets and Systems*, Vol. 84, 1996, pp. 169–185.

[254] Dubois, D., and H. Prade, "An Introduction to Fuzzy Systems," *Clin. Chim. Acta.*, Vol. 270, No. 1, 1998, pp. 1–29.

[255] Ben-Ferhat, S., D. Dubois, and H. Prade, "Practical Handling of Exception-Tainted Rules and Independence Information in Possibilistic Logic," *Applied Intelligence*, Vol. 9, 1998, pp. 101–127.

[256] Burdea, G. C., "Virtual Reality and Robotics in Medicine," *Proceedings of the 5th IEEE International Workshop on Robot and Human Communication RO-MAN'96*, Tsukuba, IEEE, New York, 1996, pp. 16–25.

[257] Steimann, F., "Fuzzy Set Theory in Medicine," *Artificial Intelligence in Medicine*, Vol. 11, 1997, pp. 1–7.

[258] Moller, D. P. F., "Fuzzy Logic in Medicine," *Fourth European Congress on Intelligent Techniques and Soft Computing Proceedings, EUFIT '96*, Verlag Mainz, Aachen, Germany, Vol. 3, 1996, pp. 2036–2045.

[259] Zimmermann, H. J., *Fuzzy Set Theory and Its Applications*, Boston, MA: Kluwer-Nijhoff Publishing, 1985.

[260] Cohen, M. E., and D. L. Hudson, "Comparative Approaches to Medical Reasoning," *Advances in Fuzzy Systems: Applications and Theory,* New York: World Scientific Publishing, Vol. 3, 1995.

[261] Teodorescu, H. N., A. Kandel, and L. C. Jain, *Fuzzy and Neuro-Fuzzy Systems in Medicine*, Boca Raton, FL: CRC Press, 1999.

CHAPTER 3

Feedback Control of Muscle Relaxation and Unconsciousness Using Predictive Control

Anesthesia is that part of medicine that ensures that the patient's body remains insensitive to pain and other stimuli during surgical operations. It includes muscle relaxation (paralysis), unconsciousness, and analgesia (pain relief). Muscle relaxation is easily measured via evoked electromyogram (EMG) responses, but depth of anesthesia (unconsciousness) is more difficult to quantify accurately. In this chapter, the pharmacokinetics and pharmacodynamics "compartmental" modeling technique applied to anesthesia is reviewed, and simulation and real-time results relating to muscle relaxation and depth of unconsciousness monitoring and control will be presented. This chapter will also highlight the problems associated with controlling systems that are nonlinear, mathematically ill defined, and characterized by uncertainties, and thus the motivation for looking for novel ways of monitoring unconsciousness more effectively.

3.1 Introduction

The major roles of anesthesia that are the concern of a clinical anesthetist are those of drug-induced unconsciousness, muscle relaxation, and analgesia. The first two roles are concentrated in the operating theater, whereas the third role is mainly concerned with postoperative conditions. Each of these roles has been researched in recent years for the possibility of automated drug-infusion via feedback strategies.

The question of measurement is a primary matter in each of the three areas of anesthesia. The measurement of pain is the hardest of all, since it is heavily subjective. However, some work has been done in this field, attempting to estimate wound healing and its associated pain level using Kalman filtering [1] and control. Although analgesia will not be the debated subject of this chapter, Chapter 6 will consider its links to unconsciousness.

The measurement of muscle relaxation (or drug-induced paralysis) is considerably easier. A common approach is the monitoring of evoked EMG signals produced at the hand via stimulation above the wrist. Supramaximal electrical pulse stimulation is applied, typically every 20 seconds. This stimulation ensures that all the nerve fibers are recruited, while suitable processing of the resultant EMG provides an analog signal inversely proportional to the level of relaxation. The signal conditioning usually entails gating, rectifying, and integrating. Commercial instruments are now available that employ this principle, and this is the measurement basis for the work described in this chapter. Early work on feedback control of drug infusion for muscle relaxation was performed on sheep [2]. The feedback controller was based on a proportional integral derivative (PID) algorithm, as was the first set of human clinical trials performed by [3], which was followed by [4], which used a proportional integral (PI) strategy for atracurium.

In contrast, depth of anesthesia is more difficult to quantify accurately. It is, in fact, agreed that there is no absolute standard for the clinical state of anesthesia against which new methods designed to measure depth of anesthesia can be calibrated [5]. Thus, one approach has been to merge a number of clinical signs and online monitored data to produce an expert system advisor for the anesthetist. This system, called RESAC, has been developed and validated in a series of clinical trials [6]. In spite of the multisensor nature of the above approach, it appears that during the majority of operating periods when no unusual emergency conditions occur, a good indication of unconsciousness can be obtained from a single online monitored variable. Thus, the use of arterial blood pressure, monitored via an inflatable cuff using a DINAMAP instrument, has been investigated for feedback control with simple PI strategies [7]. In this case the control actuation was via a stepper motor driving the dial on a gas vaporizer. This concept forms the basis for the modeling and control aspects of unconsciousness in the following work. In particular, we have used the drug isoflurane in these studies. This is an inhalational drug commonly used in modern surgery.

One problem with feedback control in biomedicine is that there are enormous patient-to-patient variations in dynamic model parameters. This is compounded by large time-varying parameters for an individual patient during the course of an operation. This makes it difficult to design a fixed-parameter PID controller that will be suitable in all cases. This has led to the need to investigate self-adaptive control strategies, and also self-organizing controllers, although, in order to facilitate the design of such advanced controllers, it is necessary to have a good mathematical model of the process.

Common to much of the work on self-tuning adaptive control is the theme of predictive control, and this is the background to the present chapter and that of Chapter 5, which investigates generalized predictive control [8] for online muscle relaxant and anesthetic drug administration. GPC has advantages over other forms of self-tuning control in that it is robust against variable and unknown time delay, overparameterization of system models, and has good disturbance rejection properties. It does, however, have a number of design “knobs” which must be adjusted carefully to suit the particular application.

The following sections will describe the first successful application of this control strategy (which has received much acclaim from academia[1]) to muscle relaxation therapy, the derivation of the first multivariable anesthesia model, and the first successful application of predictive control to simultaneous adaptive control of muscle relaxation and unconsciousness in the operating theater during surgery.

1. This in contrast to another predictive control algorithm, known as Dynamic Matrix Control (DMC), which has rather received much acclaim from industry.

3.2 The Muscle Relaxation Process and Its Physiological Background

In the chain of events that starts with the stimulation of a motor nerve and ends with the contraction of a muscle, the most crucial link is the synapse between the nerve and the muscle—the neuromuscular junction. In other words, the synaptic gap is the site that witnesses the activity leading to a muscle contraction. Therefore, the process of neuromuscular transmission can only be blocked if relaxant drugs gain access to the synaptic cleft and break the above-described chain of events, which are summarized in the diagram of Figure 3.1. Depending on their mechanism of action, muscle relaxant agents fall into two categories: depolarizing and nondepolarizing drugs. Depolarizing drugs such as suxamethonium and decamethonium are believed to act by producing a continuous depolarization of the postsynaptic membrane, rendering it unresponsive to the chemical substance Acetylcholine (Ach), and at their first application a voluntary muscle contracts; but unlike Ach, these agents are not destroyed by cholinesterase and the depolarization is maintained. Nondepolarizing agents, however, compete with Ach for the cholinoreceptors. As a result, when Ach reacts with these drugs, it fails to cause sufficient sodium pores to

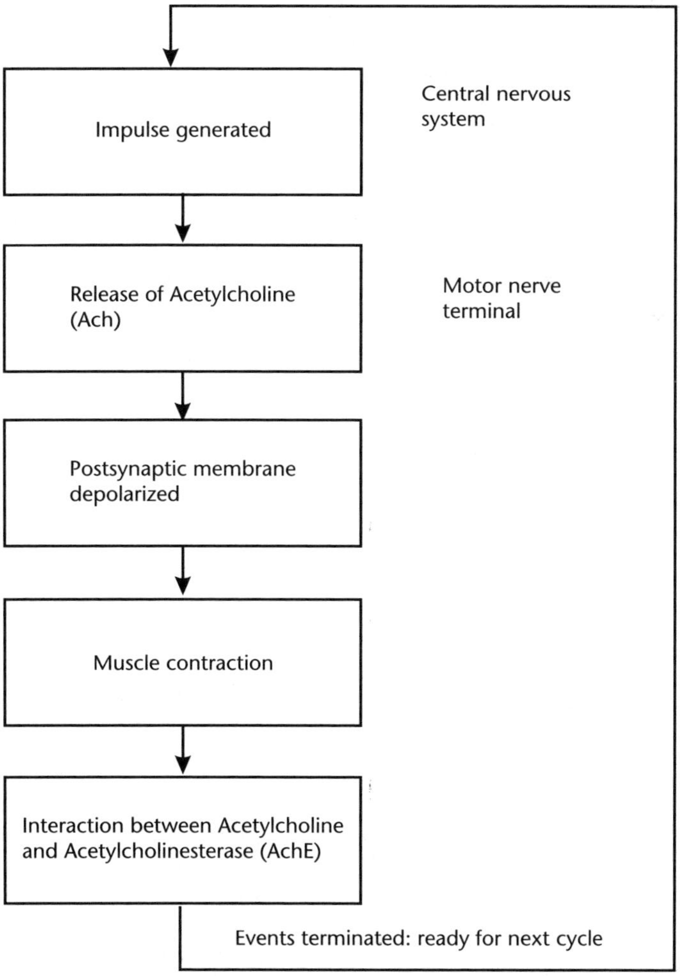

Figure 3.1 The chain of events leading to muscle contraction.

open to allow threshold depolarization to take place. These agents do not induce muscle pain following their administration, and because of this they are preferred to the other category. Atracurium, a nondepolarizing fast acting agent that has gained popularity over pancuronium and d-tubocurarine, is the subject of this research study and its associated mathematical model is presented next.

3.3 Mathematical Modeling of a Muscle Relaxant—Atracurium

In order to identify the muscle relaxation process associated with drugs, pharmacological modeling is commonly used to describe the metabolism of such drugs. As will also be highlighted in Chapter 4, pharmacological modeling comprises two main categories: pharmacokinetics and pharmacodynamics. Pharmacokinetics studies the relationship that exists between drug dose and drug concentration in the blood plasma as well as other parts of the body. Interpretation of this relationship can be given a mathematical meaning via the concept of compartmental models. Using this concept, the body consists of several compartments each representing one part of the body that involves the drug metabolism. Pharmacodynamics is concerned with the drug concentration in the blood and the effect produced.

3.3.1 Pharmacokinetics

It has been shown that after a drug injection, the plasma concentration of atracurium declines rapidly into two exponential phases corresponding to distribution and elimination [9]. Therefore, a conventional two-compartment model is used by adding an elimination path from the peripheral compartment, obeying the so-called Hofmann elimination [9, 10].

Referring to Figure 3.2, if x_i is the drug concentration at time t, $\dot{x}_i$ is its rate of change, k_{ij} is the rate of drug transfer from compartment i to compartment j, and u is the drug input, then one can write the following:

$$\begin{cases} \dot{x}_1 = -(k_{10} + k_{12})x_1 + k_{21}x_2 + u \\ \dot{x}_2 = k_{12}x_1 - (k_{20} + k_{21})x_2 \end{cases} \tag{3.1}$$

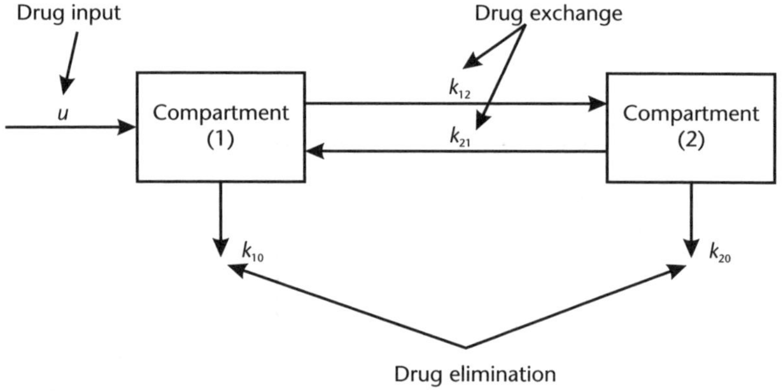

Figure 3.2 A two-compartment model for atracurium with an additional elimination path (k_{20}).

Using Laplace transforms, (3.1) can be rewritten as follows:

$$\begin{cases} sX_1 = -(k_{10} + k_{12})X_1 + k_{21}X_2 + U \\ sX_2 = k_{12}X_1 - (k_{20} + k_{21})X_2 \end{cases} \tag{3.2}$$

Hence,

$$\frac{X_1(s)}{U(s)} = \frac{s + k_{20} + k_{21}}{(s + k_{10} + k_{12})(s + k_{20} + k_{21}) - k_{12}k_{21}} \tag{3.3}$$

Using the following experimental data, which was published by Weatherley et al. [10],

$$\begin{cases} k_{12} + k_{10} = 0.26\,\text{min}^{-1} \\ k_{21} + k_{20} = 0.094\,\text{min}^{-1} \\ k_{12}\,xk_{21} = 0.015\,\text{min}^{-1} \end{cases}$$

(3.3) becomes

$$\frac{X_1(s)}{U(s)} = \frac{9.94(s+1)}{(1+3.08s)(1+34.42s)} \tag{3.4}$$

Equation (3.4) describes the pharmacokinetics of the muscle relaxation process relating to the drug atracurium.

3.3.2 Pharmacodynamics

It has been reported that for many drugs, including anesthetics, their time-course effect does not parallel their time course in the central compartment (see Figure 3.2). This phenomenon is usually referred to as *kinetic-dynamic dissociation*, and the necessary modeling extension is commonly called *effect-compartment modeling*. Effect-compartment modeling consists of including an additional compartment in the classical two-compartment structure. This so-called effect compartment is assumed to receive a vanishingly small mass of drug (so as not to affect the time course of drug disposition in the rest of the model) at a rate directly proportional to the central compartment drug concentration (k_{1E}<<) [11]. The drug exits this effect compartment in accordance with another rate constant k_{E0} (see Figure 3.3).

In this fictitious compartment (E), the drug concentration change is governed by the following equation:

$$\dot{x}_E = k_{1E}x_1 - k_{E0}x_E \tag{3.5}$$

Using Laplace transforms yields

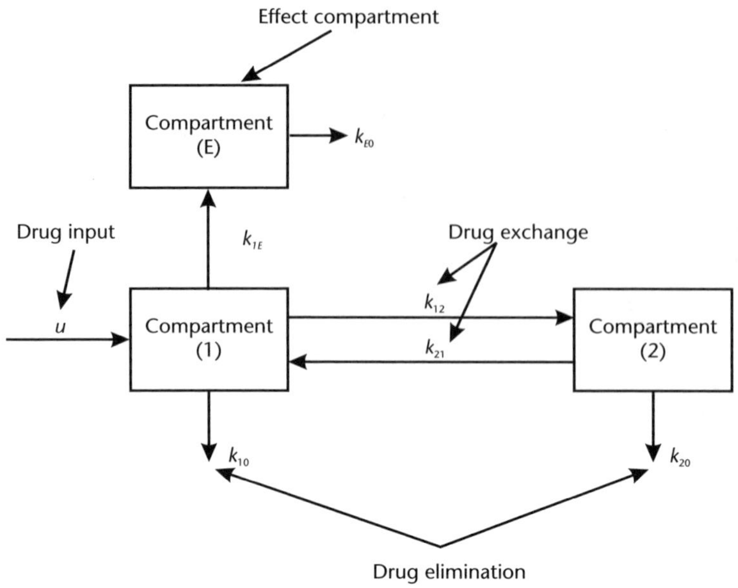

Figure 3.3 Modification of the atracurium kinetics to include the effect compartment *E*.

$$X_E(s) = \frac{k_{1E} X_1(s)}{s + k_{E0}} \tag{3.6}$$

The Hill equation [12, 13] may be used to relate the effect to a specific blood concentration of the drug, as follows:

$$E_{effect} = \frac{E_{\max}}{1 + \frac{(X_E(50))^{\alpha}}{X^{\alpha}{}_E}} \tag{3.7}$$

where X_E is the drug concentration and $X_E(50)$ is the drug concentration at the 50% effect.

One key postulate of this is that there is a considerable delay separating the first administration of a muscle relaxant drug and the onset of relaxation. This nonlinear effect is known as the *margin of safety* [12, 13] whereby no depression of twitch response can be detected until over 75% of the receptors are occluded. Also, because of saturation effects, once initiated, muscle relaxation (or paralysis) cannot increase indefinitely as the drug dosage increases.

Using the following experimental data published by Whiting and Kelman [14] and Weatherley et al. [10],

$$\begin{cases} k_{E0} = 0.208\,\text{min}^{-1} \\ X_E(50) = 0.404\ \mu\text{gml}^{-1} \\ \alpha = 2.98 \\ k_{1E} = 10^{-1}\ \text{min}^{-1} \end{cases}$$

and combining (3.4) and (3.6) and normalizing the overall open-loop gain to 1 leads to the following linear transfer function:

$$\frac{X_E(s)}{U(s)} = \frac{K(1+T_4 s)e^{-\tau s}}{(1+T_1 s)(1+T_2 s)(1+T_3 s)} \tag{3.8}$$

where

$$\begin{cases} K = 1 \\ \tau = 1 \text{ minute} \\ T_1 = 4.81 \text{ minutes} \\ T_2 = 34.42 \text{ minutes} \\ T_3 = 3.08 \text{ minutes} \\ T_4 = 10.64 \text{ minutes} \end{cases}$$

Finally, the nonlinear model can be obtained through a combination of (3.8) followed by the nonlinearity of (3.7) in a Weiner[2] structure model.

3.4 SISO Adaptive Generalized Predictive Control in Theater

Long-range prediction algorithms enjoyed considerable popularity in the late 1970s, especially with the development of a computer control algorithm called Dynamic Matrix Control (DMC) [15]. Evolving from a technique that represents process dynamics with a set of numerical coefficients together with a least-squares formulation, it promised to solve complex control problems, especially those associated with systems exhibiting large dead-times. DMC's use of a nonparametric model was later challenged by another multistep long-range predictive control algorithm: Generalized Predictive Control [8]. It is a natural successor to the Generalized Minimum Variance (GMV) design [16], and it is considered to be very robust. Its development is reviewed here.

3.4.1 Theory of SISO GPC

Consider the following locally linearized discrete model—the Controlled Auto-Regressive Integrated Moving Average (CARIMA) model—in the backward shift operator z^{-1}:

$$A(z^{-1}) \cdot \Delta y(t) = B(z^{-1}) \cdot \Delta u(t-1) + C(z^{-1})\xi(t) \tag{3.9}$$

where:

2. This refers to a nonlinear model structure which includes in series a linear dynamic equation followed by a nonlinear static equation, in contrast to the "Hammerstein" model structure where a nonlinear static function is followed in series with a linear dynamic transfer function.

$$A(z^{-1}) = 1 + a_1 z^{-1} + a_2 z^{-2} + \ldots + a_{n_a} z^{-n}$$
$$B(z^{-1}) = b_1 + b_2 z^{-1} + b_3 z^{-2} + \ldots + b_{n_b} z^{-m+1}$$
$$C(z^{-1}) = c_0 + c_1 z^{-1} + c_2 z^{-2} + \ldots + c_p z^{-p}$$

$\zeta(t)$ is an uncorrelated random sequence.

$$\Delta = 1 - z^{-1}$$

$u(t)$ represents the control input and $y(t)$ is the measured variable.

The controller computes the vector of controls using optimization of a function of the following form:

$$\begin{cases} J = E[Q_1 + Q_2] \\ Q_1 = \sum_{j=N_1}^{N_2} \left[\left(P(z^{-1})\hat{y}(t+j) - \omega(t+j) \right)^2 \right] \\ Q_2 = \sum_{j=1}^{NU} \left[\lambda(j)\left(\Delta u(t+j-1)\right)^2 \right] \end{cases} \quad (3.10)$$

where N_1 is the minimum costing (output) horizon, N_2 is the maximum costing horizon, NU is the control horizon, ω is the future set point, and $\lambda(j)$ is the control weighting sequence.

In an attempt to express the future output predictions, one can write

$$P(z^{-1}) \cdot \hat{y}(t+j) = \overline{\overline{G}}_j(z^{-1})\Delta u(t+j-1) + \Psi(t+j) \quad (3.11)$$

$$\Psi(t+j) = \overline{G}_j(z^{-1})\Delta u^f(t-1) + \overline{F}_j y^f(t) \quad (3.12)$$

$$T(z^{-1}) \cdot P(z^{-1}) = E_j(z^{-1})A(z^{-1})\Delta + z^{-j}\overline{F}_j(z^{-1})y^f(t) \quad (3.13)$$

$$P(z^{-1}) = \frac{P_n(z^{-1})}{P_d(z^{-1})} \quad (P(1) = 1) \quad (3.14)$$

$$T(z^{-1}) = 1 + t_1 z^{-1} + \ldots + t_{nt} z^{-nt} \quad (3.15)$$

$$\overline{F}_1(z^{-1}) = \frac{F_j(z^{-1})}{P_d(z^{-1})} \quad (3.16)$$

$$E_j(z^{-1}) \cdot B(z^{-1}) = \overline{\overline{G}}_j(z^{-1}) \cdot T(z^{-1}) + z^{-j} \cdot \overline{G}_j(z^{-1}) \quad (3.17)$$

$$\overline{\overline{G}}_j(z^{-1}) = \overline{\overline{g}}_0 + \overline{\overline{g}}_1 z^{-1} + \ldots + \overline{\overline{g}}_{j-1} z^{-j+1} \quad (3.18)$$

$$\overline{G}_j(z^{-1}) = \overline{g}_{j0} z^{-1} + \overline{g}_{j1} z^{-2} + \ldots + \overline{g}_{j(i-1)} z^{-i+1} \quad (3.19)$$

where:

i = from 1 to max(δB, δT)

$1 \le j \le N_2$

Superscript f denotes signals filtered by $1/T(z^{-1})$.

$P(z^{-1})$ and $T(z^{-1})$ are user-chosen polynomials in z^{-1}. $P(z^{-1})$ is referred to as the model-following polynomial, and the condition $P(1) = 1$ ensures an offset-free response [17]. $T(z^{-1})$ is the so-called "Youla" parameter or the observer polynomial for the j-step ahead predictor of (3.12).

The minimization of the cost function described in (3.10) leads to the following projected control increment:

$$\Delta u(t) = \bar{g}^{T}(\omega - \Psi) \tag{3.20}$$

where $\Psi = [\Psi(t + N_1), \ldots, \Psi(t + N_2)]$, $\bar{g}^T$ is the first row of the matrix $(\mathbf{G}_\mathrm{d}^\mathrm{T}\mathbf{G}_\mathrm{d} + \boldsymbol{\lambda}\mathbf{I})^{-1}\mathbf{G}_\mathrm{d}$, and $\mathbf{G}_\mathrm{d}$ is the dynamic (step-response) matrix of the form given in [8].

It is perhaps worth noting at this stage that this filter polynomial is an *essential* tool for ensuring the robustness properties of the closed-loop system under GPC. Indeed, as shown in Figure 3.4, this parameter is capable of reducing the feedback gain by many folds (see example below).

Consider the following plant, expressed in discrete form:

$$G_{p3}\left(z^{-1}\right) = \frac{0.1z^{-1} + 0.2z^{-2}}{1 - 0.7z^{-1}} = \frac{y(k)}{u(k)} \tag{3.21}$$

where $u(k)$ and $y(k)$ are the input and output of the system, respectively.

If we expressed the control law generated by the GPC algorithm as follows:

$$H\left(z^{-1}\right) \cdot \Delta u(t) = \omega(t) - M\left(z^{-1}\right) \cdot y(t) \tag{3.22}$$

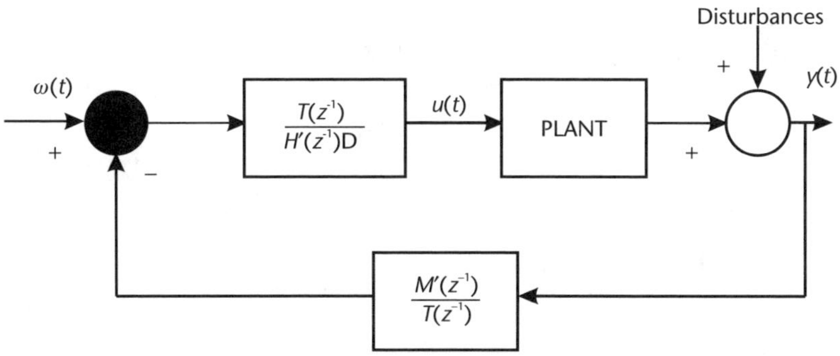

Figure 3.4 Diagram representing the GPC loop with the filter polynomial $T(z^{-1})$.

where $\omega(t)$ is the reference target and $H(z^{-1})$ and $M(z^{-1})$ are polynomials in the backward shift operator z^{-1}. Hence, for the following combination of GPC tuning factors:

$$N_1 = 1; N_2 = 10; NU - 1; \lambda = 0$$

one obtains the following GPC law:

$$\Delta u(t) = 1.231\omega - 0.701\Delta u(t-1) - 3.684y(t) + 2.453y(t-1) \quad (3.23)$$

Hence,

$$\begin{aligned} H(z^{-1}) &= (0.813 + 0.5695z^{-1}) \\ M(z^{-1}) &= (2.993 - 1.993z^{-1}) \end{aligned} \quad (3.24)$$

At low frequencies, the feedback gain is $M(1) = 1$ (0 dB). At high frequencies, however, this gain becomes $M(-1) \approx 5$ ($\approx$14 dB) (i.e., five times that at low frequencies). This only reinforces the contribution of such a filter when noise, disturbances, and unmodeled dynamics are predominant in the process under investigation.

Indeed, when a first-order polynomial $T(z^{-1}) = 1 - 0.8z^{-1}$ is included, the feedback gain becomes (see Figure 3.5, which shows the frequency response of $\frac{1}{T(z^{-1})}$ and its lowpass characteristics):

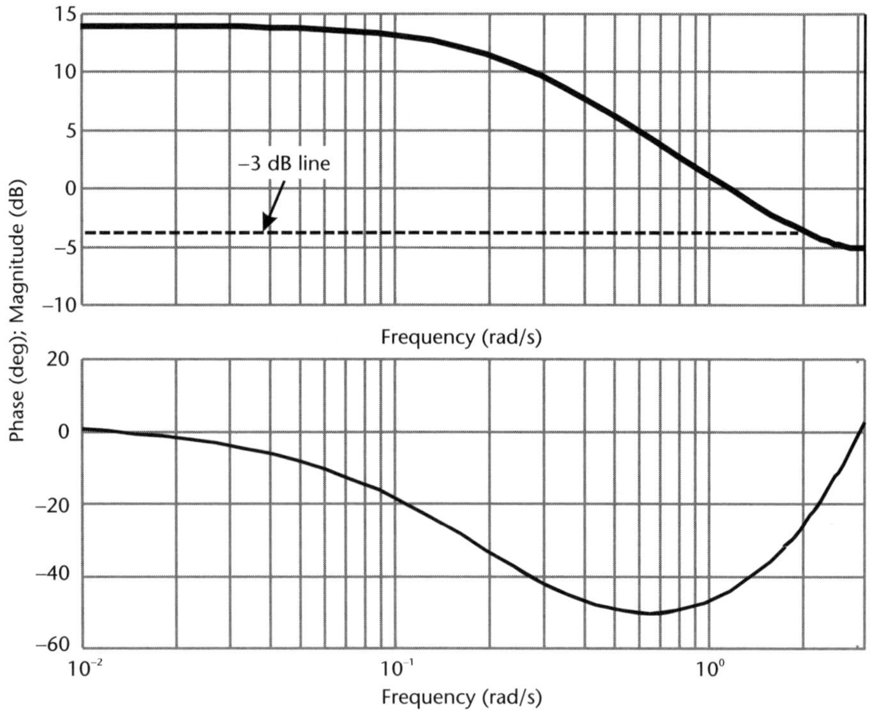

Figure 3.5 Bode diagrams relating to the lowpass filter $\frac{1}{(1-0.8z^{-1})}$.

$$\frac{M'\left(z^{-1}\right)}{T\left(z^{-1}\right)} = \frac{0.716 + 0.516z^{-1}}{1 - 0.8z^{-1}} \tag{3.25}$$

At low frequencies the gain remains equal to $M(1) = 1$ (0 dB), but at high frequencies this gain is greatly reduced to a value of $\frac{M'(-1)}{T(-1)} \approx 0.67$ (−3.5 dB).

3.4.2 Simulation Results

The overall nonlinear muscle relaxant model describing atracurium dynamics was simulated using a fourth-order Runge-Kutta method with fixed step length for numerical integration together with a sampling period of 1 minute. The pharmacokinetics of the drug are given by the three-time constant transfer function with a unit time delay of (3.8), while the pharmacodynamics are modeled using the Hill equation (3.7). The time delay in the process was made to change from 1 to 4 minutes, then back to 1 minute at every set-point change. A noise sequence of 5% was superimposed on the output signal. Parameter estimation, using incremental input and output data, takes the form of UD factorization[3] [18], which is a modified robust version of the well-known Recursive Least Square (RLS) algorithm. A third-order discrete-time model was estimated with an assumed time delay of 1 minute such that

$$G\left(z^{-1}\right) = \frac{z^{-1}\left(\hat{b}_1 z^{-1} + \hat{b}_2 z^{-2} + \hat{b}_3 z^{-3} + \hat{b}_4 z^{-4} + \hat{b}_5 z^{-5} + \hat{b}_6 z^{-6}\right)}{1 + \hat{a}_1 z^{-1} + \hat{a}_2 z^{-2} + \hat{a}_3 z^{-3}} \tag{3.26}$$

The rationale behind the choice of this discrete model structure is, by assuming a minimum value of the time delay to be 1 minute, the $B(z^{-1})$ polynomial is expanded to include three more coefficients to encompass the maximum delay of 4 minutes. Should the time delay be equal to 1 in the process, then $\hat{b}_4 = \hat{b}_5 = \hat{b}_6 \approx 0$ (although in practice this seldom is the case given noise and lack of proper excitation necessary for the RLS algorithm); however, should the time delay be equal to 4 minutes, then $\hat{b}_1 = \hat{b}_2 = \hat{b}_3 \approx 0$.

Hence, the initial covariance matrix, which expresses the level of confidence in such initial estimates, was made equal to $cov = 10^3 \cdot I$ (I being the identity matrix) and a value of $\rho = 0.975$ was adopted for the forgetting factor. Because the version of GPC that is included in this study relates to unconstrained control (see Chapter 5 for the constrained control version), the control signal as computed by the algorithm was clipped between maximum and minimum values of, respectively, 0.0 and 1.0. These limitations were reflected back to the estimator by recomputing the *actual* control sequence that was applied to the process.

The control algorithm assumed the following parameters:

3. UD factorization consists of improving the numerical stability of algorithms which handle matrices operations extensively by transforming each matrix into a combination of a diagonal matrix (D) and a triangular matrix (U).

$$N_1 = 1; N_2 = 10; NU = 1; \lambda = 0; P(z^{-1}) = \frac{1-0.7z^{-1}}{0.3}$$

$$T_{control}(z^{-1}) = T_{estimation}(z^{-1}) = (1-0.8z^{-1})^2$$

Figure 3.6(a) shows the performance of GPC, which, despite the severe changes of time delays at each set-point change, is very good in terms of target tracking and control activity. Figure 3.6(b) shows the evolution of the parameter estimates during the experiment with the appropriate $\hat{b}$ coefficients assuming small values when the delay varied from 1 to 4 and from 4 to 1.

3.4.3 Performance of SISO GPC in the Operating Theater During Surgery

The muscle relaxation control system as implemented in the operating theater can be seen in Figure 3.7, and it includes the following components:

- A microcomputer system, which includes the overall control algorithm (GPC and the interface programs).
- A DATEX RELAXOGRAPH system for measuring the degree of muscle relaxation (muscle paralysis) by electrically stimulating a peripheral nerve and displaying the resulting EMG response. Equipped with five electrodes, it employs the train-of-four principle (TOF) and features an automatic search

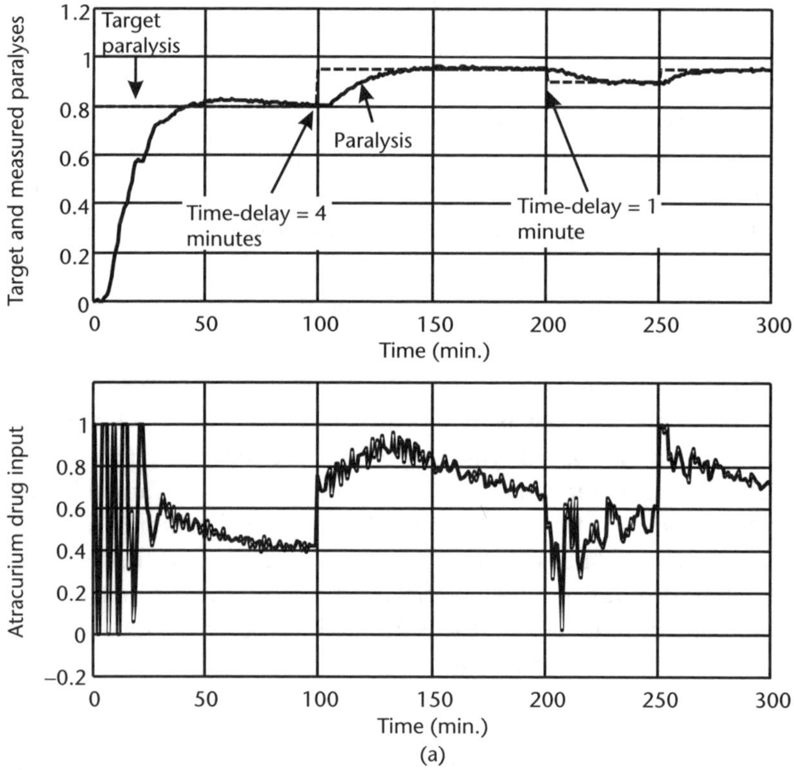

Figure 3.6 (a) Closed-loop performance under GPC with variable time delay. (b) Parameter estimates relating to the run of (a).

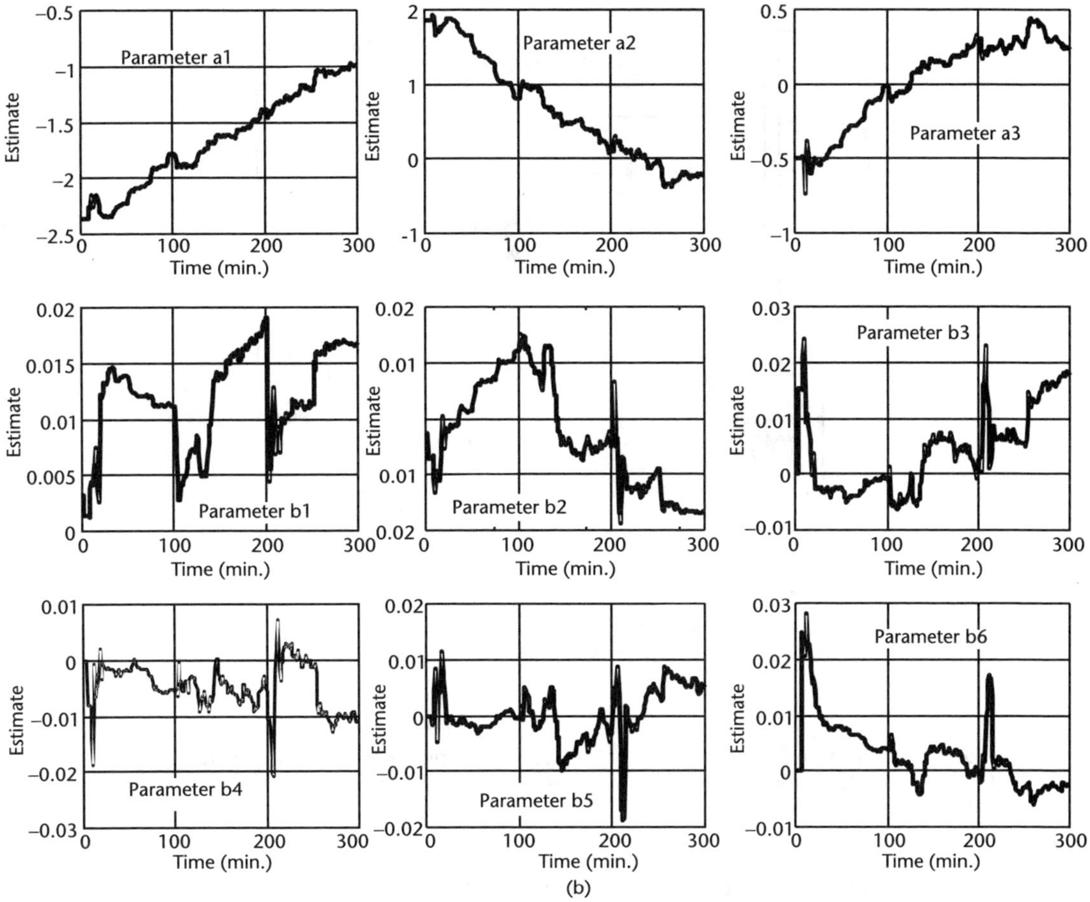

Figure 3.6 (continued.)

for supramaximal stimulation current level. Stimuli of the TOF sequence are given at a rate of two pulses per second every 20 seconds. The device displays $T1\%$ and $TR\%$ values given as follows:

$$\begin{aligned} T1 &= \frac{\textit{first twitch}}{\textit{control}} \times 100\% \\ TR &= \frac{\textit{last twitch}}{\textit{first twitch}} \times 100\% = \textit{TOF ratio} \end{aligned} \tag{3.27}$$

In our case, $T1$ will represent the value of the EMG[4] level.

- A digital pump driving a disposable 50-ml/60-ml syringe containing a solution of atracurium.

The links between the computer, the RELAXOGRAPH, and the syringe pump drive unit were via the serial RS-232 and parallel input/output ports.

4. A 100% EMG corresponds to no paralysis, whereas 0% EMG is equivalent to 100% paralysis.

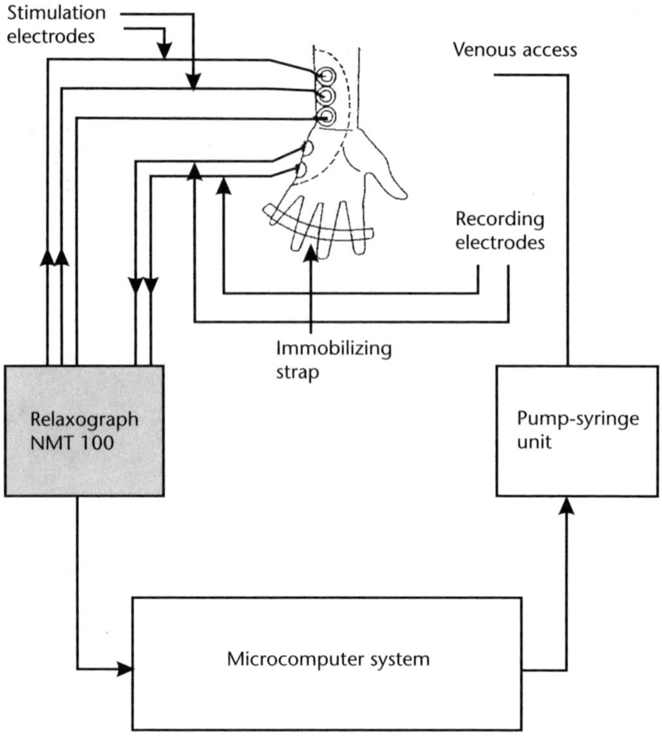

Figure 3.7 The muscle relaxation control system as implemented in the operating theater. (*From:* [19]. © 1998 Taylor & Francis Books (UK). Reprinted with permission.)

3.4.3.1 Tailoring the Measurements to the Sampling Period

The DATEX RELAXOGRAPH delivers signals at precise intervals of 20 seconds, but because it is intended to sample the closed-loop control system at a 1-minute interval, a change in the existing program had to be made to facilitate such a change. This was achieved via a three-point averaging filter of the following form:

$$G_F\left(z^{-1}\right) = \frac{1}{3}\sum_{i=0}^{2} z^{-i} \tag{3.28}$$

where

$$\begin{cases} z^{-1} = e^{-sT_f} \\ T_f = 20 \text{ seconds} \end{cases}$$

The Bode diagram of this filter is shown in Figure 3.8, where it can be seen that at a magnitude of –3 dB, the frequency is approximately 0.008 Hz. This is enough to provide lowpass filtering, which will reduce signal artifacts caused by fluctuations in the response due to diathermy[5] also to inadequate positioning of the stimulating and recording electrodes.

5. This is a severe electrical interference, very common in operating theaters.

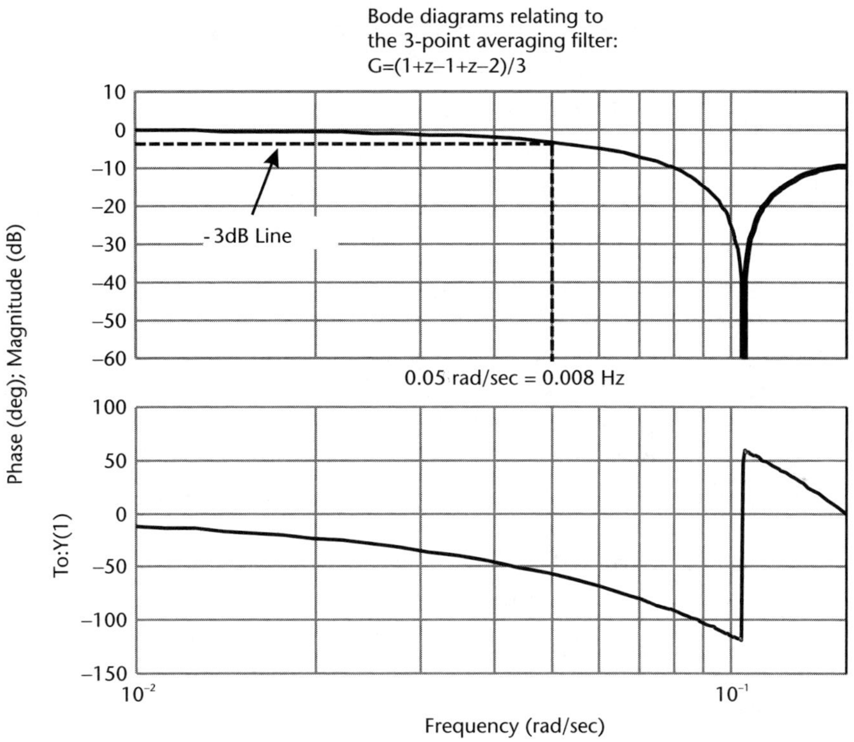

Figure 3.8 Bode diagrams relating to the three-point averaging filter $G_F(z^{-1}) = \frac{1}{3}(1 + z^{-1} + z^{-2})$ with 20-second sampling time.

3.4.3.2 Changes to the Closed-Loop Control Algorithm

When closed-loop control is established, PI controller is used to provide initial control to allow the parameter estimation routine to gather sufficient identification data. The PI parameters were obtained using Ziegler-Nichols techniques [20] applied to open-loop step responses in an off-line study. The dose of atracurium is expressed in (ml h^{-1}) and is obtained using the following formula:

$$I_{PI}\left(\text{ml h}^{-1}\right) = K_p e + K_i\left(e + PL\right) \tag{3.29}$$

where

$$\begin{cases} K_p = k_p \cdot W_t \\ K_i = k_i \cdot W_t \\ k_p = 0.002 \text{ kg}^{-1}, k_i = 0.0021 \text{ kg}^{-1} \\ PL = 150 \div 333 \end{cases}$$

I_{PI} is the atracurium infusion rate, W_t represents the actual patient's weight (in kilograms), PL is the initial preloaded dose of atracurium, and e is the difference between the actual $T1\%$ and the target $T1\%$.

3.4.3.3 Clinical Preparation of Patients Before Surgery

The patients were selected on the basis that they did not suffer from any known sensitivity to anesthetic drugs or myoneural disorders and had not been taking drugs known to affect neuromuscular transmission. All patients underwent either abdominal or orthopedic surgery normally requiring muscle relaxation. Approximately 60 minutes before surgery, they were premedicated with temazepam by mouth. Anesthesia was induced with methohexitane 1 mg kg^{-1}. The trachea was intubated when $T1$ reached a value between 10% and 15%. The lungs were inflated with 30% oxygen, 70% nitrous oxide, and 1% enflurane. During surgery, enflurane anesthesia was supplemented with boluses of fentanyl at 1 μg kg^{-1}. While the patient was still in the anesthetic room, the DATEX RELAXOGRAPH electrodes were carefully placed on the patient's arm, and then the calibration proceeded. Once transferred to the theater, the patient, already connected to the control system, was given an intravenous bolus[6] dose of atracurium of 0.15 to 0.25 mg kg^{-1}.

The online controlled infusion was started when $T1$ (induced by the initial bolus dose) reached a level judged adequate by the anesthetist (usually 10% to 15% of the baseline value). The level of muscle relaxation was monitored until the surgeon requested cessation of paralysis. The control was then switched off immediately, and residual muscle blockade was reversed using antagonist agents such as neostigmine (2.5 mg) and atropine (0.8 mg).

3.4.3.4 Results and Discussions

After local Ethics Committee approval, 10 patients (eight females and two males) were selected as being suitable for the experiments. Information relating to the patients is presented in Table 3.1. The atracurium concentrations used were all 1 mg ml^{-1} unless otherwise specified.

Table 3.1 Summary of Patients' Personal Details

Patient	*Sex*	*Age (years)*	*Weight (kg)*	*Duration of Procedure (min)*	*Duration of Control (min)*
1	F	68	50	107	67
2	F	33	60	56	30
3	F	21	68	60	45
4	F	69	50	120	69
5	F	65	58	66	33
6	F	37	60	58	32
7	F	17	56	165	130
8	M	32	69	90	52
9	M	46	73	177	106
10	F	41	71	63	22

Source: [19].

6. This is a large atracurium dose initially given by anesthetists to patients to obtain a high level of relaxation in a relatively short time.

All 10 trials were conducted using a sampling-time interval of 1 minute. Control and parameter estimation were performed at 1-minute intervals, while EMG readings were obtained every 20 seconds. Parameter estimation was based on the UD-factorization algorithm and was triggered at the same time as the closed-loop control was established with an initial covariance matrix $cov = 10 \cdot I$ and forgetting factor $\rho = 0.999$, respectively. A 20% reference EMG level ($T1\%$) was required by the operating surgeon in all trials except for the last case, corresponding to Patient 10, where a 15% EMG reference level was targeted. Results corresponding to each patient are presented in two parts. The first part consists of two traces: the upper trace representing the recorded EMG level ($T1\%$) in the form of bars[7]; and the lower trace shows the variations of the infusion rate of atracurium in (ml hr^{-1}). The time axis is labeled in samples of 20 seconds, and a zero-order hold is applied to the infusion rate sampled data. The second part of the results includes the parameter estimate variations plotted at every 1-minute interval. Only 5 of the 10 experiments will be described here, illustrating lessons learned from the trials.

Patient 3

For this experiment, GPC embodied a combination of (1, 30, 1, 0) for (N_1, N_2, NU, λ) and a second-order observer polynomial $T(z^{-1}) = (1 - 0.7z^{-1})^2$ to compensate for any unmodeled dynamics. The idea was to use a faster root to quickly reject any disturbances and also to double to roll-off[8] to enhance robustness. The parameter estimation routine, using incremental filtered data, assumed the third-order discrete model of (3.25). Parameter estimates were initialized at 0.0 except estimate $\hat{b}_1$, which was taken to be 1. Also, for this experiment, whose results are shown in Figure 3.9(a), the atracurium concentration was taken to be 500 μg ml^{-1}. Marks 1 and 2 on the upper trace of the same figure represent the times at which the anesthetist administered bolus doses of 7.5 mg and 2.5 mg, respectively, to bring the EMG level down to approximately 15%. At mark 3, the closed-loop control mode was entered with a fixed PI allowed to run for 30 minutes, after which the self-adaptive GPC took over at mark 4. Both control modes succeeded in keeping a remarkably steady level of paralysis with hardly any fluctuations, but a closer look at the trace reveals that the period corresponding to GPC protocol was steadier. Finally, at mark 5 the controller was switched off and the blockade reversed at mark 6. Notice the return to a 100% baseline, suggesting that no unnecessary drug has been administered and that only a little drift in the relaxation calibration level has occurred. Figure 3.9(b) illustrates the variations of the parameter estimates, which were studier during the trial. These finally converged to the following values:

$$\begin{cases} \hat{a}_1 = -1.1926 & \hat{a}_2 = 0.3059 & \hat{a}_3 = 0.0745 \\ \hat{b}_1 = 0.8109 & \hat{b}_2 = -0.1358 & \hat{b}_3 = -0.074 \end{cases}$$

These, in fact, correspond to the following stable pole/zero positions in the z-plane:

7. Bars of light intensity represent the relevant EMG values, while those of darker intensity represent the TOF ratios).
8. This is the frequency point which corresponds to the slope magnitude-frequency plot becoming steeper.

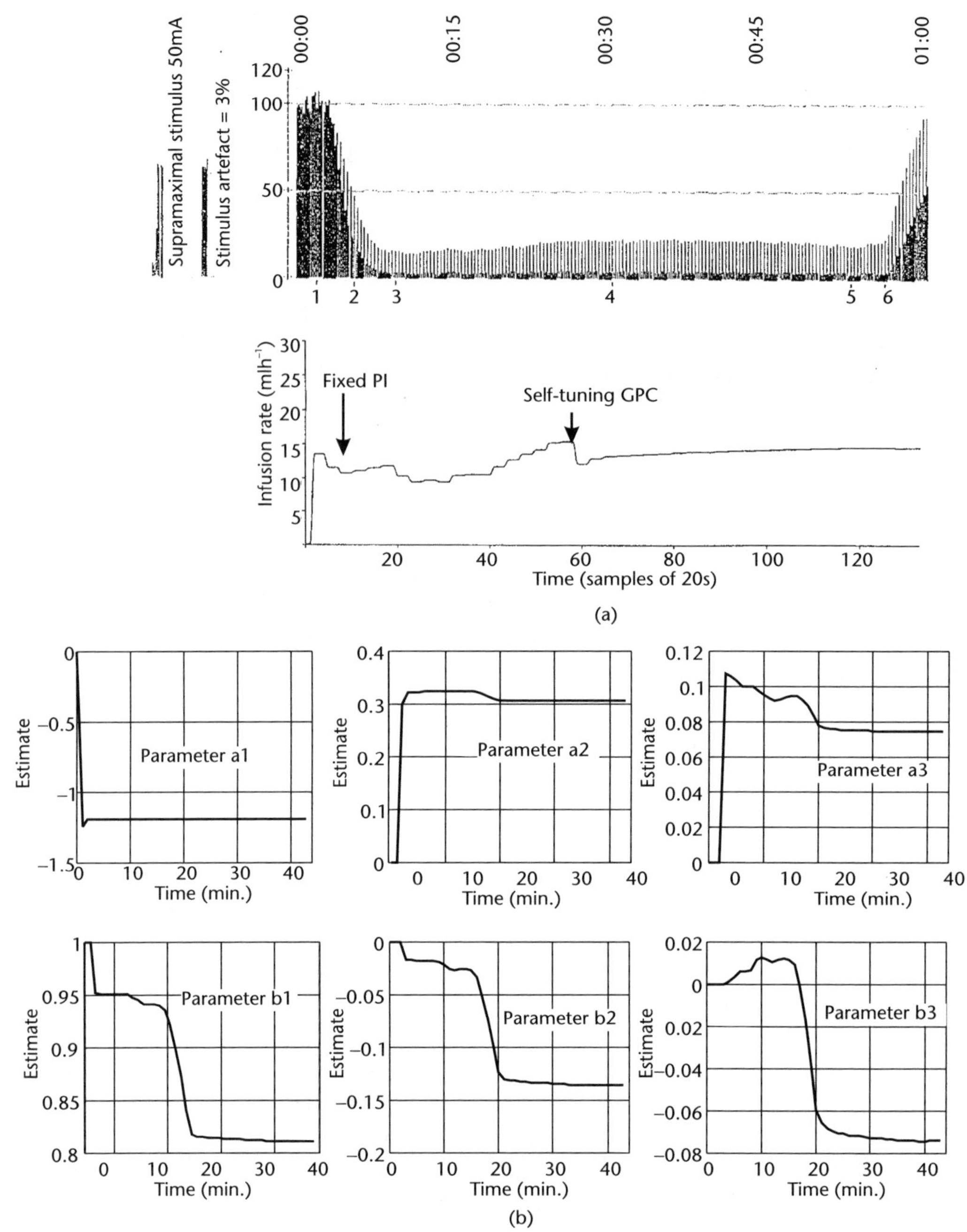

Figure 3.9 (a) Patient 3: recorded EMG and pump infusion rate during surgery. (*From:* [19]. © 1998 Taylor & Francis Books (UK). Reprinted with permission.) (b) System parameter estimates corresponding to (a).

Zeros: 0.3972; –0.2297.

Poles: $(0.6702 \pm 0.2342j)$; –0.1478.

with an estimated open-loop gain of 3.20. It is worth noting that the presence of a pair of complex poles, which does not reflect the compartmental idea, is probably due to the lack of proper excitation necessary for the identification routine.

Patient 4

During this particular experiment, the subject, a young female, demonstrated very high resistivity to the muscle relaxant drug. Indeed, as the EMG recording in Figure 3.10(a) shows, the patient was insensitive to the first bolus dose of 9 mg intravenously administered at mark 2. Another 2 mg and then 3 mg were given at mark 3 and mark 4, respectively. At this stage, the anesthetist decided to commence automatic control of the infusion at mark 5 with the fixed PI, which was allowed to run for 5 minutes only. The infusion rate of atracurium at 500 μg ml^{-1} began at approxi-

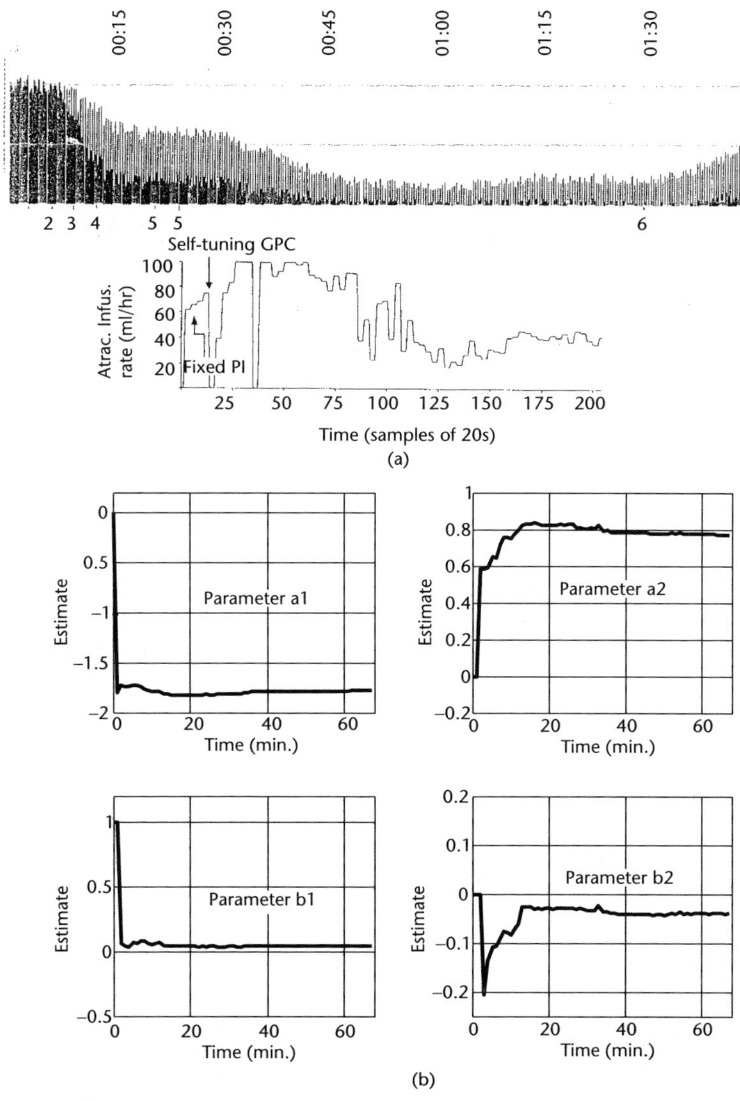

Figure 3.10 (a) Patient 4: recorded EMG and pump infusion rate during surgery. (*From:* [19]. © 1998 Taylor & Francis Books (UK). Reprinted with permission.) (b) System parameter estimates corresponding to (a).

mately 60 ml hr^{-1} and then increased gradually to 80 ml h^{-1}, but that did not cause the EMG to drop below the 50% line. However, when GPC took over at mark $\overline{5}$, with a combination of (1, 10, 1, 0) for (N_1, N_2, NU, λ) and $T(z^{-1}) = (1 - 0.95z^{-1})^2$, it was quick to drive the EMG level to the target in spite of the noise level, estimated at ±3%, persistently exciting the output. The controller behaved rather well by rejecting these disturbances and produced a control signal, illustrated on the lower trace of the same figure, which was reasonably active. Undoubtedly, the use of a slower root in the $T(z^{-1})$ polynomial made the controller more robust. The parameter estimation routine, which in this case assumed a second-order model with a 1-minute time delay, used filtered incremental data for the measurement vector. The variations of the parameter estimates are shown in Figure 3.10(b) and they converged to the following values:

$$\begin{cases} \hat{a}_1 = -1.7674 & \hat{a}_2 = 0.7717 \\ \hat{b}_1 = 0.0471 & \hat{b}_2 = -0.0386 \end{cases}$$

This is equivalent to a continuous second-order system with the following gain and time constants:

$$G\hat{a}in = 1.97 \quad \hat{T}_{c_1} = 4.19 \text{ minutes} \quad \hat{T}_{c_2} = 48.88 \text{ minutes}$$

At this stage it is worth noting that the for the remainder of these trials, the RLS parameter estimation assumed full valued data rather than incremental data in its information vector, which, it is recognized here, contravenes the CARIMA model idea, but this was found to provide good parameters and good control as described below

Patient 7

The experiment conducted with this subject is interesting in that it allowed one to test the robustness of the algorithm with respect to online variation of the drug concentration. Figure 3.11(a) shows the EMG recording as well as the infusion rate variations. Mark 1 represents the time at which a 12-mg bolus dose of muscle relaxant was intravenously administered, and mark 2 is when automatic control (PI) was initiated using a drug dilution of 500 μg ml^{-1}. Mark $\overline{2}$ shows when the self-adaptive GPC took over (5 minutes later) with a combination of (1, 30, 1, 5) for (N_1, N_2, NU, λ). The atracurium concentration was doubled to 1 mg ml^{-1} at mark 3. The controller was later switched off at mark 5, and the blockade reversed at mark 6. The 22% overshoot induced by the PI was quickly countered via the GPC by generating abrupt control actions, which, because of the low concentration, reached a mean level of 40 ml h^{-1}. Control activity, which was high during the first 100 samples, decreased slightly between samples 100 and 200, as a result of which the EMG level was kept steady with 2% fluctuations around the 20% target. When the change in atracurium concentration occurred at mark 3, and because of the time delay, the EMG level still remained at 20% for a few samples and then dropped 5% below the target, while the controller tried to overcome this by driving the pump sometimes at 0.1 ml h^{-1} (minimum speed) and other times at 20 ml h^{-1}. Figure 3.11(b) shows how the parameter estimates were affected by this change. As illustrated in the same fig-

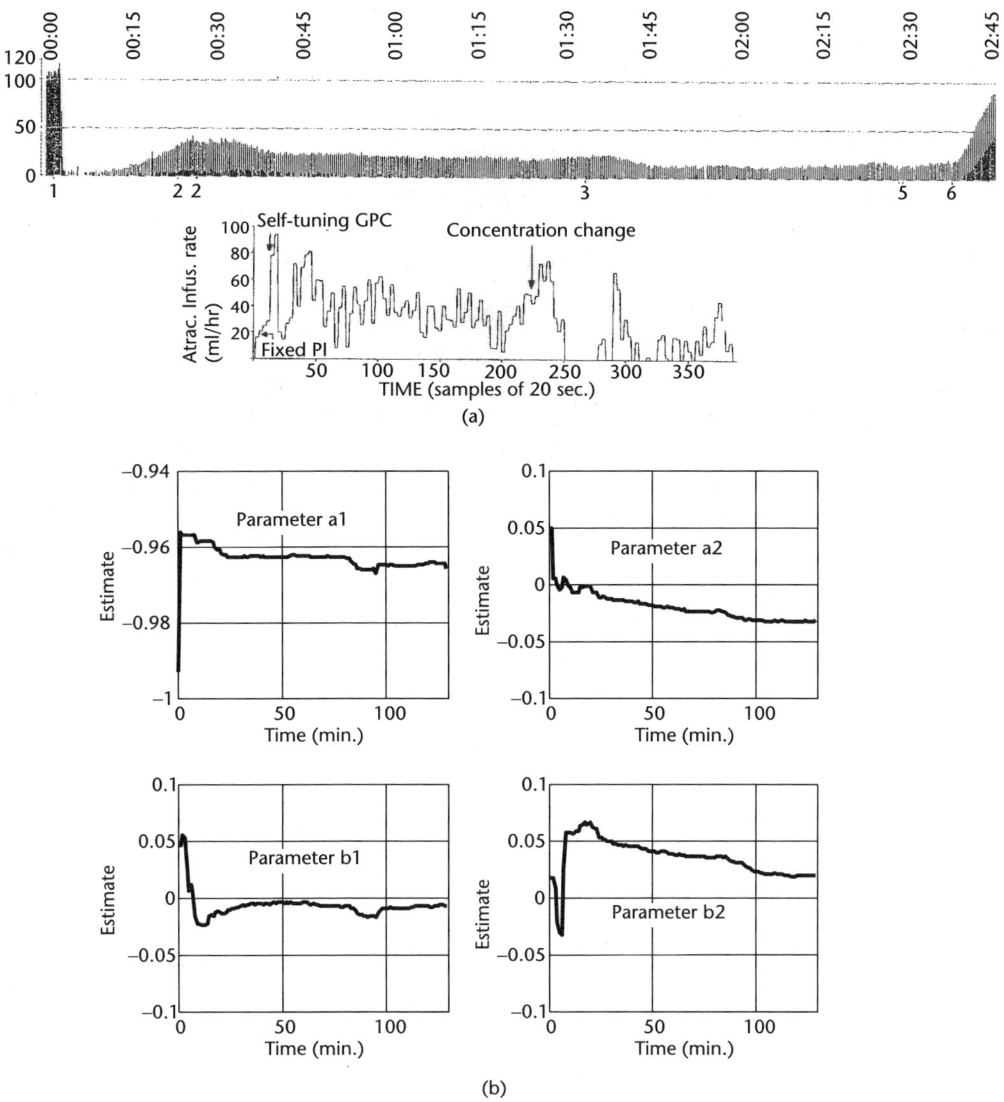

Figure 3.11 (a) Patient 7: recorded EMG and pump infusion rate during surgery. (*From:* [19]. © 1998 Taylor & Francis Books (UK). Reprinted with permission.) (b) System parameter estimates corresponding to (a).

ure, these parameters did not settle at all, and this may be due to the wrong assumption of the time delay and the absence of the observer polynomial $T(z^{-1})$ to compensate for any unmodeled dynamics. In fact, at the end of the run, the estimate's final values suggested a nonminimum phase system, as the following values demonstrate:

$$\begin{cases} \hat{a}_1 = -0.9655 & \hat{a}_2 = -0.0316 \\ \hat{b}_1 = -0.0072 & \hat{b}_2 = -0.0195 \end{cases}$$

corresponding to the following pole/zero positions in the z-plane:

Zero: 2.7083

Poles: 0.9972; –0.0317

The patient's estimated open-loop gain had the following values at the end of each of the two phases corresponding to the different concentrations:

$$\hat{Gain} = 2.15 \text{ for concentration } 500\ \mu\text{g ml}^{-1}$$

$$\hat{Gain} = 4.24 \text{ for concentration } 1{,}000\ \mu\text{g ml}^{-1}$$

Patient 8

After bolus doses of 8 mg then 3 mg administered at mark 1 and mark 2, respectively, on the upper trace of Figure 3.12 (again suggesting a low-sensitivity patient), the loop was closed at mark 3 when the EMG level reached approximately 28%. The PI was allowed to run for 5 minutes and produced an overshoot of 12%. When GPC took over at mark $\bar{3}$ assuming the same controller parameters as before, it was quick to reduce the overshoot by making the EMG efficiently track the 20% target. The control signal, whose variations are shown on the lower trace of the same figure, was good and reasonably active.

Patient 10

In contrast to the nine previous trials, a reference EMG level of 15% was targeted for this patient, thus moving the operating point closer to the nonlinear region. Hence, after a bolus dose given at mark 1 of Figure 3.13, automatic control was

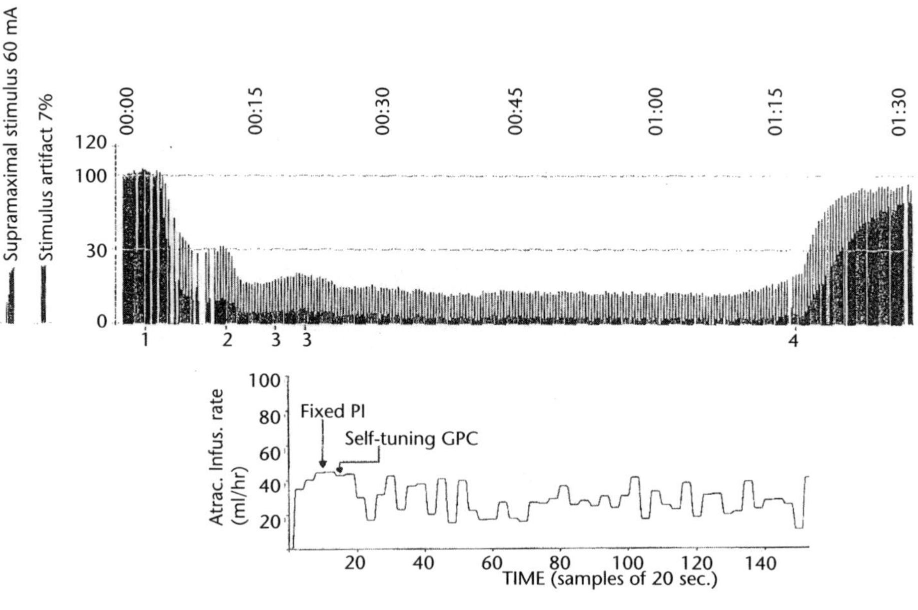

Figure 3.12 Patient 8: recorded EMG and pump infusion rate during surgery. (*From:* [19]. © 1998 Taylor & Francis Books (UK). Reprinted with permission.)

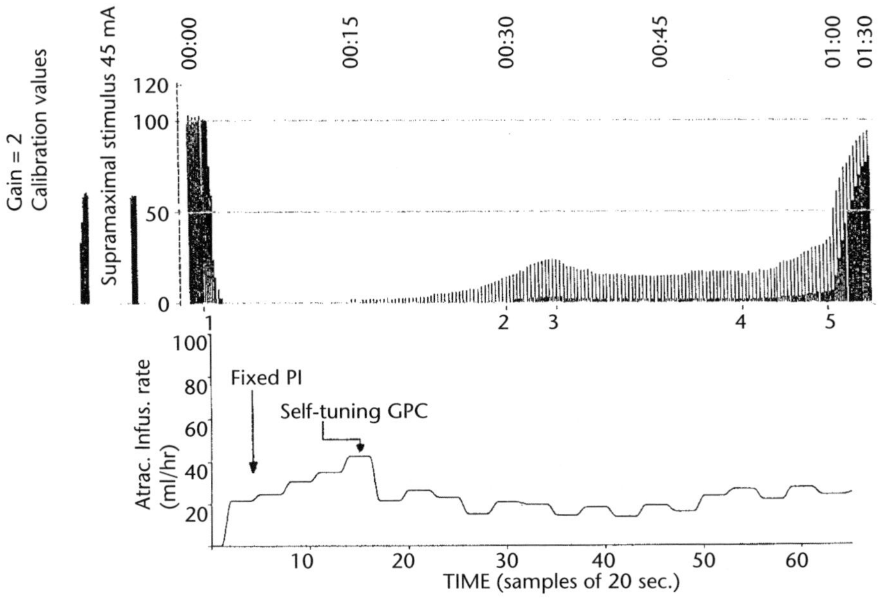

Figure 3.13 Patient 10: recorded EMG and pump infusion rate during surgery. (*From:* [19]. © 1998 Taylor & Francis Books (UK). Reprinted with permission.)

switched on at mark 2 with the PI providing initial control for only 5 minutes. After this, self-tuning control under the same conditions as before was initiated to counteract the overshoot, which was induced previously. A remarkably steady EMG level at 15% was achieved together with a well-behaved control signal, until mark 4, where the controller was switched off prior to reversing the blockade at mark 5. Parameter estimates converged to the following values:

$$\begin{cases} \hat{a}_1 = -0.9852 & \hat{a}_2 = -0.0153 \\ \hat{b}_1 = -0.0418 & \hat{b}_2 = -0.0695 \end{cases}$$

This is equivalent to a continuous second-order system with the following gain and time constants:

$$G\hat{a}in = 3.70 \quad \hat{T}_{c_1} = 0.24 \text{ minute} \quad \hat{T}_{c_2} = 32.20 \text{ minutes}$$

3.4.3.5 Analysis of the Data

To analyze the data, three indices were used: the mean value, the standard deviation (SD), and the root-mean square deviation (RMSD), the last two indices being commonly used to give an indication of the spread of a set of values around the mean value and the target value, respectively. The indices are defined by the following expressions:

$$\overline{X} = \frac{1}{N}\sum_{i=1}^{N} X_i \tag{3.30}$$

$$SD = \sqrt{\frac{1}{N-1}\sum_{i=1}^{N}\left(X_i - \overline{X}\right)^2} \tag{3.31}$$

where X_i, $\overline{X}$, and N are the current measurement, the mean value, and the total number of points considered, respectively.

$$RMSD = \sqrt{\frac{1}{N-1}\sum_{i=1}^{N}\left(X_i - X_{Trgt}\right)^2} \tag{3.32}$$

Here, X_{Trgt} is the EMG reference level.

A summary of the first nine patients' results is given in Table 3.2, where the mean and SD indices are stated for the duration of the automatic control, the mean drug dose consumption, the mean, the RMSD, and the SD. The last of the 10 experiments conducted was excluded since the 15% target was different from the other trials.

The value of 3.12% for the mean of the standard deviation of $T1\%$ indicates that, generally, a steady level of blockade was obtained. Moreover, the fact that the value was so close to the value of the mean of the root square deviation of $T1\%$ (3.41%) implies that the degree of neuromuscular blockade was close to the target. The mean value of the mean of $T1\%$ (19.74%) reinforces the argument that the individual $T1\%$ values were also close to the target. Moreover, it was observed that for the three control modes used (manual, fixed controller, self-adaptive GPC), the self-adaptive scheme proved more robust and efficient. Also, the mean dose of the muscle relaxant drug (5.05 μg kg^{-1} min^{-1}) was far lower than that which would be obtained by the anesthetist using bolus doses, and certainly lower that the range suggested by the manufacturers of atracurium.

3.5 Review of the Multivariable Anesthesia Control System

The application of multivariable predictive control, particularly multivariable GPC, has been demonstrated, first, on a multivariable model for two aspects of anesthesia comprising muscle relaxation as well as unconsciousness. The necessary transfer function components for the identified model have been obtained in various ways. The two drugs considered in the model for humans are atracurium (for muscle relax-

Table 3.2 Summary of Patients' Data (n = 9)

Parameter	*Mean*	*SD*	*Range*
Automatic control duration (min)	62.33	33.04	30–130
Dose (μg/kg/min)	5.05	1.80	1.57–6.81
Mean of $T1$ (%)	19.74	1.37	17.50–21.85
RMSD of $T1$ (%)	3.41	1.69	1.07–6.89
SD of $T1$ (%)	3.12	1.68	1.02–6.89

Source: [19].

ation) and isoflurane (for inducing unconsciousness).[9] This section reviews the identified multivariable anesthesia model.

3.5.1 Identification of the Multivariable Anesthesia Model

3.5.1.1 The Atracurium Mathematical Model

This is described by (3.7) and (3.8) in Section 3.3.

3.5.1.2 Isoflurane Unconsciousness Model

As already stated, there is no direct method of measuring depth of anesthesia. Previous research work namely by Schwilden [21, 22] and Savege [23] used quantitative EEG analysis in humans to give an indication of their anesthetic state. However, the interpretation of the tracings is a difficult and subjective task. The information proved unreliable even when interpreted by experienced staff, since the characteristic patterns were often disturbed by factors such as anoxia, surgical stimulations, and the anesthetic agents used [24].

Consequently, anesthetists have to resort to the merger of several clinical signs such as blood pressure and respiration to obtain the closest possible indication of how well the patient is anesthetized. Indeed, in a study conducted by Asbury [unpublished correspondence letter, 1990], anesthetists were asked to rank the relative importance of 10 clinical signs ranging from movement and response to surgery to capillary refill. These signs were ranked on a scale of 1 to 10 based on the mean values provided by these anesthetists. From these surveys, it emerged that arterial blood pressure can be selected as one variable to give a good indication of depth of anesthesia when no emergency conditions occur. In fact, previously published work by Gray and Asbury [25] used systolic arterial pressure (SAP) as the variable for a closed-loop control system. Also, Schills [26] used mean arterial pressure (MAP) and a measure of EEG frequency to control anesthesia via halothane in an on-off control strategy.

In other studies [27, 28], in which step response trials to isoflurane administration were carried out on several patients, it emerged that a first-order linear model with dead time can be adopted, having a time constant of 1 to 2 minutes. In order to estimate the steady-state gain, it is assumed that a relatively sensitive patient needs 2% isoflurane for a 30-mmHg reduction of the mean arterial pressure. Therefore, the model describing the variation of blood pressure to changes in inhaled isoflurane concentrations can written as follows:

$$G_{22}(s) = \frac{\Delta MAP}{U_2(s)} = \frac{K_2 e^{-\tau_2 s}}{(1+T_5 S)} \tag{3.33}$$

where ΔMAP is the change in arterial blood pressure and $U_2(s)$ is the isoflurane drug input, and

9. These may vary from a clinical hospital to another.

$$\begin{cases} \tau_2 &= 25\,s \\ T_5 &= 2 \text{ minutes} \\ K_2 &= -15 \end{cases}$$

3.5.1.3 Atracurium to MAP Interaction

This interaction has been investigated in humans and was found to be clinically insignificant [unpublished correspondence letter, 1990].

3.5.1.4 Isoflurane to Muscle Relaxation Interaction

In order to identify this type of interaction, which is small but significant, an experiment was performed by Asbury [unpublished correspondence letter, 1990], in which a 47-year-old patient without a kidney but having a renal transplant, had to be anesthetized. An infusion of atracurium of 5 mg h^{-1} was initiated until a steady state was achieved in the muscle relaxation level at which time a step change of isoflurane concentration from 0% to 1% was introduced (see Figure 3.14). Once the changes were observed, the isoflurane was switched off and a new equilibrium was achieved. The isoflurane was again switched on to 1% and then switched off once the expected changes were observed (see Figure 3.15).

In order to analyze the results, it was necessary to divide the experiment into the on-phase and off-phase parts to establish a certain symmetry in the model. Using the identification methods described in [29], the following second-order transfer functions were obtained:

$$\begin{aligned} G_{on}(s) &= \frac{0.24e^{-0.84s}}{(1+2.33s)(1+1.17s)} \\ G_{off}(s) &= \frac{0.29e^{-s}}{(1+3.33s)(1+1.33s)} \end{aligned} \tag{3.34}$$

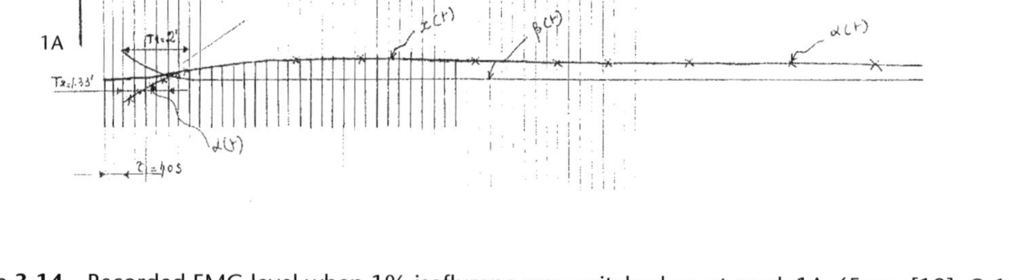

Figure 3.14 Recorded EMG level when 1% isoflurane was switched on at mark 1A. (*From:* [19]. © 1998 Taylor & Francis Books (UK). Reprinted with permission.)

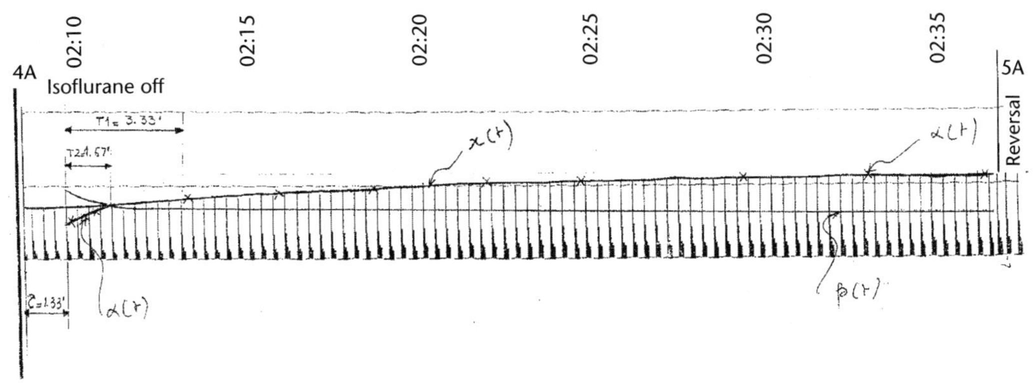

Figure 3.15 Recorded EMG level when isoflurane was switched off at mark 4A and the blockade reversed at mark 5A.

A quick analysis of these two transfer functions does indeed suggest that the model is more or less symmetrical. Hence, taking the mean values of the values of (3.33) gives the following overall interactive model, which describes the effect of the inhaled agent isoflurane on muscle relaxation during surgery:

$$G_{12}(s) = \frac{K_4 e^{-\tau_3 s}}{(1+T_6 s)(1+T_7 s)} \tag{3.35}$$

where

$$\begin{cases} K_4 = 0.27 \\ \tau_3 = 1 \text{ minute} \\ T_6 = 1.25 \text{ minutes} \\ T_7 = 2.83 \text{ minutes} \end{cases}$$

3.5.1.5 Overall Multivariable Anesthetic Model

The overall linear multivariable system combining muscle relaxation together with unconsciousness (via blood pressure measurements) is illustrated in Figure 3.16, and its components are described by the following P-canonical triangular representation:

$$\begin{bmatrix} Paralysis \\ \Delta MAP \end{bmatrix} = \begin{bmatrix} G_{11}(s) & G_{12}(s) \\ 0 & G_{22}(s) \end{bmatrix} \begin{bmatrix} U_1 \\ U_2 \end{bmatrix}$$

$$G_{11}(s) = \frac{(1+10.64s)e^{-s}}{(1+3.08s)(1+4.81s)(1+34.42s)} \tag{3.36}$$

$$G_{12}(s) = \frac{0.27e^{-s}}{(1+2.83s)(1+1.25s)}$$

$$G_{22}(s) = \frac{-15e^{-0.42s}}{(1+2s)}$$

To represent the nonlinearity of (3.7) in the muscle relaxation channel, changes in the overall gain were incorporated to reflect the parameters of this Hill equation.

3.5.2 Extension of GPC to the Multivariable Case

Consider the following m-input m-output linear discrete-time system[10]:

$$A\left(z^{-1}\right)y(t) - z^{-k_{ij}} B\left(z^{-1}\right)u(t-1) + \frac{C\left(z^{-1}\right)}{\Delta}\zeta(t) \tag{3.37}$$

where:

$$A(z^{-1}) = I + A_1(z^{-1})z^{-1} + A_2z^{-2} + \ldots + A_nz^{-n}$$
$$B(z^{-1}) = B_1 + B_2z^{-1} + B_3z^{-2} + \ldots + B_mz^{-m+1}$$
$$C(z^{-1}) = C_0 + C_1z^{-1} + C_2z^{-2} + \ldots + C_pz^{-p}$$
$$y(t) = [y_1(t), y_2(t), \ldots y_m(t)]$$
$$u(t) = [u_1(t), u_2(t), \ldots u_m(t)]$$
$$\Delta = 1 - z^{-1}$$

$y(t)$ and $u(t)$ are vectors of m measurable outputs and m measurable inputs, respectively; k_{ij} is the integral time delay of the ijth element of $B(z^{-1})$, and $\zeta(t)$ denotes a vector of m uncorrelated sequences of random variables with zero mean and covariance σ. I is the identity matrix.

In order to derive a j-step-ahead predictor of $y(t + j)$ based on (3.37), consider the following Diophantine equation:

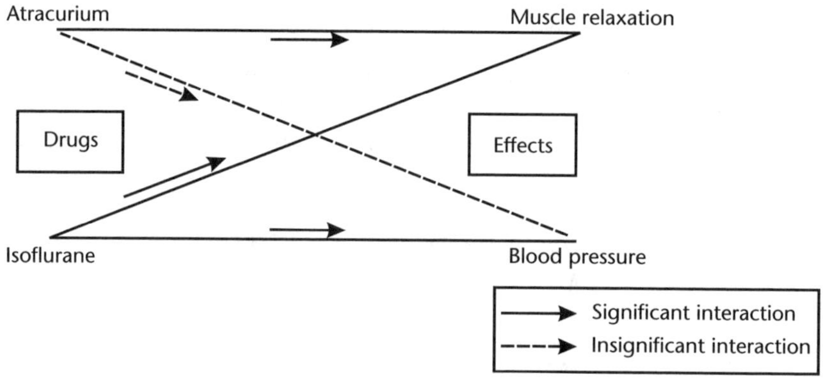

Figure 3.16 Schematic diagram representing the various components forming the multivariate anesthesia model. (*From:* [19]. © 1998 Taylor & Francis Books (UK). Reprinted with permission.)

10. The consideration of a "square" system will not imply any loss of generality here.

$$T(z^{-1})P_N(z^{-1})(P_D(z^{-1}))^{-1} = E_j(z^{-1})A(z^{-1})\Delta + z^{-j}F_j(z^{-1})(P_D(z^{-1}))^{-1} \tag{3.38}$$
$$j = 1,2, \ldots, N_2$$

$E_j(z^{-1})$ and $F_j(z^{-1})$ are matrix polynomials defined given the matrix polynomial $B(z^{-1}$ $A(z^{-1})$ and the prediction interval j · $P(z^{-1}) = P_N(z^{-1})(P_D(z^{-1}))^{-1}$ is the so-called model-following matrix polynomial which creates an auxiliary output $\Psi(t)$ such that

$$\Psi(t) = P(z^{-1})y(t) \tag{3.39}$$

Following the same procedure as in the SISO case of Section 3.3, it can be shown that the predictor becomes

$$\begin{cases} \Psi(t+j) = (T(z^{-1}))^{-1}\begin{bmatrix} G_j(z^{-1})\Delta u(t+j-1) + F_j(z^{-1})(P_D(z^{-1}))^{-1} y(t) + \\ E_j(z^{-1})x(t+j) \end{bmatrix} \\ G_j(z^{-1}) = E_j(z^{-1})B(z^{-1}) \end{cases} \tag{3.40}$$

For the controller to be optimal, the residual $(T(z^{-1}))^{-1}$ $E_j(z^{-1})x(t+j)$ must be orthogonal to data at time t, thus suggesting that $T(z^{-1})$ and $x(t+j)$ are uncorrelated, which is not always the case, leading, therefore, to suboptimal predictions. However, the choice of $T(z^{-1})$ as diagonal and with identical elements in each row, implies commutativity of the product of matrices in (3.40), reducing considerably the computational burden. Hence, by taking $(T(z^{-1}))^{-1}E_j(z^{-1})x(t+j) = 0$, the following set of prediction equations is finally obtained:

$$\begin{cases} \Psi(t+j) = G_j(z^{-1})\Delta u^f(t+j-1) + F_j(z^{-1})(P_D(z^{-1}))^{-1} y^f(t) \\ y^f(t) = T(z^{-1})y(t) \\ u^f(t) = T(z^{-1})u(t) \end{cases} \tag{3.41}$$

Because the minimization of the cost function should be performed in terms of Δu instead of Δu^f, the following equation is also considered:

$$G_j(z^{-1}) = G'_j(z^{-1})T(z^{-1}) + z^{-j}\Gamma_j(z^{-1}) \tag{3.42}$$

where the matrix polynomial $G'_j(z^{-1})$ is equivalent to the matrix polynomial $G_j(z^{-1})$, which would have been obtained if $T(z^{-1}) = I$ (the identity matrix). Consequently, (3.40) can be rewritten as follows:

$$\begin{aligned} \Psi(t+j) = {} & G'_j(z^{-1})\Delta u(t+j-1) + \Gamma_j(z^{-1})\Delta u^f(t-1) + \\ & F_j(z^{-1})(P_D(z^{-1}))^{-1} y^f(t) \end{aligned} \tag{3.43}$$

Similarly to the SISO case, the controller minimizes the following cost function, which is, in fact, the expectation subject to data available up to time t:

$$J(N_1, N_2, NU) = \sum_{j=N_1}^{j=N_2} \left[\Psi(t+j) - \Omega(t+j)\right]^2 + \sum_{j=1}^{j=NU} \Lambda(j)\left[\Delta u(t+j-1)\right]^2 \qquad (3.44)$$

where Ω is the vector of set points relating to the m loop channels, and Λ is the control weighting sequence, which is a diagonal matrix of the following form:

$$\Lambda = \begin{bmatrix} \lambda_{11} & 0 \ldots 0 \\ 0 \; \lambda_{22} & \ldots 0 \\ 0 \;\; 0 \ldots & \lambda_{mm} \end{bmatrix}$$

It is worth noting that in (3.44) it is assumed that all loop channels acquire the same tuning factors N_1, N_2, NU, but in general different settings can be given to these factors across these loop channels depending on the known channels' respective dynamics.

Since only the first increment is considered, the solution for the minimization of the previous cost function can be summarized as follows:

$$\Delta u(t) = (I, 0, \ldots, 0)\left(G^T G + \Lambda I\right)^{-1} G^T \left(\Omega - \Psi_f\right) \qquad (3.45)$$

where Ψ_f denotes the signals in (3.42) which are "known" up to time t.

3.5.3 Simulation Results

Several studies have involved the simulation of the model described in Section 3.4 using a fourth-order Runge-Kutta method with a fixed-step integration interval of 0.1 and a sampling interval of 1 minute. Command signals of 80% then 70% for relaxation, and 130 mmHg then 140 mmHg for blood pressure were assumed throughout. Initial conditions were 0% relaxation and 150 mmHg mean arterial pressure. At time $t = 0$, a bolus dose of atracurium was administered and the loop remained open until the muscle relaxation reached a safe level for the loop to be closed with GPC. The input signal was clipped between 0 and 1.0 for the atracurium drug input, and between 0% and 5% for the isoflurane input. For parameter estimation, a UDU factorization method was used on incremental data, with an initial covariance matrix and forgetting factor given by $cov = 10^2 \cdot I$ and $\rho = 0.975$, respectively. A discrete multivariable model of five diagonal $A(z^{-1})$ matrices and five upper triangular $B(z^{-1})$ matrices of the form given below was estimated with an assumed time delay of 1 sample:

$$B(z^{-1}) = \begin{bmatrix} b_{11} & b_{12} \\ 0 & b_{22} \end{bmatrix} + \begin{bmatrix} b_{13}z^{-1} & b_{14}z^{-1} \\ 0 & b_{24}z^{-1} \end{bmatrix} + \ldots + \begin{bmatrix} b_{1(2m+1)}z^{-m} & b_{1(2m+2)}z^{-m} \\ 0 & b_{2(2m+2)}z^{-m} \end{bmatrix}$$

and

$$A(z^{-1}) = I_t \begin{bmatrix} a_{13}z^{-1} & 0 \\ 0 & a_{24}z^{-1} \end{bmatrix} + \begin{bmatrix} a_{15}z^{-2} & 0 \\ 0 & a_{26}z^{-2} \end{bmatrix} + \ldots + \begin{bmatrix} a_{1(2n+1)}z^{-n} & 0 \\ 0 & a_{2(2n+2)}z^{-n} \end{bmatrix}$$

It is worth noting that in this version of the algorithm, the future set-point values for muscle relaxation and blood pressure were known to the control algorithm (i.e., preprogrammed in advance). This is done in order to allow the algorithm to react to those changes early enough to avoid undesirable switching transients and sudden reactions.

Figure 3.17 is a screen shot of the user-interactive multivariable anesthesia simulator built for the purpose of training medical staff, including junior doctors and nurses, to understand the interactions behind such mechanisms.

Figure 3.18 shows the performance of the multivariable control algorithm when the following tuning parameters were assumed:

$$N_2(\text{Channel}\,1) = 10; \quad N_2(\text{Channel}2) = 20;$$
$$NU(\text{Channel}\,1) = 2; \quad NU(\text{Channel}2) = 1;$$
$$T_{control}(z^{-1}) = T_{estimation}(z^{-1}) = (1 - 0.8z^{-1}) \cdot I$$
$$\Lambda = I; \quad P = I$$

As seen in this figure, the closed-loop control performance was very good despite the stochastic activity and the various disturbances, which were simulated in both channels.

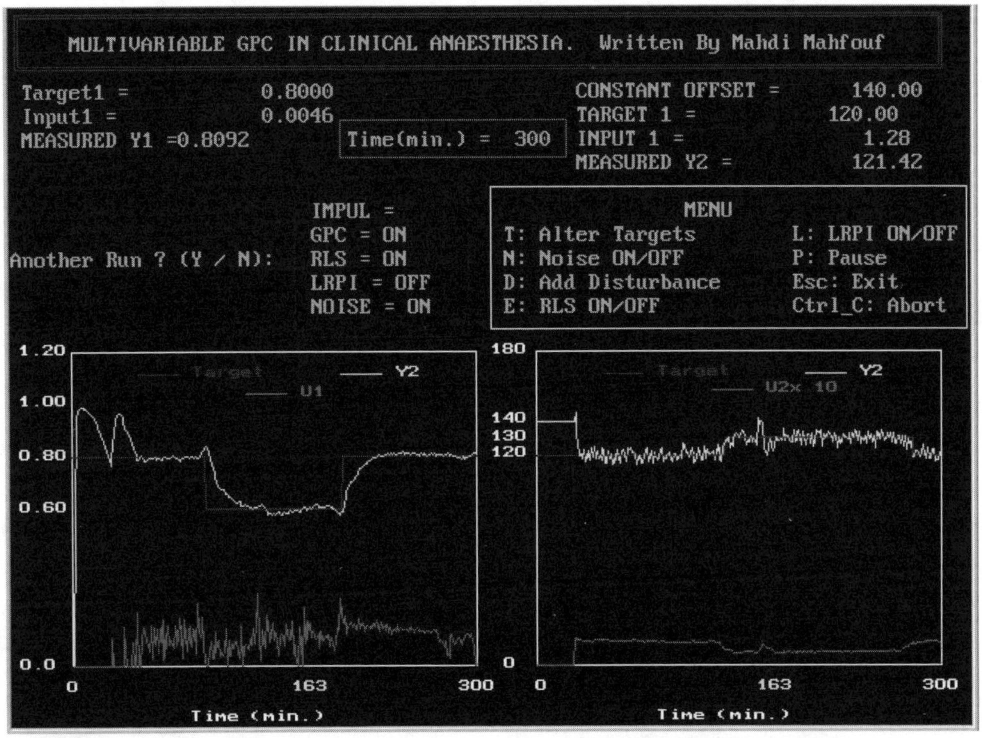

Figure 3.17 Screen shot of the interactive software relating to the multivariable anesthesia simulator. (RLS: recursive least squares; LRPI: long-range predictive identification.)

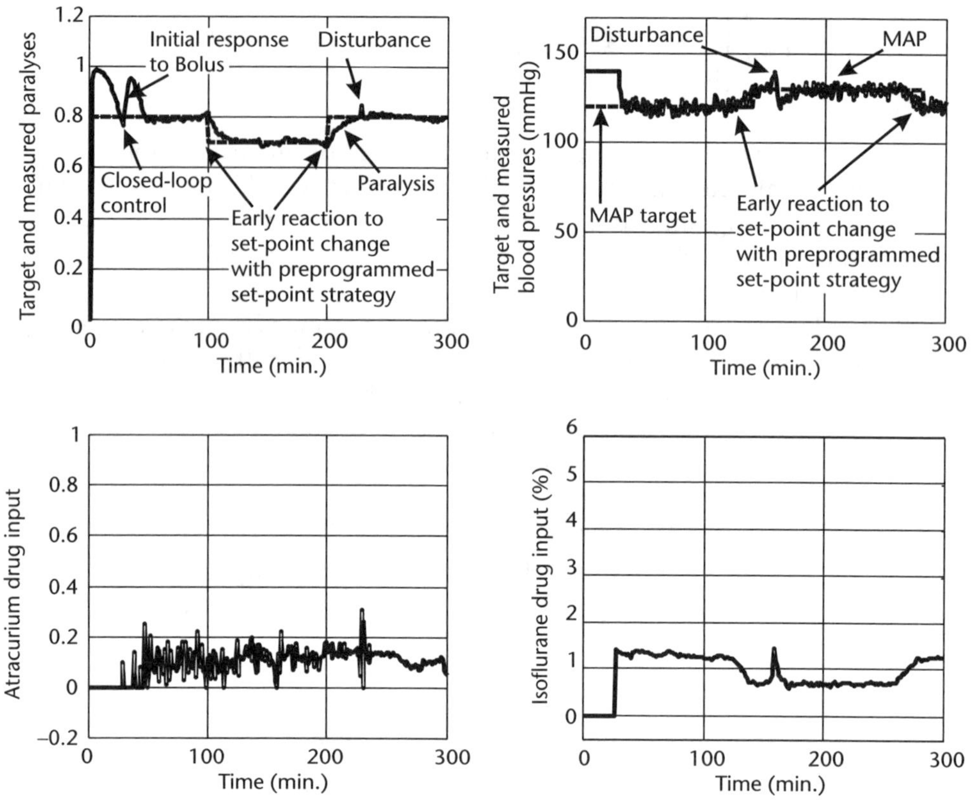

Figure 3.18 Performance of multivariable GPC using the nonlinear multivariable anesthesia model.

3.5.4 Real-Time Experiments

The multivariable control system used in the operating theater is shown in Figure 3.19 and consists of the following:

- An IBM-compatible microcomputer that incorporates the control algorithm;
- A DATEX RELAXOGRAPH system for measuring the degree of muscle relaxation (paralysis);
- A BRAUN PERFUSOR SECURA digital pump driving a disposable 50-ml/60-ml syringe containing a solution of atracurium calibrated between a minimum of 0 ml h^{-1} to a maximum of 99.99 ml h^{-1};
- A DATASCOPE-3000 instrument, which provides systolic arterial pressure (SAP) measurements every minute;
- An Enfluratec3 vaporizer (Cyprane), for delivering isoflurane, modified to allow it to be driven via a stepper motor under computer control and calibrated between a minimum of 0% and a maximum of 5% with increments of 0.5%. The system for controlling the vaporizer contains several features that guarantee its safe use [5].

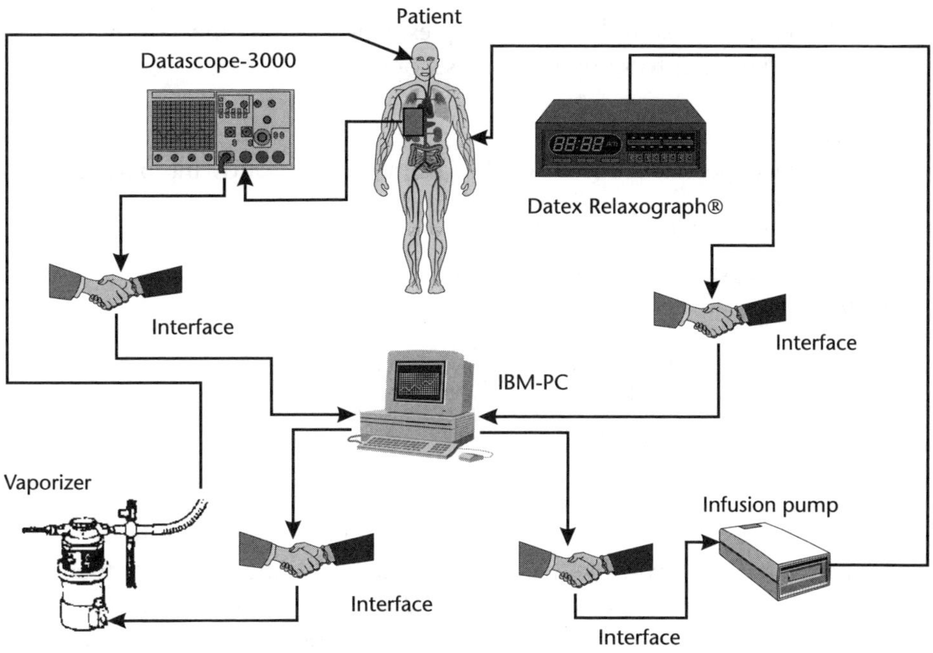

Figure 3.19 The multivariable control system set-up in the operating theater. (*From:* [19]. © 1998 Taylor & Francis Books (UK). Reprinted with permission.)

The links between the DATEX RELAXOGRAPH, the DATASCOPE-3000, and the BRAUN PERFUSOR-SECURA pump are via three RS-232 serial communication ports, whereas the communication with the vaporizer is via the parallel port.

After local Ethics Committee approval, three patients were selected for the clinical trials, and they all underwent surgery, which required muscle relaxation as well as anesthesia. The online controlled infusion was started with the same conditions as in Section 3.3 except that the fixed PI controller, which was operational for 29 minutes, was replaced by the multivariable GPC with fixed parameters derived from average population dynamics. EMG measurements were taken every 20 seconds, and three data measurements were averaged over a 1-minute interval with a 20% EMG target, whereas blood pressure measurements were taken at 1-minute intervals with the target being taken as a drop of 10% from the baseline systolic pressure, which was measured prior to surgery.

For parameter estimation, a UD-factorization method was used on incremental data, with an initial covariance matrix and forgetting factor given by $cov = 10 \cdot I$ and $\rho = 0.999$, respectively. A multivariable P-canonical model of five diagonal time-shifted $A(z^{-1})$ matrices together with five upper triangular time-shifted $B(z^{-1})$ matrices was used. This structure was implemented to absorb any additional time delays within the system. It is worth noting that for the "muscle relaxation" channel EMG readings are obtained and updated every 20 seconds and averaged over three samples to obtain one EMG value over 1 minute. The *same* SAP readings are transmitted through the RS-232 port every 10 seconds but *updated* every minute. Hence, in the following data plots, the time axis in the muscle relaxation channel will be subdivided into samples of 20 samples, whereas the time axis in the SAP channel will be subdivided into samples of 10 seconds.

Patient PXF

For this experiment the GPC algorithm had the following parameters:

$$N_1 = 1 \quad NU(\text{Channel1}) = 1$$
$$N_2 = 10 \quad NU(\text{Channel2}) = 1$$
$$\lambda = 0 \; P(z^{-1}) = \begin{bmatrix} \dfrac{1-0.9z^{-1}}{0.1} & 0 \\ 0 & 1 \end{bmatrix}$$
$$T_{estimation}(z^{-1}) = T(z^{-1})_{control} = (1-0.8z^{-1}) \cdot I$$

The idea behind the choice of a slow $P(z^{-1})$ pole was to force the EMG signal to track the target cautiously without inducing an undershoot, which is undesirable in muscle relaxation therapy. Figure 3.20(a, b), which includes the system's outputs and inputs, respectively, shows the closed-loop control performance under these conditions. The muscle relaxation channel was rather slow to converge to the target due to the choice of $P(z^{-1})$, whereas the SAP channel shows that, despite an isoflurane input of approximately 2%, the SAP would not converge to the 100-

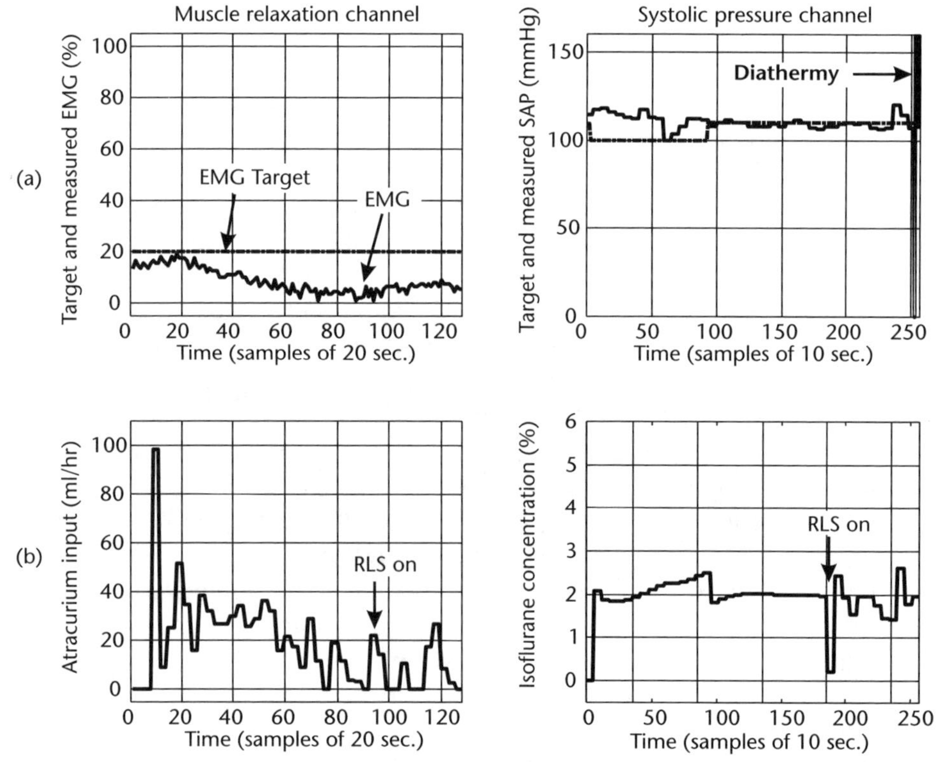

Figure 3.20 Patient PXF: performance of multivariable GPC during surgery: (a) system outputs; and (b) system inputs. (*From:* [19]. © 1998 Taylor & Francis Books (UK). Reprinted with permission.)

mmHg target; this is believed to be because at that time the surgeon was in the process of cutting tissues, which made the blood pressure jump to almost 120 mmHg for approximately 16 minutes, at which time the anesthetist decided that a 110 mmHg target would be more appropriate. This made the output track the target very efficiently with minimum activity even when the adaptation was switched on. The whole trial lasted 40 minutes, after which the surgeon requested the administration of both drugs to be ceased.

Patient JXR

Using the same controller and estimator parameters, a second experiment involving another patient was carried out. Figure 3.21(a, b) shows the result of the trial where it can be seen that the EMG signal was much closer to the 20% target than in the previous trial, except that at time 44 minutes the RELAXOGRAPH device started giving wrong readings due to heavy electrical interference, and as a result the closed-loop control was interrupted. Note that in this case the SAP signal was less settled than in the previous case, although the anesthetist was satisfied with the overall performance.

Patient GXD

For this experiment the GPC algorithm had the following parameters:

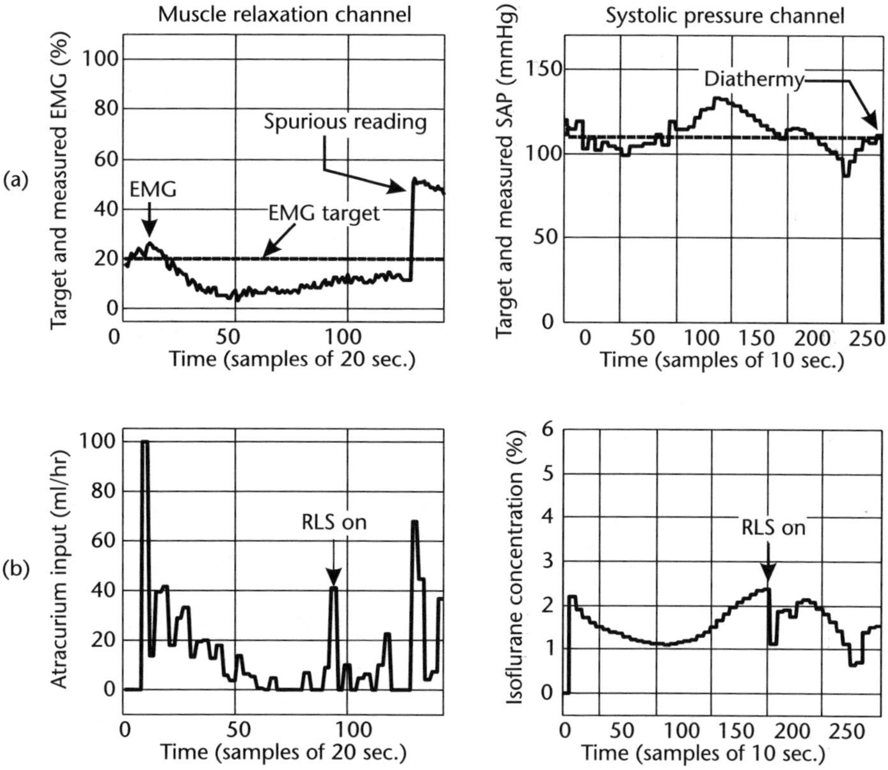

Figure 3.21 Patient JXR: performance of multivariable GPC during surgery: (a) system outputs; (b) system inputs. (*From:* [19]. © 1998 Taylor & Francis Books (UK). Reprinted with permission.)

$$N_1 = 1 \quad NU(\text{Channel1}) = 2$$
$$N_2 = 10 \quad NU(\text{Channel2}) = 1$$
$$\lambda = 0 \quad P(z^{-1}) = \begin{bmatrix} \frac{1-0.7z^{-1}}{0.3} & 0 \\ 0 & \frac{1-0.9z^{-1}}{0.1} \end{bmatrix} I$$

For control a $T_{control} = (1 - 0.85z^{-1})^2 I$ was used, whereas for estimation the filter took the form of $T_{estimation} = (1 - 0.8z^{-1})^2 I$.

This time, the idea behind the choice of the above tuning "knobs" was to make the muscle relaxation channel relatively fast by increasing the number of degrees of freedom (with the obvious risk of a more activated control signal) and choosing a $P(z^{-1})$ such that the muscle relaxation channel is fast, but the blood pressure, which is the interactor channel, is slow. A form of *jacketing* was devised according to the diagram in Figure 3.22, where the limits shown were considered to represent the safe regions of adaptive controller operation. As shown in Figure 3.23(a, b), the outputs in both channels tracked the set-points reasonably well until the 30th minute, where the adaptation was switched on. At that time, the blood pressure rose drastically due to surgical stimuli, and the input responded with a high isoflurane input, which, in turn, produced the interaction in the muscle relaxation channel that caused the switching off of the adaptation according to the jacketing policy of Figure 3.22.

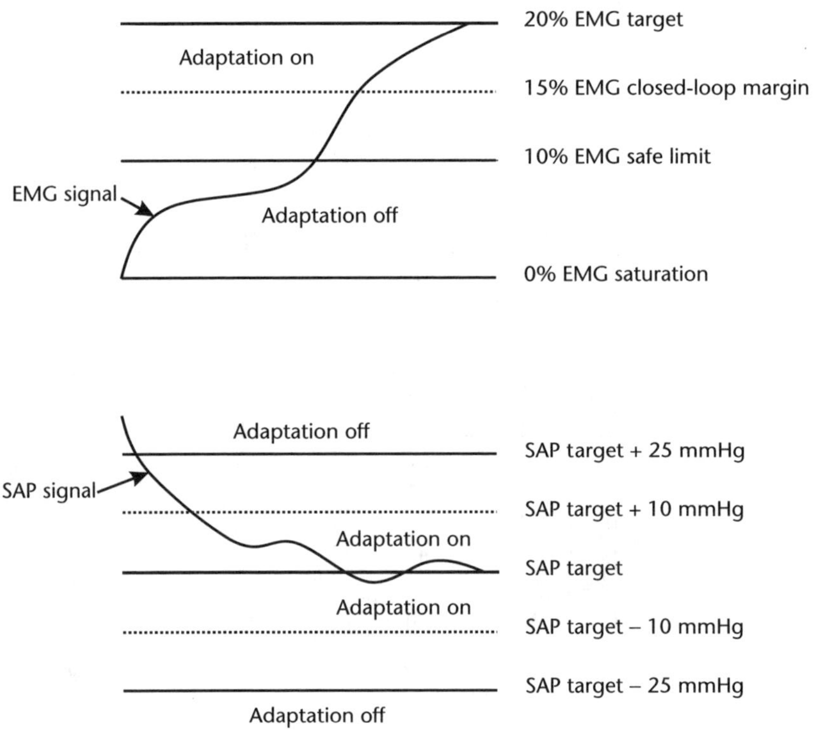

Figure 3.22 Diagram representing the EMG limits used within the jacketing procedure. (*From:* [19]. © 1998 Taylor & Francis Books (UK). Reprinted with permission.)

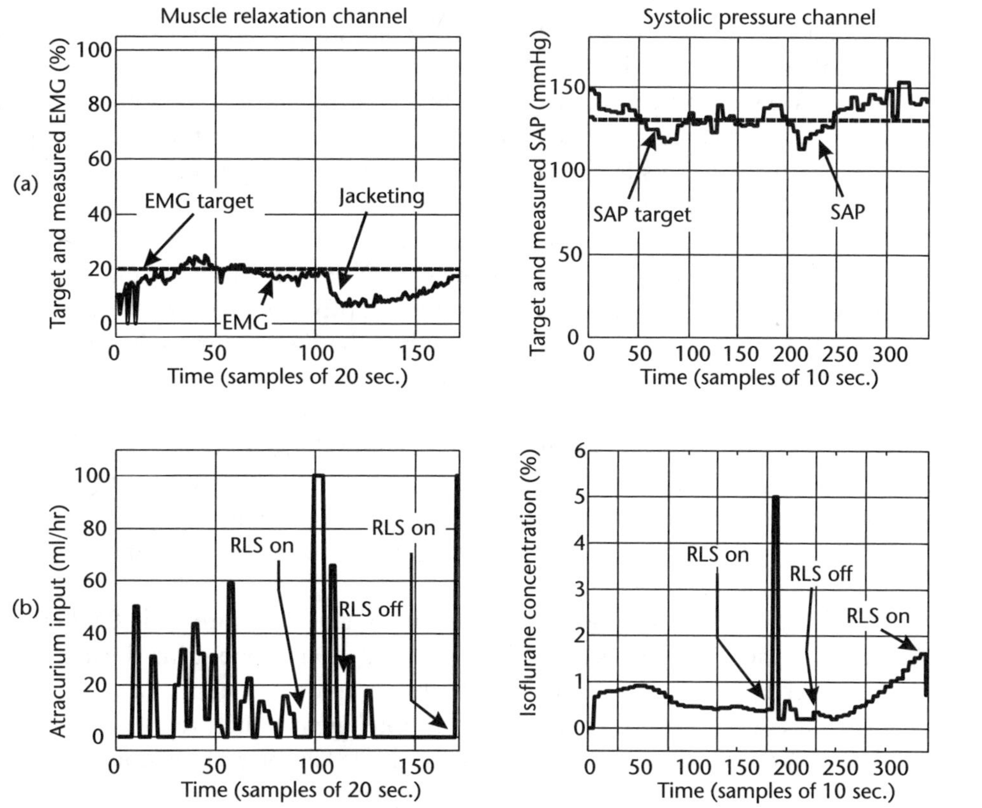

Figure 3.23 Patient GXD: performance of multivariable GPC during surgery: (a) system outputs; and (b) system inputs. (*From:* [19]. © 1998 Taylor & Francis Books (UK). Reprinted with permission.)

However, in the clinical validation report sent to us by the anesthetist, the performance was considered to be good.

3.6 Conclusions

The application of adaptive GPC to online control of drug infusion in anesthesia has been reviewed. The study focused on two system configurations: SISO GPC and multivariable GPC. In the first case, GPC has been successful in achieving two primary goals in the operating theater. These are the maintaining of a steady level of paralysis, crucial to the surgeon performing the operation, together with minimum deviation from the target set point and reduction of total muscle relaxant dosage. Perhaps one of the most important aspects explored during these trials is that of data filtering. Indeed, it has been shown that the inclusion of the filter within the control and estimation algorithms leads to a less active control signal. In this case, a second-order filter was found to be adequate.

The second case considered the extension of the above work to simultaneous control of muscle relaxation as well as unconsciousness (via blood pressure measurements). The multivariable control system proved a very challenging process, since we had to make the computer communicate with four devices at each sampling time by making sure that it (the computer) read two signals and deliver two signals successfully. In this unique work we succeeded in conducting three trials where we were able to control muscle relaxation as well as systolic arterial pressure by online infusions of atracurium and isoflurane. From these encouraging preliminary results, various lessons can be learned: (1) as far as the muscle relaxation channel is concerned, we cannot afford to have an overactive channel by choosing a larger control horizon *NU* and a fast pole for the model-following polynomial; (2) a smoother approach to the set point should be encouraged; and (3) some form of jacketing is necessary to switch from the adaptive mode to the fixed mode, especially near the nonlinear zone (in our case when the EMG approaches 10%).

As far as blood pressure is concerned, it is widely agreed that the SAP variable is not easy to control due to various conditions independent of the feedback loop, such as blood loss and the patient's state of mind. Suffice to say that GPC managed to maintain the SAP signal within reasonable bounds acceptable to anesthetists; it is worth noting here that when controlling blood pressure as an inferential variable to monitor unconsciousness, it is not always necessary to obtain tight control; a tolerance range is sufficient.

In the trial with Patient JXR, the phenomenon of unreliable EMG measurements was experienced. This is a common phenomenon in operating theaters where electrical interference from surgical diathermy is quite frequent. To alleviate such problems, it is necessary to superimpose an intelligent structure by adding a supervisory layer that monitors the variables involved (see Chapter 5). By doing so, the controller will be able to successfully accomplish three tasks: control, adaptation, and supervision. Chapter 5 will consider certain aspects relating to fault tolerant systems.

Finally, it is worth mentioning that "measurement" is the *one* aspect that came out strongly in this research work; that is, in relation to the *integrity* of the various measurements obtained (are we actually measuring what we think we are measuring?), but also in relation to what useful (if any) information are these variables carrying. More investigations into other meaningful variables and "whiter" models will continue in the following chapters.

References

[1] Jacobs, O. L. R., et al., "Modeling Estimation and Control in the Relief of Post-Operative Pain," *Automatica*, Vol. 21, 1985, pp. 349–360.

[2] Linkens, D. A., S. G. Greenhow, and A. J. Asbury, "Clinical Trials with the Anaesthetic Expert Adviser RESAC," *Expert Systems in Medicine* 6, London, U.K., June 11–18, 1990.

[3] Cass, N. M., et al., "Computer-Controlled Muscle Relaxation: A Comparison of Four Muscle Relaxants in Sheep," *Anesthesia in Intensive Care*, Vol. 4, 1976, pp. 16–22.

[4] Brown, B. H., et al., "Closed-Loop Control of Muscle Relaxation During Surgery," *Clinical Physics and Physiological Measurement*, Vol. 1, 1980, pp. 203–210.

[5] Robb, H. M., et al., "Towards Automatic Control of General Anaesthesia," presentation at the *Conference on Medical Informatics '88*, Nottingham, U.K., 1988, pp. 121–126.

[6] MacLeod, A. D., et al., "Automatic Control of Neuromuscular Block with Atracurium," *British Journal of Anaesthesia*, Vol. 63, 1989, pp. 31–35.

[7] Robb, H. M., "Towards a Standardized Anaesthetic State Using Isoflurane and Morphine," *British Journal of Anaesthesia*, Vol. 71, 1993, pp. 366–369.

[8] Clarke, D. W., C. Mohtadi, and P. S. Tuffs, "Generalized Predictive Control: Part I and II," *Automatica*, Vol. 23, No. 2, 1987, pp. 137–160.

[9] Ward, S., "Pharmacokinetics of Atracurium Besylate in Healthy Patients (After a Single IV Bolus Dose)," *British Journal of Anaesthesia*, Vol. 55, 1983, p. 113.

[10] Weatherley, B. C., S. G. Williams, and E. A. M. Neil, "Pharmacokinetics, Pharmacodynamics and Dose Response Relationship of Atracurium Administered i.v.," *British Journal of Anaesthesia*, Vol. 55, 1983, p. 39S.

[11] Sheiner, L. B., "Simultaneous Modeling of Pharmacokinetics and Pharmacodynamics: Application to d-tubocurarine," *Clinical Pharmacology and Therapeutics*, Vol. 25, 1979, p. 358.

[12] Paton, W. D. M., and D. R. Waud, "The Margin of Safety of Neuromuscular Blockade Transmission," *Journal of Physiology (London)*, Vol. 191, 1967, p. 59.

[13] Waud, D. R., and B. E. Waud, "The Relation Between Tetanic Fade and Receptor Occlusion in the Presence of Competitive Neuromuscular Block," *Anesthesiology*, Vol. 35, 1971, p. 456.

[14] Whiting, B., and A. W. Kelman, "The Modeling of Drug Response," *Clinical Science*, Vol. 59, 1980, pp. 311–314.

[15] Cutler, C. R., and B. L. Ramaker, "Dynamic Matrix Control: A Computer Control Algorithm," Paper WP5-B, *Proceedings of the Joint Automatic Control Conference*, San Francisco, CA, 1980.

[16] Clarke, D. W., and P. J. Gawthrop, "Self-Tuning Controller," *Institution of Electrical Engineers Proceedings, PtD, Control Theory and Applications*, Vol. 122, No. 9, 1975, pp. 929–934.

[17] Clarke, D. W., "Implementation of Self-Tuning Controllers," in C. J. Harris and S. A. Billings, (eds.), *Self-Tuning and Adaptive Control: Theory and Application*, London, U.K.: Peter Peregrinus, Vol. 15, 1985, pp. 146–147.

[18] Bierman, G. J., *Factorization Methods for Discrete Sequential Estimation*, New York: Academic Press, 1977.

[19] Mahfouf, M., and D. A. Linkens, *Generalised Predictive Control and Bioengineering*, London, U.K.: Taylor & Francis Publishers, 1998.

[20] Ziegler, J. G., and N. B. Nichols, "Optimum Settings for Automatic Controllers," *Transactions of the American Society of Mechanical Engineers*, Vol. 64, 1942, p. 759.

[21] Schwilden, H., H. Stoeckel, and J. Schuttler, "Closed-Loop Feedback Control of Propofol Anaesthesia by Quantitative EEG Analysis in Humans," *British Journal of Anaesthesia*, Vol. 62, 1989, pp. 290–296.

[22] Schwilden, H., J. Scuttler, and H. Stoeckel, "Closed-Loop Feedback Control of Methohexital Anesthesia by Quantitative EEG Analysis in Humans," *Anesthesiology*, Vol. 67, 1987, pp. 341–347.

[23] Savege, T. M., et al., "Preliminary Investigation into a New Method of Assessing the Quality of Anaesthesia: The Cardiovascular Response to a Measured Noxious Stimulus," *British Journal of Anaesthesia*, Vol. 50, 1978, pp. 481–487.

[24] Breckenridge, J. L., and A. R. Aitkenhead, "Awareness During Anaesthesia: A Review," *Annals of the Royal College of Surgeons of England*, Vol. 6, 1983, pp. 93–96.

[25] Gray, W. M., and A. J. Asbury, "Measurement and Control of Depth of Anaesthesia in Surgical Patients," *3rd IMEKO Conference on Measurement in Clinical Medicine*, Edinburgh, U.K., 1986, pp. 167–172.

[26] Schills, G. F., F. G. Sasse, and V. C. Rideout, "Automatic Control of Anesthesia Using Two Feedback Variables," *Annals of Biomedical Engineering*, Vol. 15, 1987, pp. 19–34.

[27] Millard R. K., et al., "Controlled Hypotension During ENT Surgery Using Self-Tuners," *Biomedical Measurement, Informatics, and Control*, Vol. 2, No. 2, 1988a, pp. 59–72.

[28] Millard R. K., C. R. Monk, and C. Prys-Roberts, "Self-Tuning Control of Hypotension During ENT Surgery Using a Volatile Anesthetic," *Institution of Electrical Engineers Proceedings, PtD, Control Theory and Applications*, 125, 1988b, pp. 95–105.

[29] Graupe, D., *Identification of Systems*, Malabar, FL: Robert Krieger Publishing Company, 1976.

CHAPTER 4

A New Generic Approach to Model Reduction for Complex Physiologically Based Drug Models

As was shown in the previous chapter, compartmental modeling is a rather crude representation of the dynamics associated with drugs kinetics as they lump the whole human body into a very limited number of sites where it is believed these drugs act. In this chapter, a rather detailed approach of representing the body is reviewed via physiological modeling. As a result, the well-known Mapleson model is analyzed and formulated for anesthetic drugs. The identification problem relating to this type of model is explored and a new *generic* approach is proposed that bridges the gap between the *white* box and the popular *black* box model representations. Finally, the modeling approach is extended to include an architecture deemed more amenable to systematic control design.

4.1 Introduction

In pharmacology the understanding and exploration of the interrelationships between drug dose, concentration, effect and time is crucial. While pharmacokinetics allows one to describe and to predict the time course of drug concentration, pharmacodynamics considers the relationships between drug concentration and its effects (see Chapter 3). Figure 4.1 depicts pharmacokinetics and pharmacodynamics as the fundamental elements of pharmacology.

There has been an increasing interest in pharmacokinetics modeling in recent years, and many concepts and trial models have been proposed, and some of them validated, as a result. Among the proposed models, one can cite empirical models [1], compartmental models [2–4], and physiological models [5]. Tucker [6] reviewed the evolution of various concepts in pharmacokinetics and pharmacodynamics, but the paragraphs below give a brief summary relating to these three different categories of models.

Empirical models relate the drug dose to the drug concentration using analytical expressions. In the case of a drug displaying linear kinetics, such drug concentration (output of the system) can be described using the sum of exponential functions, each exponential representing a particular time constant.

Compartmental modeling consists of lumping the whole body into a series of "compartments" that adequately fit the experimental data, which are related to the drug concentration history in the blood or plasma; although in a good number of

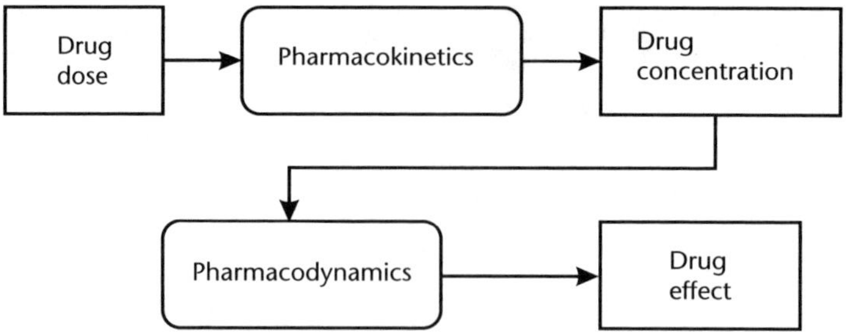

Figure 4.1 The bridge between pharmacokinetics and pharmacodynamics.

cases such a structure can represent a reasonable approximation to the underlying physiology of the system under investigation. Such conceptual compartments may be assigned separate locations where drug resides, such as plasma, urine, tissue, and drug metabolism. The precondition for a compartmental model to be valid is that the drug concentration within each compartment must be *homogenous*. This means that the drug behaves identically and equilibrates[1] instantly within any of the compartments. Should a particular model include more than two compartments, these can be arranged in either mammilliary or catenary form as shown in Figure 4.2.

Physiological modeling is a branch of pharmacology that saw a drastic resurgence in the late 1970s. It consists of defining structures whose mathematical characteristics mirror, as closely as possible, those real tissues and organs by providing an insight into the actual processes occurring. Unlike compartmental modeling,

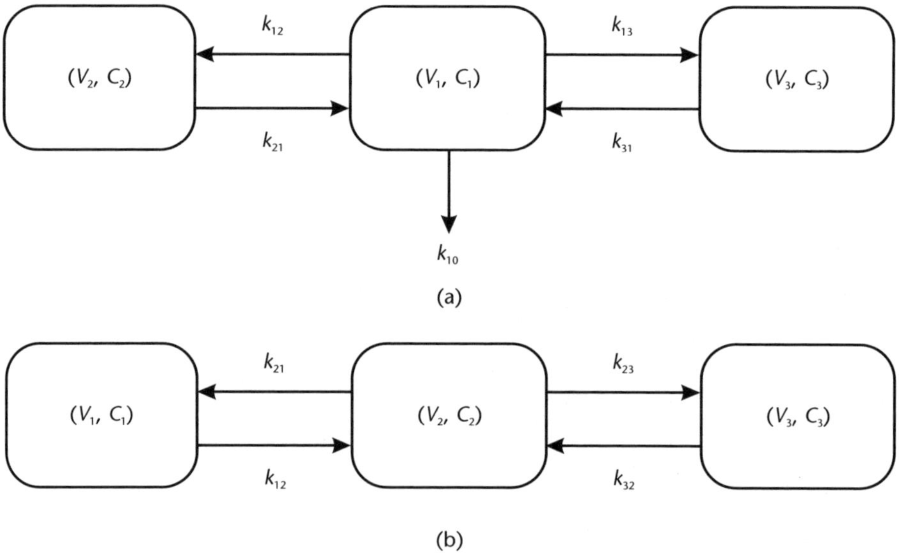

Figure 4.2 Two configurations for a three-compartment model: (a) mammilliary form; and (b) catenary form. (V_i, C_i refer to volumes and concentrations, respectively. K_{ij} refers to the rate of distribution.)

1. Equilibration (or equilibrium) is a steady-state phase when the tissue acquires the same drug concentration as the blood washing (perfusing) it.

which proposes, in some cases, a structure after measurement of the experimental response to be simulated, physiological modeling consists of models that are developed a priori in that experimental data are used to propose a model before any experimental response is available [5].

Physiological models possess advantages as well as disadvantages; the advantages can be summarized as follows:

- Physiological models provide a heuristic insight into processes that determine drug concentration under various physiological conditions.
- Physiological models have the ability to incorporate feedback (e.g., drug concentration) which influences the model.
- Physiological models can be tested by comparing the predicted values of drug concentrations in various parts of the model with those taken from observation.
- Physiological models allow scaling from animal (e.g., rat) to human [7].
- Physiological models lead to better accuracy in determining the drug levels that allow one to achieve adequate effect and hence give the opportunity to alter drug administration regimes.

Conversely, physiological models have several disadvantages:

- Physiological models rely on precise information, which in some cases is difficult or impossible to measure. Hence, various assumptions are made with respect to drug distribution to organs such as *the perfusion limitation* concept.
- Mathematical equations describing the structures involved in the model do not always lead to analytical solutions. These are obtained via digital numerical computations.
- It is difficult to prove that a particular physiological representation of a model is valid. Indeed, as Jarvis noted in his paper [5], the fact that the results of a model (predicted) match those measured does not prove in any way its validity. In fact, two different models may produce the same results, which match the experimental data.

To the question of how can one obtain two different representations of a physiological model associated with a particular drug, the answer would be that it very much depends on the assumptions that are made prior to the building of such a model. Among these assumptions one can mention the inclusion of a blood flow heterogeneity, or *diffusion-limited* distribution, of the drug, and also, the lumping of various organs into one believed to be a faithful representative of them all.

Perhaps one of the biggest challenges facing physiological models is their ability to predict drug concentration levels at time intervals less than a minute (i.e., the first few seconds that follow drug administration.) In order to assemble model structures that are capable of doing just that, it is vital to increase our knowledge of the exchange processes occurring between capillaries and tissues [8], including the movement of drug through membranes and the mechanisms behind blood flow.

4.1.1 Drug Movement Through Membranes

Transport Processes

The transport of drugs can be described as the movement across a series of membranes and spaces. Each of the interposing cellular membranes and spaces hampers drug transport to varying degrees, and any one of them can rate-limit the overall process. It is this complexity of structure that makes quantitative predictions of drug transport difficult.

Membrane transport processes can be classed as passive or active. Passive processes do not require any energy expenditure or activity of the cell membranes. The particles move by using energy that they already have. In contrast, active processes do require the expenditure of metabolic energy by the cell. Hence, the transported particles are actively "pulled" across the membrane.

Active transport is distinguished from passive transport in that the rate of transport in the latter increases linearly with concentration, whereas the rate of transport in the former approaches a maximum value at high concentrations [9].

4.1.2 Blood Flow

Blood consists of a fluid, cells, and specialized cell fragments called platelets. The fluid portion of blood (plasma) is one of the three major body fluids (interstitial and intracellular fluids are the other two). Blood is a complex transport medium that performs vital pickup and delivery services for the body. Accordingly, viewing any tissue as a whole, the movement of a drug through membranes is always associated with the concept of *perfusion*, which is usually expressed in units of millimeter per minute per volume (or mass) of tissue.

In lipophilic drugs, or drugs that pass freely through the tissue pores and channels (such as analgesics and certain anesthetics), typical membranes offer virtually no barrier to drug movement. However, in general, the limiting step controlling the rate of movement of drug across a membrane from blood to tissue, or the inverse, varies. As Figure 4.3 shows, if the membrane offers no resistance, drug in the blood leaving the capillary and in the tissue is virtually in equilibrium; blood and tissue may be viewed as one. Here, movement of the drug is limited by blood flow; the process is said to be *perfusion-limited*. As membrane resistance to the drug increases, the rate limitation moves away from one of perfusion to one of *permeability*. Nearly all physiological models have assumed that the distribution of drug to tissues is flow-limited. Tissues are individually modeled as a simple compartment whereby the concentration of the active drug in blood perfusing a tissue bed comes to equilib-

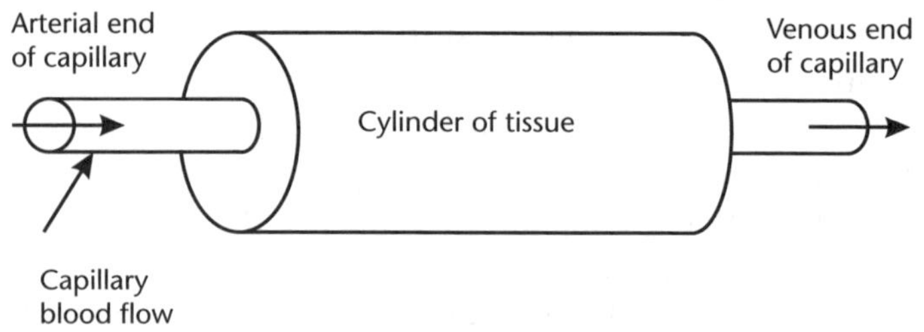

Figure 4.3 The idealized concept of the Krogh cylinder model.

rium with it leading to the concentration in the effluent blood equaling the concentration in the tissue. This assumption, however, can only hold under near steady-state conditions and not under conditions where tissue transients are significant. For instance, Wada et al. took into account the fact that with some model tissues the impediment to drug diffusion was so high that they could not be modeled by a perfusion-limited model. Bjorkman [7] observed that fentanyl and alfentanil pharmacokinetics could not be described by a flow-limited model due to the existence of diffusion barriers within the tissue. Because of the existence of several assumptions regarding the blood distribution in capillaries and tissues, the next section looks at the various postulated capillary-tissue models.

4.2 Models Associated with Capillary-Tissue Exchange

One of the earliest attempts at modeling capillary-tissue exchange was carried out by Krogh [10], whose concept of modeling the region of muscle tissue deriving nutrients from a single capillary was later to be known as the Krogh cylinder model. Later, Renkin [11] used the same concept but slightly modified to derive an analytical equation with respect to the fraction of drug extracted during its passage through the capillary.

In contrast to the Krogh cylinder model, Lutz [12] modeled the organ as a three-compartment model each separated by membrane barriers. It is the aim of this section to review the theory behind these concepts.

4.2.1 Krogh's Cylinder Model

An ideal capillary-tissue unit is shown in Figure 4.4, and this is known as the Krogh cylinder model since Krogh's use of it in his calculations on oxygen delivery. In this arrangement, exchange occurs only in capillary-tissue regions and not in large vessels. A fundamental prerequisite is that flow in adjacent capillaries and volumes in adjacent capillary-tissue regions are similar. Although the liver and heart microvasculatures can be described by parallel capillary-tissue units, the capillaries of bone, brain, and mesentery have much more variable arrangements [13–15].

Although it is not the objective of this chapter, and indeed this book, to review comprehensively this particular model and others in general, suffice it to say that Rose et al. [16] developed the first analytical model for a single capillary-interstitial fluid cell using a series of nontrivial differential equations whose solutions can be found in Bassingthtwaighte and Goresky [17].

4.2.2 Crone and Renkin's Idea

These authors [11, 18] assumed that diffusible material, which had left the capillary, was immediately and completely isolated, so that none of it could return to the capillary lumen and that the rate of loss of this material was proportional to the remaining concentration at each point [19].

Consider the situation in which drug transfer across a capillary wall results in an homogeneous interstitial fluid drug concentration C_t, and that the concentration

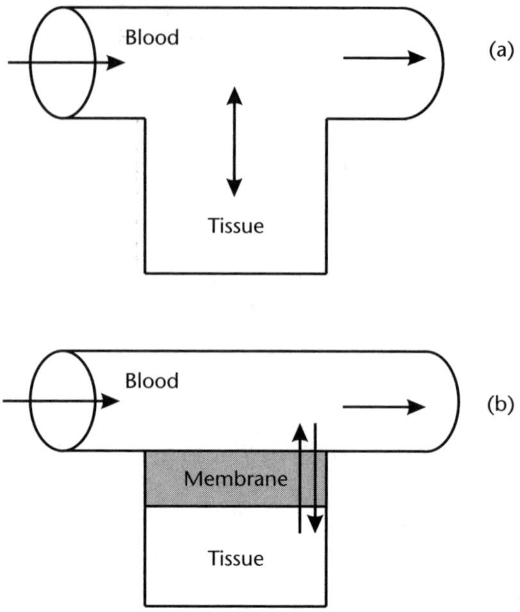

Figure 4.4 Diagram illustrating the two causes to consider when dealing with drug movement across a membrane: (a) perfusion-rate limitation; and (b) permeability-rate limitation.

of drug in blood leaving the capillary is equal to C_t. If the drug concentration entering an infinitesimal capillary segment of length dl, thickness b, and surface area dA is given by C, and X is the total amount of drug in the capillary segment, then by Fick's First Law of Diffusion,

$$\frac{dX}{dt} = -\frac{D \cdot dA}{\delta}(C - C_t) \tag{4.1}$$

For a cylindrical capillary of surface area A and a volume V, one can write that

$$dA = \frac{dl}{L} A \tag{4.2}$$

$$dV = \frac{dl}{L} dV \tag{4.3}$$

The change in drug concentration within the capillary segment is given by

$$dC = \frac{dX}{dV} \tag{4.4}$$

Following lengthy transformations (beyond the scope of this chapter), the following expression for the total amount of drug in the capillary segment is obtained:

$$\frac{dX}{dt} = P \cdot \overline{\Delta} = Q \cdot \left(1 - e^{-(P/Q)}\right) \cdot (C_b - C_t) \tag{4.5}$$

where P is the permeability coefficient of the drug, Q is the flow through the capillary, C_b is the drug concentration in blood entering the capillary, and $\overline{\Delta}$ is the mean gradient which determines the rate of drug transfer across the capillary segment.

Hence, drug clearance across the capillary wall is given as follows:

$$CL = q \cdot \left(1 - e^{-(P/Q)}\right) \tag{4.6}$$

Equation (4.6), which is known as the Crone-Renkin equation, can be used to determine whether the distribution of a drug is flow-limited or diffusion-limited. Indeed, such a decision can be made once the power in the exponent is evaluated; that is,

If $P >> Q$, then $e^{-P/Q} \rightarrow 0$ and $CL = Q \Rightarrow$ the distribution is flow-limited.

If $P << Q$, then $e^{-(P/Q)} = 1 - \frac{P}{Q} + \left(\frac{P}{Q}\right)^2 - \ldots$ and $CL = P \Rightarrow$ the distribution is diffusion-limited.

At this stage it is worth noting that many forms of the Renkin-Crone equation exist, but all are equivalent. For instance, an expression of the extraction ratio $E = 1 - e^{(P/Q)}$ is often used to make the above diagnosis (i.e., if $E \approx 1$, then the distribution is flow-limited) [7].

4.2.3 Kety's Model

Kety observed in his 1951 paper [20] that Fick's principle may be stated as follows:

> The amount of drug taken up by the tissue per unit of time is equal to the quantity brought to the tissue by the arterial blood minus the quantity carried away in the venous blood, i.e.,

$$\frac{dX_t}{dt} = Q_t \cdot (C_a - C_v) \tag{4.7}$$

where X_t is the amount of drug in the tissue, C_a is the arterial drug concentration, C_v is the venous drug concentration, and Q_t is the blood flow, which is assumed to be constant.

Equation (4.7) can also be rewritten as follows:

$$\frac{dC_t}{dt} = \frac{1}{V_t}\frac{dX_t}{dt} = \frac{Q_t}{V_t} \cdot (C_a - C_v) \tag{4.8}$$

where V_t is the tissue volume.

Kety borrowed one of Zunt's basic assumptions [21], which is that venous blood from a tissue is in equilibrium with the tissue itself (nowhere in the tissue will there be concentration gradients); that is,

$$C_t = \lambda_{tb} \cdot C_v \tag{4.9}$$

with λ_{tb} being a partition coefficient between tissue and blood.

Substituting (4.9) into (4.8) gives

$$\frac{dC_t}{dt} = -\frac{Q_t}{\lambda_{tb} \cdot V_t} \cdot (C_v - (\lambda \cdot C_a)) \tag{4.10}$$

If the venous concentration is sought, the above substitution leads to

$$\frac{dC_v}{dt} = -\frac{Q_t}{\lambda_{tb} \cdot V_t} \cdot (C_v - C_a) \tag{4.11}$$

For a saturation process (i.e., $C_a \neq 0$), one can write

$$C_t = \lambda \cdot C_a \cdot \left(1 - e^{-kt}\right) \tag{4.12}$$

$$\begin{aligned} C_v &= C_a\left(1 - e^{-kt}\right) \\ k &= \frac{Q_t}{\lambda \cdot V_t} \end{aligned} \tag{4.13}$$

However, for a desaturation process, also known as washout (i.e., $C_a = 0$), the following equations can be obtained:

$$C_t = C_{t0} \cdot e^{-kt} \tag{4.14}$$

$$C_v = C_{v0} \cdot e^{-kt} \tag{4.15}$$

An attempt was made by Kety [20] to take the diffusion process into consideration by introducing a factor m into (4.8) such that

$$\frac{dC_t}{dt} = \frac{m}{\lambda}\frac{Q_t}{V_t}(\lambda C_a - C_v) \tag{4.16}$$

where $m = 1 - e^{-D \cdot S/Q_t}$ with D being the diffusion coefficient and S the surface area.

However, this attempt seems to have attracted many criticisms, such as that of Goresky [19], because, from their point of view, the way in which a permeability phenomenon is introduced is internally inconsistent with the implication of (4.9) that diffusion is everywhere instantaneous. They concluded that use of the constant m in the model cannot change Kety's model in a significant way. They also concluded that when the implications of restricted capillary permeability are considered in detail, it is apparent that diffusion equilibration cannot everywhere be instantaneous and that to achieve a complete description of the system, one must use a spatially distributed model.

4.2.4 The Concept of the In Vivo Approach to Membrane Transport

Still in the domain of lumped compartments, Lutz et al. [12] developed a simple organ compartment model whereby each tissue is represented by three entities: tissue plasma, interstitial space, and intracellular space, separated between each other by membrane barriers as shown in Figure 4.5. In this representation, the drug enters the compartment's plasma space with the arterial flow at some flow rate Q, and at the arterial concentration C_p. The drug moves from the plasma to the interstitial fluid at some net rate $Flux_1$, and it is transferred from the interstitial fluid to the intracellular compartment at another net rate $Flux_2$. It is worth noting that the drug may bind to soluble proteins, cell membranes, or subcellular particles, so that the drug may be characterized by its total concentration or free concentration. The drug exits the plasma compartment at the venous concentration, which equals that of the plasma concentration; in other words, blood equilibrates with the plasma compartment. Let us take each compartment and write the corresponding mass balance equations of the following general form:

Accumulation = sum of influx of drug − sum of efflux of drug

For plasma:

$$V_1 \frac{dx_1}{dt} = Q \cdot x_p - Q \cdot x_1 - Flux_1 \tag{4.17}$$

For interstitial fluid:

$$V_2 \frac{dx_2}{dt} = Flux_1 - Flux_2 \tag{4.18}$$

For intracellular space:

$$V_3 \frac{dx_3}{dt} = Flux_3 \tag{4.19}$$

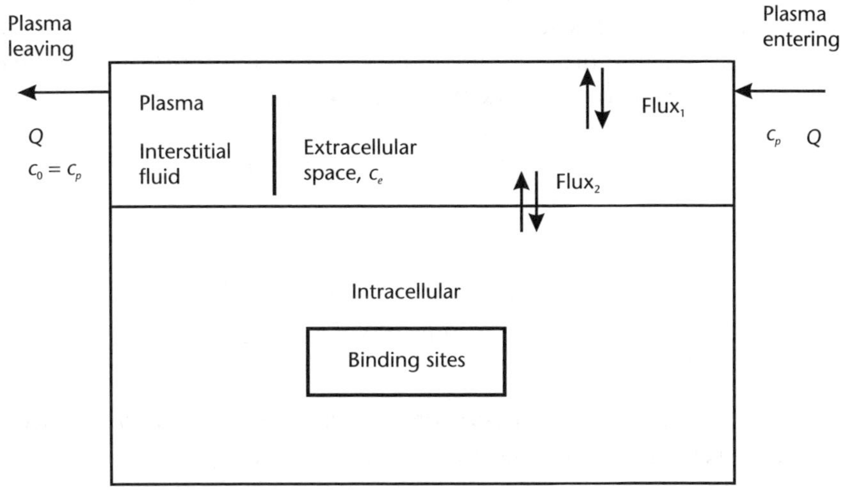

Figure 4.5 A Schematic representation of a lumped compartment as described by Lutz. (*From:* [12]. © 1980 Elsevier. Reprinted with permission.)

where:

V is the compartment volume.

Q is the blood flow.

x_i is the total concentration (free and bound) in compartment i.

t is time.

In order to solve (4.17) through (4.19), four types of information must be available:

1. Anatomical information such as compartment volumes V_1, V_2, V_3. These can be determined from literature. Plasma or extracellular volumes can be determined from experiments using marker molecules.
2. Physiological information such as blood flow rate Q, clearance and rates of metabolism.
3. Thermodynamic information such as the relationship between the total concentration x_i (free and bound) and the free concentration C. Hence, three cases may emerge:

$$x = C(\text{total} = \text{free}) \tag{4.20}$$

$$x = R \cdot C(\text{total concentration proportional to free concentration}) \tag{4.21}$$

$$x \frac{a \cdot C}{E + C} \tag{4.22}$$

(nonlinear Michaelis-Menten binding with occurrence of saturation of binding sites; a: the maximum binding capacity; E: the dissociation constant).

4. Transport information that describes the transfer of drug across the capillary membrane $Flux_1$ and across the cell membrane $Flux_2$. As we saw in the previous sections, Pick's first principle involves a simple first-order exponential process where the net flux is proportional to the concentration gradient across the membrane; that is,

$$Flux = P \cdot A \cdot (C_1 - C_2) \tag{4.23}$$

where:

$P \cdot A$ is the permeability-area product for the membrane.

C_1 and C_2 are the free concentrations on each side of the membrane.

In the case of cell transport being facilitated by a carrier process as described previously, the net flux is best described by an equation that exhibits saturable transport; that is,

$$Flux = \frac{k_1 V_1 C_1}{K_1 + C_1} - \frac{k_2 V_2 C_2}{K_2 + C_2} \tag{4.24}$$

where the first term on the right side of (4.24) is the trans-membrane influx, the second term is the trans-membrane efflux, k_1 and k_2 representing the maximum transport rates.

Using proper assumptions, the plasma and interstitial fluid extracellular compartments of Figure 4.5 can be lumped to form one extracellular compartment at equilibrium, where it is assumed that the transport of drug across the capillary membrane is so rapid that both compartments are at equilibrium. Hence, the previous three mass-balance equations can be reduced to only two equations, such that

$$V_e \frac{dx_e}{dt} = Q \cdot x_p - Q \cdot x_3 - flux \tag{4.25}$$

and

$$V_i \frac{dx_i}{dt} = flux \tag{4.26}$$

where the e and i subscripts refer ton the extracellular and intracellular spaces, respectively. Here again, Lutz et al. [12] considered the two cases of flow-limited and membrane-limited, or diffusion-limited, models.

When the flux across the cell membrane is faster that the perfusion rate (i.e., $P > Q$), then the intracellular compartment is considered to be in equilibrium with the extracellular concentration leading to the reduction of the system of (4.25) and (4.26) to one single equation of the following form:

$$V_t \frac{dx_t}{dt} = Q \cdot \left(x_p - x_e\right) \tag{4.27}$$

where x_p is the total tissue concentration and x_e is the venous concentration exiting the compartment and is assumed to be in equilibrium with that of the tissue as in (4.9).

Conversely, when the flux across the cell membrane is slower than the tissue perfusion (i.e., $P \cdot A < Q$), then the uptake into the intracellular compartment is governed by the cell membrane permeability and the membrane area. Therefore, the mass balance equations (4.25) and (4.26) are valid in this case.

4.2.5 Discussions

This study focused on the theoretical background relating to capillary-tissue exchange in physiological modeling. First, a section was devoted to the drug movement across the membrane where we distinguished between passive transport, which is a spontaneous process not caused by any external interference, and active transport, which is a process helped by cellular energy. All processes described in this study were assumed to be passive. The movement of the drug through membranes can be limited by blood flow or membrane permeability. In the former case the process is said to be perfusion-limited, and in the latter case it is said to be diffusion-limited. The majority of physiological models are built around the assumption of instantaneous equilibration between blood and tissue drug concentration (perfusion limitation process), which afforded a tremendous ease on the computation side, on one hand, but also caused the hence-built models to be inaccurate as far as predicting drug behavior in various tissues, especially in the first few seconds of drug administration. This has prompted various researchers to reopen the case for allowing the diffusion processes to be included in any modeling step. The report focused on four major ideas ranging from the most complicated to the simplest idea: the Krogh model (complicated), Crone and Renkin's idea (moderate), Lutz's model (moderate to simple), and Kety's model (simple). Kety's model was the simplest of them all in that instant equilibration between blood and tissue is assumed through Fick's First Law of Diffusion. Lutz's [12] model has been used extensively by Wada et al. [22], Ebling [23], and Bjorkman [7]; it consisted of defining a representation of a lumped compartment comprising plasma, interstitial fluid, and intracellular space where drug exchanges between them occurs according to well-defined rate constants. The differential equations are straightforward and easy to solve. Wada reported that using his models, he successfully matched predicted and measured concentration data provided that an appropriate pulmonary pharmacokinetic model is also considered. Crone and Renkin's idea consisted of using the 12 assumptions used by Krogh to define his model plus two more to simplify it even further. This resulted in a structure that depended on the permeability surface; the equations hence involved are easy to solve and the final expression relating the clearance to the blood flow and permeability allows one to declare whether the process is flow-limited or diffusion-limited. This model has also been criticized for being incomplete since the return of material to the lumen has been neglected. Finally, the distributed model of Krogh, whose equations were developed by Bassingthwaighte and Goresky [17] and Goresky [19], is the most complete and also the most involved compared to the others. The analytical solutions were developed for the plasma-interstitial fluid exchanges but also for the plasma-interstitial fluid-cell exchanges, which are even more complicated. The equations are said to be valid for the majority of organs but definitely not for others such as bone, brain, and mesentery, which have much more variable arrangements.

In the light of the above considerations that relate to the question of *which is the most appropriate model that can reproduce most physiological phenomena that occur in the human body once a drug has been administered as faithfully as possible and concomitantly is simple enough to allow its easy quantification, implementation as well as it exploitation*, it is of the opinion of this author that the structure proposed by Mapleson [24] and Davis [25, 26] represents the best compromise in the sense of Einstein's principle of parsimony.[2] It is in this spirit that the next study, which is described in the next sections, has been conducted. The study shows how to bridge the gap between a "white" model description to a "grayish" model without the loss of vital information as far as a particular exploitation of this model is concerned. The drug fentanyl was considered to be the agent under investigation without loss of generality, as this has been chosen purely on the basis of available data.

2. This keeps in mind the fact that a model is just as good as what it is intended to be used for.

4.3 The Mapleson-Higgins Flow-Limited Model for Fentanyl

4.3.1 Fentanyl Pharmacokinetics

Fentanyl is a potent synthetic opioid that is considered to be a prototype of a series of narcotic analgesics which include alfentanil and sufentanil. Its attractive properties include the relative lack of cardiovascular side effects. Fentanyl is a high lipid soluble agent that should enable it to equilibrate quickly across biological membranes. At a pH of 7.4, fentanyl is about 80% bound to plasma proteins. Protein binding is decreased and comparative partitioning into red cells is increased at low plasma protein concentrations. After administration of a large dose of fentanyl, during the period of falling plasma concentrations, the plasma concentration curve may spontaneously rise again before continuing to fall; the release of fentanyl from muscle, stomach, or lungs have been suggested as a possible cause. These secondary rises can be associated with respiratory depression appropriate to the higher concentrations of fentanyl. Clearance of fentanyl occurs mainly by metabolism in the liver to hydrophilic metabolites, which are excreted by the kidney.

4.3.2 Structure of the Model

The structure of the model developed by Higgins [27] is based on Mapleson's Model P [24, 25]. Its features include a division of the body mass into various discrete blocks and a representation of the blood circulation that simulates both the rates of blood volume flow and the mean transition time (MTT) through these blocks.

Other than the gut and spleen, kidney, liver, and lung blocks, which correspond to their anatomical counterparts, the remainder of the body mass is grouped into four compartments: unperfused tissue, other viscera, lean, and fat. In addition, sample brain, intramuscular injection, and nasal mucosal sites are of negligible mass (the bulk of the brain tissue being absorbed by the "other viscera" compartment). It is worth noting that each organ is represented by one single compartment with the important assumption of blood-flow homogeneity (i.e., the blood-washed tissue acquires the same concentration as the incoming blood—with the drug—almost instantaneously).

4.3.3 Quantification of the Model

The composition of cell membranes is, in general, made up of lipid material. As a result, a particular drug needs to be mostly lipid soluble in order to diffuse across such a membrane, for lipid soluble drugs establish high concentrations in the membrane, which in turn result in high concentration gradients as well as rapid diffusion. For such lipid soluble drugs, the buildup of such concentrations is dependent upon the delivery rate to the diffusion site. Because it is argued that concentration levels depend on the arterial perfusion, which represents a percentage of cardiac output, the equilibration of drug concentrations follows the laws regulating a perfusion-limited model (see Section 4.1.2). Conversely, if the limiting step in the concentration gradient buildup is the diffusion rate, then equilibration is diffusion-limited.

In the original physiological model proposed by Davis and Mapleson [26] on an anesthetic drug called pethidine, the drug is assumed to ionize, dissolve in fat, and bind to protein. This is also true of fentanyl [24], an analgesic drug which has also some anesthetic effect. Hence, according to the previous authors, the drug concentrations exist in four different forms in the body:

- Unionized, dissolved in water;
- Ionized, dissolved in water;
- Unionized, dissolved in lipid;
- Bound to protein.

Furthermore, the concentration of the drug in the plasma, the erythrocytes (red blood cells), the tissue cells, and the tissue interstitial fluid will involve equilibration between the above four forms of the drug.

Equilibration

The process of equilibration where only the unionized form of the drug diffuses through the cell membranes and dissolves in fat [26]. The unionized, ionized, and the total drug concentration in water can bind to proteins. At equilibrium, the total drug concentration in the system remains unchanged (i.e., the membranes store a negligible amount of drug and equilibration is total).

Ionization

Ionization involves equilibrium between the unionized for a drug in water. The equilibrium point depends upon the physiochemical properties of the drug and the pH of the medium. The Henderson-Hasselbach equation describes the relationship between the ionized drug concentration in water C_i, and the standard drug concentration (unionized dissolved in water), C_{std}:

$$C_i = 10^{b[pK_a - pH]} \cdot C_{std} \tag{4.28}$$

where $b = \begin{cases} +1 \text{ If drug is acidic} \\ -1 \text{ If drug is basic} \end{cases}$

Partition into Fat

As far as lipid solubility is concerned, only the ionized form is dissolved in lipid, and when the concentration of pH of the aqueous phase changes, the concentration of the drug in lipid changes in proportion to the unionized concentration in water according to the following law:

$$C_l = C_s \cdot \lambda_{ls} \tag{4.29}$$

where C_l is the concentration dissolved in lipid, C_s is the concentration of the standard form (unionized dissolved in water), and λ_{ls} is the lipid/water partition coefficient. In fact, λ_{ls} can be calculated using the following extended Henderson Hasselbach equation:

$$\lambda_{ls} = \lambda_{lw} \cdot \left(1 + 10^{(pK_a - pH)}\right)$$
$$\lambda_{lw} = \frac{C_l}{C_w} \tag{4.30}$$
C_w: total concentration of drug in aqueous phase

Protein Binding

In this model configuration, drug concentrations exist in the body according to four different forms: unionized dissolved in water, ionized dissolved in water, unionized dissolved in lipid, and bound to proteins. Concentration of a drug in the plasma, red cells, tissue cells, and the tissue interstitial fluid involves equilibration between these four forms of the drug.

In order to describe protein binding, Higgins expanded the Scatchard equation to accommodate different binding constants for the ionized and unionized forms as follows:

$$C_p = C_s\left(10^{(pK_a - pH)} \cdot k_i + k_s\right) \tag{4.31}$$

where C_p is the aqueous concentration bound to protein, C_s is the aqueous concentration of unionized drug, k_i and k_s are constants for the ionized and unionized species, and pK_a is the negative logarithm of the acid dissociation constant.

Distribution in Blood

For the distribution of the drug in the blood, Higgins asserted that the simplest and most robust way of describing drug distribution between plasma and erythrocytes is by a ratio factor that will still hold even if plasma pH changes:

$$eprat = \frac{\text{total fentanyl concentration in red cells}}{\text{total fentanyl concentration in plasma}} \tag{4.32}$$

Elimination

Finally, Higgins divided elimination into hepatic metabolism and renal extraction. Fentanyl has a high extraction fraction and is mainly removed from the body by hepatic metabolism with a small amount excreted unchanged in the urine. Hence, metabolism is expressed as an *extraction fraction*:

$$extraction = \frac{clearance}{perfusion} \tag{4.33}$$

In addition, and in order to allow for renal elimination, Higgins introduced a concentration factor, which operates by decreasing the mass of the kidney, and perfusing stroke volume. This is after the total mass of the drug in blood and kidneys have been calculated.

Blood-Tissue Distribution

The total amount of fentanyl in the blood and tissue of a particular organ can be found using the following expression that makes use of the corresponding concentration of unionized drug C_s:

$$\begin{aligned} A_{f_{bt}} &= A_{f_p}(\text{amount of plasma}) + A_{f_r}(\text{amount in red blood cells}) + \\ & A_{f_c}(\text{amount in tissue cells}) + A_{f_e}(\text{amount in extracellular fluid}) \end{aligned} \tag{4.34}$$

where

$$A_{f_e} = wt_p \cdot C_s \cdot \left(\left(10^{(pK_a - pH_p)} k_i + k_s\right) \cdot gp_p + fw_p + 10^{(pK_a - pH_p)} \cdot fw_p\right) \tag{4.35}$$

$$A_{f_r} = wt_r \cdot eprat \cdot C_s \cdot \left(\left(10^{(pK_a - pH_p)} k_i + k_s\right) \cdot gp_p + fw_p + 10^{(pK_a - pH_p)} \cdot fw_p\right) \tag{4.36}$$

$$A_{f_c} = wt_c \cdot C_s \cdot \left(\left(10^{(pK_a - pH_p)} k_i + k_s\right) \cdot gp_c + \lambda \cdot fl_c + fw_c + 10^{(pK_a - pH_p)} \cdot fw_c\right) \tag{4.37}$$

$$A_{f_e} = wt_e \cdot C_s \cdot \left(\left(10^{(pK_a - pH_p)} k_i + k_s\right) \cdot gp_e + fw_e + 10^{(pK_a - pH_p)} \cdot fw_e\right) \tag{4.38}$$

and

wt is compartment mass.
n is the number of biding sites per protein molecule (constant).
k is the binding constant.
gp is the protein concentration.
λ is the lipid solubility coefficient.
fl is the lipid concentration.
fw is the water concentration.

It is worth noting that in the cases of the nasal and kidneys blocks, calculations need to include the nasal mucus as well the urine for the extra compartments in these blocks, respectively. Hence, for the nasal block, the total amount of fentanyl can be derived using the following expression:

$$\begin{aligned} A_{f_{btm}} &= A_{f_{bt}} + A_{f_m} \\ A_{f_m} &= wt_m \cdot \left(C_s \cdot \left(10^{(pKa - pH_m)} \cdot k_i + k_s\right) \cdot gp_m + C_s \cdot fw_m \cdot \left(1 + 10^{(pKa - pH_m)}\right)\right) \\ pH_m &: \text{mucus pH} \end{aligned} \tag{4.39}$$

and for the kidney block such amount can be expressed as follows:

$$A_{f_{btu}} = A_{f_{bt}} + A_{f_u}$$
$$A_{f_u} = wt_u \cdot C_s \cdot \left(10^{(pKa - pH_u)} \cdot k_i + k_s\right) \tag{4.40}$$
$$pH_u\text{: urine } pH$$

Calculation Sequence

Since the model is recirculatory, the sequence of calculations in one cycle (corresponding to one heartbeat) is as follows.

Stroke volumes of blood are taken out of the arterial mixed venous, and injection pools and fractional stroke volumes are removed from the portal, muscle, and fat pools. The stroke volume from the arterial pool is divided into fractional stroke volumes, and, after mixing the fractional stroke volume from the portal pool with the hepatic fraction, each is equilibrated with the corresponding compartment tissue. The amount of drug metabolized is removed from the combined liver fractional stroke volume according to the extraction fraction. The stroke volume from the injection pool is equilibrated with the lungs compartment at the same time.

After equilibration, the fractional stroke volumes from the fastest tissue compartments (peripheral shunt, kidneys, liver, other viscera, sample brain, nasal mucus, and injection site) and from the muscle and fat pools are combined into a mixed venous stroke volume. The whole drug in the urine is excreted.

Finally, the injection pool, the intramuscular injection site, the lung, and the nasal mucosal site can serve as sites for any additional administered drug input.

4.4 A Dynamic Representation of the Mapleson-Higgins Model for Fentanyl

Referring to Figure 4.6, it can be seen that some of the submodels can be identified directly using a SISO structure assumption. It is worth noting that only the drug flow from one organ (tissue or pool) to another can be described, and the concentration of drugs in individual tissues cannot be determined. Hence, we suggest referring back to this model structure as the *drug-flow model* structure. We will describe later in this section how the drug concentration in an individual tissue can also be modeled. Let us outline a few considerations that will form the basis for the modeling operation:

- Each tissue is perfused with a blood fraction K_i of the arterial blood.
- A dynamic model is appended to the peripheral shunt and also to the intravenous injection (i.v.) pool.
- Because the intramuscular injection site will not be used in this study, it was removed from the overall model.

The advantages of modeling the Mapleson-Higgins structure as a dynamic model can be summarized as follows:

- A dynamic model is simpler to manipulate for extensions, reductions, or substitutions.

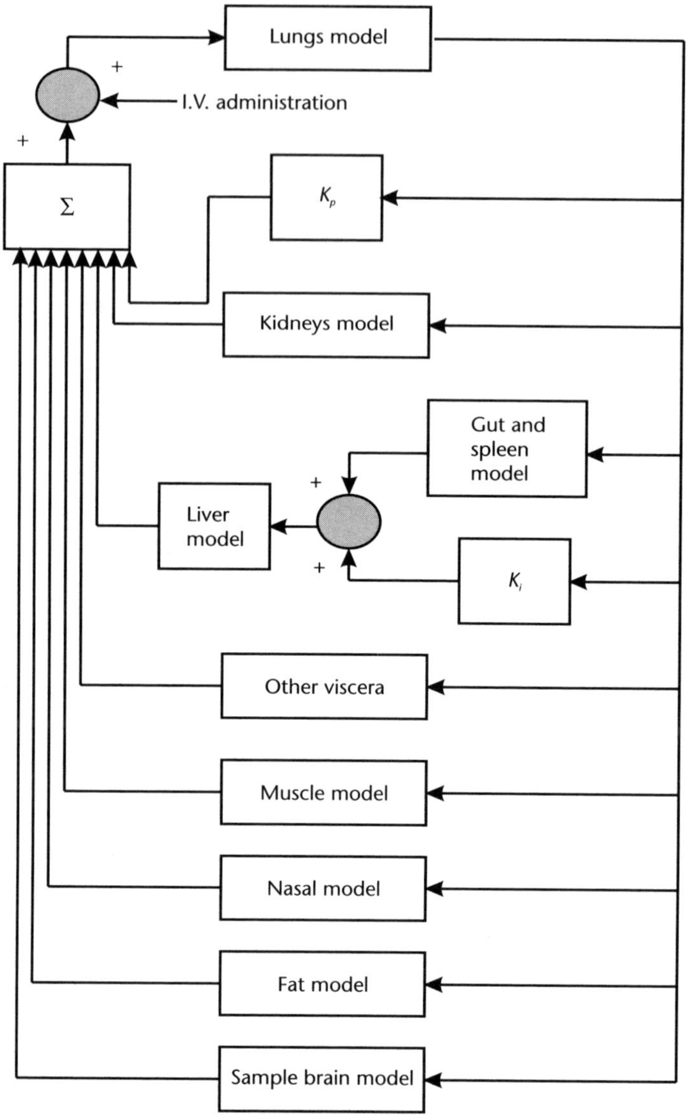

Figure 4.6 The proposed Mapleson-Higgins model structure for fentanyl.

- A dynamic structure is more amenable to control design using the transfer function or state-space forms.
- A dynamic structure is independent of the injection scheme adopted, which usually has a complex coding structure when algebraic equations are used; this was mainly due to the fact that time in the original model was almost added as an ad-hoc variable rather than being included in an integration process.

4.4.1 Individual Organs' Model Fitting

In a different study [28], it has been shown that compartmental modeling is not suitable for representing such a drug-flow structure; instead, a companion format fitting using the arx ( ) function[3] in the MATLAB Identification Toolbox has been shown to be more appropriate. Indeed, in the model fitting study below, a body weight of 70 kg, a cardiac output of 6.48 liters/minute, a default injection value of 100 μg over 60 seconds together with a simulation time of 21,600 seconds and a calculation step of 1 second were assumed. In the following, and throughout, we will be referring to the *original data* as that data which was obtained using the Mapleson-Higgins algebraic model via the *cmex-dll*[4] module. In the following we shall find linear approximations to the various organs and tissues as well as pools. Space prohibits a description of all tissue models that were obtained using this technique; it will suffice to mention only typical organs, with a chosen model structure as shown in Figure 4.6.

Having obtained the drug flow in each organ, one can use the data generated to identify the associated models, with the input signal being the arterial pool drug and the output being the drug amount coming out of the organ itself. Although the identification study concerned all 11 organs and the blood pools, only a few strategic cases will be described in this section that illustrate the most valuable lessons learned.

Fat Model Fitting

The Identification Toolbox in the MATLAB package is used to identify the discrete models, which are then transformed back into their continuous forms. For the fat model, first and second-order models will be identified.

Figure 4.7 shows the curve fitting for first and second-order models using logarithmic and linear scales, respectively. As can be seen from the figures, the second-order model for fat seems to suggest a better fit, and hence, it will be the adopted model that is represented by the following transfer function:

$$G_{Fat}(s) = \frac{-0.0058s + 0.6994}{s^2 + 0.6592s + +0.0008} \times 10^{-5} \tag{4.41}$$

Peripheral Shunt

Repeating the same procedure as before, the following first-order model for the Peripheral Shunt was obtained:

$$G_{\text{Shunter}}(s) = \frac{51.5622}{s + 2136.1} \tag{4.42}$$

Equation (4.42) shows that this first-order lag includes a time constant that is so fast that its corresponding transient cannot be seen. In fact, further investigations into the relationship between the input and output signals reveal that a gain factor is sufficient to represent such a linear relationship. Hence, by defining an error (cost) function of the following form:

3. Auto-Regressive Exogenous (ARX).
4. cmex is a MATLAB term for an executable that can be run from within the MATLAB environment, and a dll is a dynamic link library, which is a code that is loaded at runtime by one or more programs as opposed to static programs that are made with ll library code, which is built in.

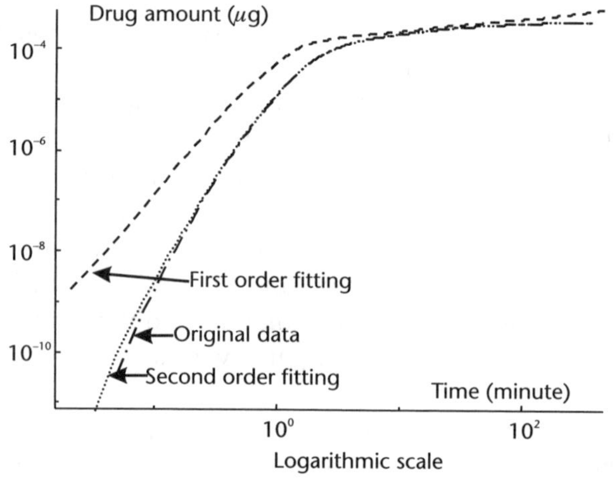

Figure 4.7 Fitting the fat model via first and second-order continuous transfer functions.

$$J_p = \sum_{i=1}^{N} \left(y_i - u_i K_p\right)^2 \tag{4.43}$$

where (u_i, y_i) is the input/output pair, respectively, N is the total number of input/output pairs, and K_p is the proportional gain to be identified.

Equation (4.43) possesses a minimum at

$$K_p = \sum_{i=1}^{N} y_i u_i \Big/ \sum_{i=1}^{N} u_i^2 \tag{4.44}$$

A parameter value $K_p = 0.0241$ was found. Figure 4.8 shows the new fit, which uses only this gain factor between the input and the output.

Lungs Model

By selecting the input signal as the injected drug and the amount of drug coming out of the venous pool, three fits were attempted: a first-order model with a transmission zero, a first-order model without a transmission zero, and a second-order model, as shown in Figure 4.9. The second-order model was retained as it corresponded to the

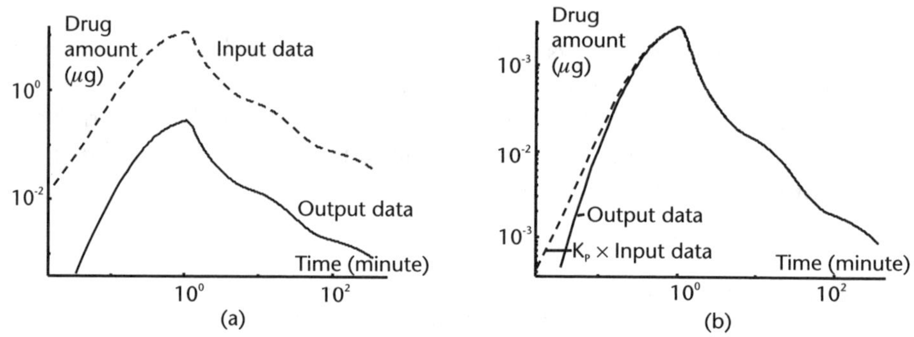

Figure 4.8 Fitting the Peripheral Shunt model.

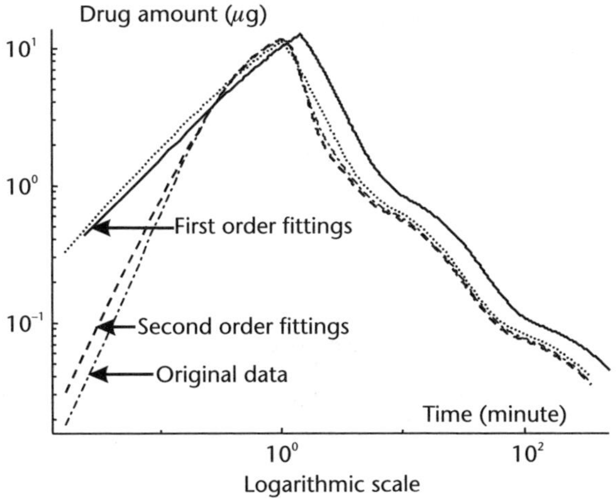

Figure 4.9 Fitting the lungs model via first- and second-order continuous transfer functions.

best fit (higher order models did not improve the fit). Hence, the following transfer function relating to the Lungs compartment was obtained:

$$G_{\text{Lungs}}(s) = \frac{0.1786s + 116.2879}{s^2 + 7.6121s + 19.4950} \tag{4.45}$$

Brain Model

To identify the dynamics relating to the brain, the input signal was selected as the arterial drug amount. Figure 4.10 shows how a first-order model was sufficient to describe such dynamics. The following transfer function was obtained:

$$G_{\text{Brain}}(s) = \frac{4.0981}{s + 0.2441} \times 10^{-5} \tag{4.46}$$

4.4.2 Simulation of the Overall System

Having identified the linear models relating to each organ, the overall model structure of Figure 4.11 is obtained. Selecting the output signal to be the arterial pool drug amount, Figure 4.12 shows how well the data from the original model (algebraic Mapleson-Higgins model) fit the identified drug-flow model simulated using the SIMULINK representation (a special method for stiff equations is used for integration).

4.5 Model Parameters' Sensitivity Study

Recall that so far in all our experiments for model fitting, default values for body weight and cardiac output of 70 kg and 6.48 l/min were used, respectively. In order to test the validity of our model outside these values, a study was conducted in which these parameters were varied inside a realistic range and the repercussions on

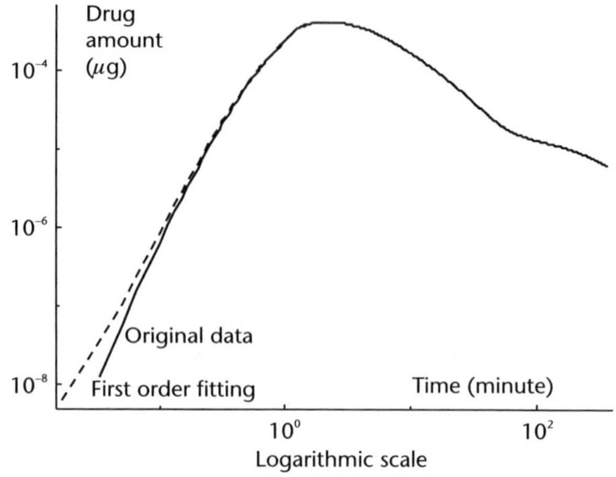

Figure 4.10 Fitting the brain model.

the derived nominal models were observed. The following sections present and discuss the results obtained.

4.5.1 Model Parameters' Sensitivity with Respect to Body-Weight Variations

Most first-order and second-order submodels have been expressed generically as follows:

$$G^1(s) = \frac{b_1}{s + a_1}, \quad G^2(s)\frac{b_1 s + b_2}{s^2 + a_1 s + a_2} \tag{4.47}$$

When the body weight was varied from 50 kg to 110 kg, the model parameters a_i and b_i varied linearly for all organs as Figure 4.13(a, b) shows for the instances of fat and kidneys. To cater for such variations, a logarithmic-based linear interpolation formula relating to the parameters in the fitted submodels can be derived, as shown in Figure 4.14, where the model parameter Ψ_w is displayed as a function of the weight w only. Hence the slope be evaluated as follows:

$$k_w = \frac{\log(\Psi_{w_M}) - \log(\Psi_{w_m})}{\log(w_M) - \log(w_m)} = \frac{\log(\Psi_{w_M}/\Psi_{w_m})}{\log(w_M/w_m)} \tag{4.48}$$

where $w_M = 110$ (maximum weight) and $w_m = 50$ (minimum weight).

Using the nominal point (W_0, Ψ_{W_0}) and with $W_0 = 70$, one can rewrite the linear equation (4.48) as follows:

$$\log(\Psi_w) - \log(\Psi_{W_0}) = k_w\left[\log(w) - \log(W_0)\right] \tag{4.49}$$

which can be simplified further as follows:

$$\log(\Psi_w/\Psi_{w_0}) = k_w \log(w/W_0) \tag{4.50}$$

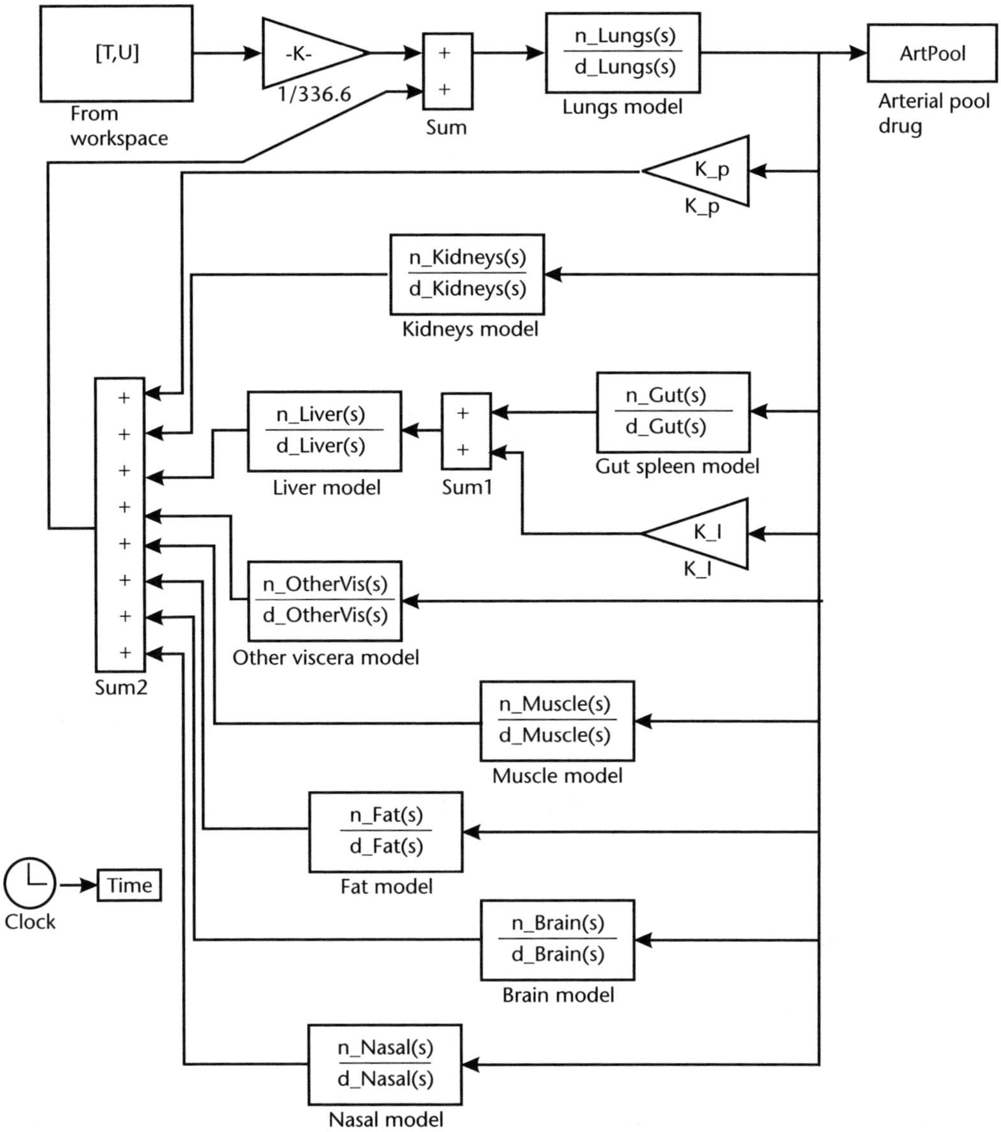

Figure 4.11 SIMULINK representation of the fentanyl model structure.

and with a simple rearrangement, one can finally write the interpolation formula as follows:

$$\Psi_w = \left(\frac{w}{W_0}\right)^{k_w} \Psi_{W_0} \tag{4.51}$$

By building this interpolation formula inside the MATLAB-SIMULINK environment, one can safely simulate the model for any weight, as shown in Figure 4.15, where a body weight of 82 kg was considered and the fit was still accurate.

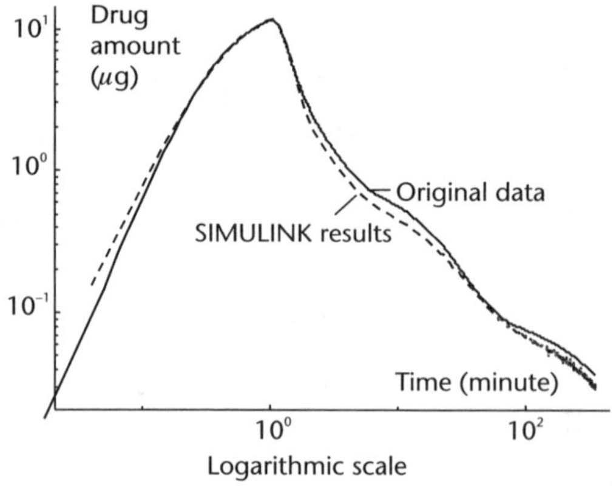

Figure 4.12 Simulated drug amounts in the arterial pool using the overall SIMULINK model.

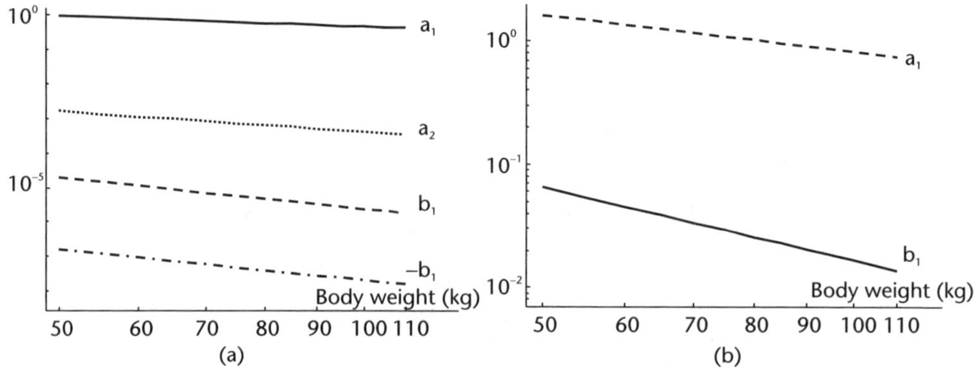

Figure 4.13 Variations of model parameters with respect to body weight: (a) fat and (b) kidneys model parameters.

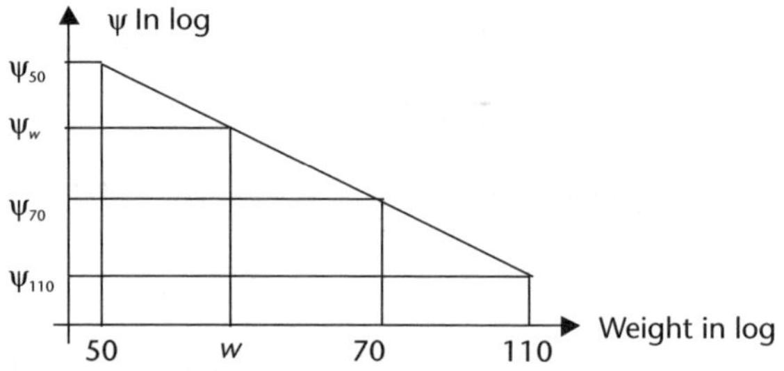

Figure 4.14 Illustration of the logarithmic linear interpolation.

4.5.2 Model Parameters' Sensitivity with Respect to Cardiac Output Variations

Repeating the same procedure as above, one can write the interpolation formula with respect to cardiac output as follows:

$$\phi_c = \left(\frac{c}{C_0}\right)^{k_c} \phi_{C_0} \tag{4.52}$$

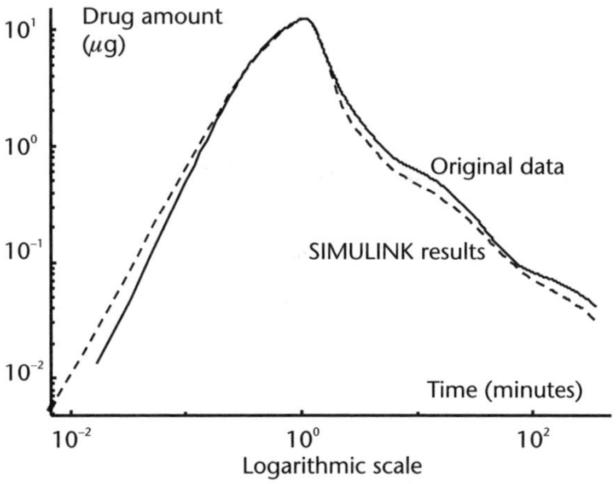

Figure 4.15 Simulated arterial pool drug amount for a patient with a body weight of 82 kg.

where $C_0 = 6.48$ l/min, ϕ_{C_0} is the nominal parameter identified using C_0, and k_c is the slope relating to the linear interpolation.

Using (4.52), the model was simulated for a cardiac output of 8 l/min and the result is shown in Figure 4.16 where it can be seen that the fit thus obtained is still accurate.

4.5.3 Model Parameters' Sensitivity with Respect to Simultaneous Variations of Cardiac Output and Body Weight

In the previous sections it was shown that for a fixed cardiac output value, the logarithmic value of the parameters is linearly dependent on the logarithmic value of the body weight; and conversely, for a fixed body weight the logarithmic value of the parameters are linearly dependent on the logarithmic value of the cardiac output. Hence, if both body weight and cardiac output vary simultaneously, then a surface plane can represent the logarithmic values of the model parameters.

Using (4.52), one can write

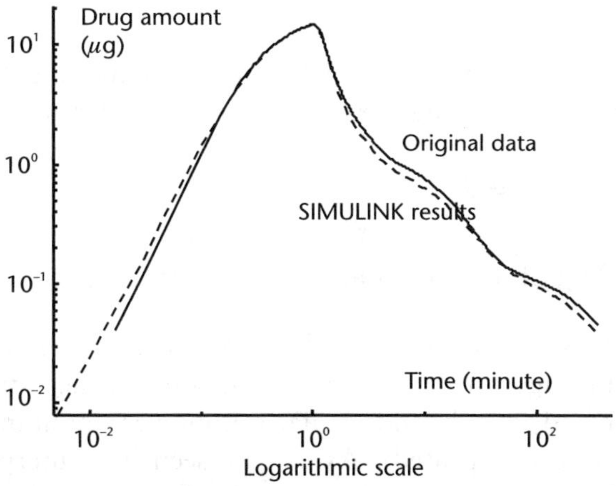

Figure 4.16 Simulated arterial pool drug amount for a cardiac output of 8 l/min.

$$\phi_{c,w} = \left(\frac{c}{C_0}\right)^{k_{c,w}} \phi_{C_0,w} \tag{4.53}$$

where the slope can be expressed as follows:

$$k_{c,w} = \frac{\log(\phi_{c_M,w}/\phi_{c_m,w})}{\log(c_M/c_m)} \tag{4.54}$$

and

$$\phi_{c_M,w} = \left(\frac{w}{W_0}\right)^{k_{C_M,w}} \phi_{c_M,W_0} = \left(\frac{w}{W_0}\right)^{k_{C_m,w}} \phi_{c_m,W_0} \tag{4.55}$$

with the following definitions:

$$k_{c_M,w} = \frac{\log(\phi_{c_M,w_M}/\phi_{c_M,w_m})}{\log(w_M/w_m)}, k_{c_m,w} = \frac{\log(\phi_{c_m,w_M}/\phi_{c_m,w_m})}{\log(w_M/w_m)} \tag{4.56}$$

The slope $k_{c,w}$ can be determined if the sets of model parameters at the four corners of the fitting region are obtained. Having obtained $k_{c,w}$, the linear interpolation of $\varphi_{c,w}$ can be obtained as follows:

$$\phi_{c,w} = \left(\frac{c}{C_0}\right)^{k_{c,w}} \phi_{C_0,w} = \left(\frac{c}{C_0}\right)^{k_{c,w}} \left(\frac{w}{W_0}\right)^{k_{C_0,w}} \phi_{C_0,W_0} \tag{4.57}$$

where

$$k_{C_0,w} = \frac{\log(\phi_{C_0,w_M}/\phi_{C_0,w_m})}{\log(W_M/w_m)} \tag{4.58}$$

Using the above interpolation formula, an experiment was conducted in which a patient with the following parameters was considered:

- Body weight = 93 kg.
- Cardiac output = 5.4 l/min.
- A 100 μg of fentanyl is administered over a period of 60 seconds.

Table 4.1 shows the submodel parameters that were obtained using the interpolation formula against those that were obtained using a direct identification method (the specific body weight and cardiac output were input into the model and the identification exercise repeated). As can be seen, the interpolated model parameters proved to be comparable to those obtained using the direct identification method.

Table 4.1 Comparing Interpolation and Direct Identification Models as Body Weight and Cardiac Output Vary Simultaneously

Tissue	*Interpolation Results*	*Direct Identification*
Fat	$\frac{-1.437.10^{-8}s+1.722.10^{-6}}{s^2+0.4126s+0.000324}$	$\frac{-1.438.10^{-8}s+1.722.10^{-6}}{s^2+0.4126s+0.000324}$
Lungs	$\frac{64.78}{s^2+4.17s+6.97}$	$\frac{-0.0318s+65.77}{s^2+4.19s+7.039}$
Gut spleen	$\frac{-1.34.10^{-5}s+0.0016}{s^2+0.906s+0.0936}$	$\frac{-1.34.10^{-5}s+0.0016}{s^2+0.906s+0.0936}$
Kidneys	$\frac{0.0132}{s+0.744}$	$\frac{0.0132}{s+0.743}$
Liver	$\frac{0.00173}{s+0.09}$	$\frac{0.00173}{s+0.09}$
Other viscera	$\frac{0.00173}{s+0.09}$	$\frac{0.00173}{s+0.09}$
Muscle	$\frac{-2.764.10^{-7}s+3.31.10^{-5}}{s^2+0.415s+0.00187}$	$\frac{-2.763.10^{-7}s+3.31.10^{-5}}{s^2+0.415s+0.00187}$
Brain	$\frac{1.614.10^{-5}}{s+0.153}$	$\frac{1.614.10^{-5}}{s+0.153}$

4.6 Model Fitting for Drug Concentrations in Tissues and Blood Pools

The submodels identified in Section 4.4 do not represent the drug amounts in the tissue but in the outgoing blood from the tissue. Hence, while the drug-flow model was useful in computing the recirculatory modes, we may need to determine the drug concentrations within each tissue. For that, we can use the arterial pool drug amount as the input signal and the drug concentration from each tissue obtained from the original algebraic Mapleson-Higgins model as the output signal, as schematically represented in Figure 4.17.

4.6.1 Concentrations in Tissues

The drug concentration in the various tissues, such as brain and lungs, is defined as the drug amount in the tissue itself divided by the tissue weight. Because the original Mapleson-Higgins algebraic model for fentanyl can provide data relating to concentrations as well as drug amounts, they were used for our direct identification study. Taking the brain as an example led to the following transfer function:

$$G_{\text{Brain}}(s) = \frac{1.4358}{s+0.2441} \tag{4.59}$$

Comparing (4.59) with (4.46) suggests that a gain factor is enough to represent the drug concentration in the brain. Figure 4.18 shows the brain concentration curve fitting. Repeating the same procedure with all other tissues led to the same conclusion: gain factors are necessary for obtaining concentrations from outputs of submodels that were identified in Section 4.4.

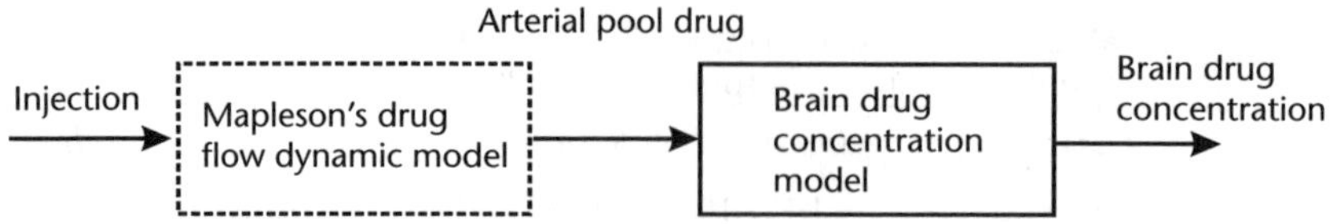

Figure 4.17 The new model structure for determining the drug concentration in tissues.

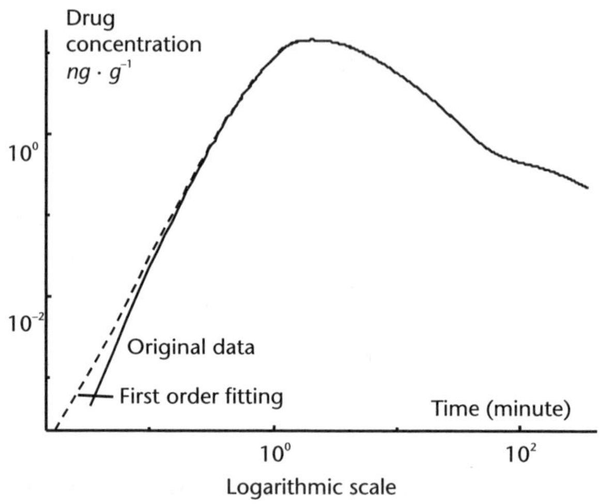

Figure 4.18 Fitting the brain concentration model.

4.6.2 Concentrations in Blood Pools

In the original Mapleson-Higgins model, blood pools have been included to reflect the transport delay of the drug traveling through the main parts. They comprise:

- The arterial pool;
- The venous pool;
- The gut-spleen pool;
- The muscle pool;
- The fat pool.

Two types of concentrations can be obtained from such pools: the blood concentration and the plasma concentration. Figure 4.19 shows the fat pool blood and plasma concentrations from the original algebraic model (which are close) drawn against the new model fit, which can be considered to be a good fit.

4.7 Model Reduction Analysis

The structure of the model shown in Figure 4.6 can be further reduced by analyzing the signal from the i.v. injection to the arterial pool blocks. In this case, the overall model obtained is of the thirteenth order and as follows:

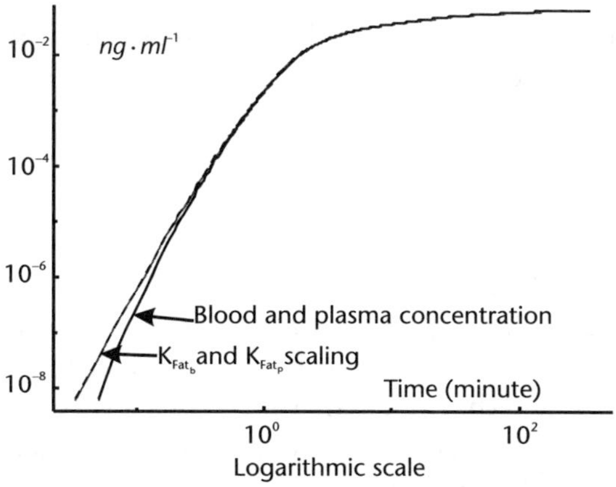

Figure 4.19 Blood and plasma concentrations relating to the fat pool.

Gain: 0.1786

Zeros: −654.7; −1.26; −1.169; −0.6579; −0.6562; −0.2442; −0.1894; −0.1436; −0.1338; −0.08033; −0.007249; −0.001255.

Poles: −1.26; −0.89388; −0.6608; −0.6568; −0.2438; −0.1939; −0.1339; −0.1178; −0.05358; −0.00412; −0.0009189; −3.9925 ±1.8303*i*.

The presence of a pair of imaginary poles, which cannot be justified in pharmacokinetics, is due to lack of proper excitation. One way to eliminate these modes is to carry out a model reduction technique based on *dominant pole* or *balanced realization* techniques [29]. The latter was chosen for this particular study. The following gives a brief introduction to this chosen technique:

Suppose that the state-space representation of the system is given by

$$\begin{cases} \dot{x} = Ax + Bu \\ y = Cx + Du \end{cases} \tag{4.60}$$

where (A,B,C,D) are compatible matrices. However, from the above system equation, it is very difficult to determine which state is important to the input-output relationship. The Controllability and Observability Gramians can be defined as follows:

$$G_c = \int_{t_0}^{\infty} e^{A(t_0-t)} BB^T e^{A^T(t_0-t)} dt \tag{4.61}$$

$$G_o = \int_{t_0}^{\infty} e^{A^T(t_0-t)} C^T C e^{A(t_0-t)} dt \tag{4.62}$$

In fact, the Gramians can also be obtained by solving the following two Lyapunov equations:

$$AG_c + G_c A^T = -BB^T \text{ and } A^T G_o + G_o A = -C^T C \tag{4.63}$$

If we introduce an internal state linear transformation (i.e., introduce a new state vector $z = Tx$), we can obtain a model under a new framework as follows:

$$\begin{cases} \dot{z} = A_b z + B_b u \\ y = C_b z + D_b u \end{cases} \tag{4.64}$$

where $A_b = T^{-1}AT$, $B_b = T^{-1}B$, and $C_b = CT$.

If adequately chosen, the system under the new framework can guarantee that the two Gramians are equal. In this case, the system is called *balance realized*, and there are algorithms that calculate the balanced realization from a given state-space model. The common Gramian matrix is diagonal, and the diagonal elements are the singular values of the matrix.

Hence, using this technique with a criterion of $\varepsilon = 10^{-3}$ for pole/zero cancellation, the following fourth and fifth-order models were obtained.

Fourth-Order Model

Gain: 0.037188

Zeros: –3309; –0.1764; –0.009554.

Poles: –5.365; –3.087; –0.005012; –0.09477.

Fifth-Order Model

Gain: 0.1737

Zeros: –676.3; –1.123; –0.1053; –0.00746.

Poles: –0.00416; –0.8092; –0.0597; –3.9895 ±1.8177*i*.

Figure 4.20 compares the fourth and fifth-order models with the full-order model relating to the arterial pool drug amount. It is worth noting that before the model reduction technique was applied, all near enough pole-zero cancellations were carried out at the above precision point.

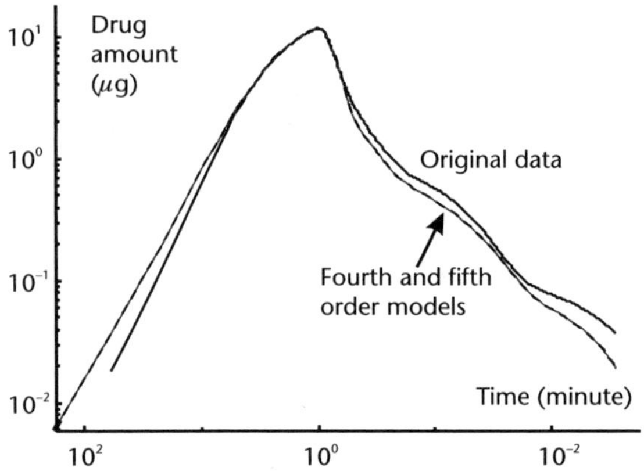

Figure 4.20 Comparing the overall fourth- and fifth-order models to the original data.

4.8 Wada's Model

It is believed that the propagation and distribution of drug in a human body, especially for intravenous administration, is mainly via blood circulation.

All blood pumped by the right side of heart passes through the lungs for the O_2 pickup and CO_2 removal. The blood pumped by the left side of the heart is transported out in various proportions to the systemic organs through a parallel arrangement of vessels that branch from the aorta. Arteries progressively branch as they carry blood from the heart to the tissues. A separate small arterial branch delivers blood to each of the various organs. As a small artery enters the organ, it branches into arterioles, which further branch into an extensive network of capillaries. The exchanges of oxygen, as well as the drug components to cells are mainly done via these capillaries. The capillaries then merge to form venules, which further unite to form small veins that leave the organ. The small veins progressively merge as they carry blood back to the heart. According to Sherwood [30], the nominal blood distribution to various organs under resting conditions is given in Table 4.2. With respect of drug disposition, an organ can be considered to be made up of blood in the capillary network, interstitial water, and tissue cells. In order to establish a physiological model, it is vital to increase understanding and knowledge of the exchange process occurring between capillaries and tissues.

Conceptually, an organ can be represented by blood in capillary network, interstitial water, and tissue cells. In Wada's model, an organ is represented by a multicompartment model where the drug is assumed to equilibrate instantaneously within each compartment. The impedance of the capillary endothelium or the cellular membranes against drug movement determines a diffusion barrier between compartments. When the impedance of all diffusion barriers becomes very small (i.e., the drug can move across all the components more freely), the multicomponent model collapses to a single-compartment model. The number of compartments required for a specific organ model will depend on the characteristics of the drug movement through the organ's different tissue structures. In Wada's model, the most general organ has three compartments. The physiological interpretation of the first component is blood in capillary, the second is interstitial fluid, and the third is

Table 4.2 Nominal Blood Distribution to Various Organs Under Resting Conditions

Organ	%
Digestive system	21
Skeletal muscle	15
Heart muscle	3
Kidneys	20
Skin	9
Bone	8
Brain	13
Liver	6
Other	8

intracellular fluid. The rate of removal of drug from component "i" to compartment "j" is denoted by k_{ij}, and the elimination rate, which represents the metabolic process assumed to happen in the intracellular compartment (i.e., compartment 3), is represented by k_{30}.

Assume that the mass flux of drug to the organ is $J_{in}(t)$, and the volume flux of blood to the organ is $Q_{in}(t)$. Correspondingly, the mass flux of drug and the volume flux of blood out of the organ are $J_{out}(t)$ and $Q_{out}(t)$, respectively. Since the drug flux is very small compared with the blood flux, it is reasonable to assume that blood inflow $Q_{in}(t)$ equals the blood outflow $Q_{out}(t)$ for any specific organ. One can express the mathematical equation for a three-compartmental model as follows:

$$\begin{aligned}\dot{x}_1(t) &= -\left(\frac{Q_{out}(t)}{V_1 \cdot w} + k_{12}\right)x_1(t) + k_{21}x_2(t) + J_{in}(t)\\ \dot{x}_2(t) &= k_{12}x_1(t) - (k_{21} + k_{23})x_2(t) + k_{32}x_3(t)\\ \dot{x}_3(t) &= k_{23}x_2(t) - (k_{32} + k_{30})x_3(t)\end{aligned} \tag{4.65}$$

$$J_{out}(t) = Q_{out}(t) \cdot \frac{x_1(t)}{V_1 \cdot w} \tag{4.66}$$

where x_i is the mass of drug residing in compartment i, V_1 is the volume of blood in compartment 1 per organ weight, and w is the total weight of the organ.

Table 4.3 shows the nominal parameter values used for simulation [private communication with Dr. Wada, 1996].

For the purpose of this study, similar parameters to the Higgins' model simulation relative to the dose (100 μg), the infusion duration (1 minute), and body weight (70 kg) were used to simulate Wada's model for fentanyl whose MATLAB-SIMULINK representation is shown in Figure 4.21.

Figure 4.22 shows the drug concentrations in the lungs, liver, gut, and fat for the full period of simulation (360 minutes) and the reduced period (12 minutes) in order to visualize the early stages of drug uptake. As shown in the above figure, the various curve trends do not seem to differ considerably. Indeed, during the first minute, the model used by Higgins seems to predict higher concentrations than the model used by Wada in almost all tissues. Obviously, this cannot be used as a major criterion that either model is more accurate but merely to underline the fact that because Wada's model includes the phenomenon of diffusion within certain tissues, the corresponding data could be more credible. However, Higgins' choice to model fentanyl disposition using Mapleson's approach, which considers only the flow-limited concept, can be fully justified since fentanyl is a lipophilic drug and therefore more capable of crossing biological membranes. However, Wada found that some tissues had lower extraction ratios, suggesting poor extraction and hence presence of the diffusion phenomenon. Peak concentrations in the lungs and arterial and venous pools seem to occur at almost the same time but have different magnitudes, whereas for fat tissue, the predictions are almost identical in the steady state. The predicted peak concentrations in the lungs differ considerably, although their respective times of occurrence are similar. This could be due to the parameters Wada et al. took with respect to the volume and the weight (21 ml/kg and 0.87 kg,

Table 4.3 Nominal Parameter Values for Simulation of Wada Model for Fentanyl

Parameter	*Value*	*Parameter*	*Value*	*Parameter*	*Value*
Cardiac output	6.25 l/min	Dose rate	100 μg/min	Dose period	[0, 1] min
Body weight[1]	66.5	Brain weight	1.34	Heart weight	0.41
Lung weight	0.87	Liver weight	2.19	SPC weight	0.28
Kidneys weight	0.34	Fat weight	15.29	Muscle weight	26.64
Skin weight	2.85	Gut weight	1.21	Crc weight	14.64
Blood weight	5	Brain flow[2]	0.54 w_{brain}	Heart flow	0.66 w_{heart}
Lung weight	0.19 w_{lung}	Liver flow	0.18 w_{liver}	SPC flow	0.9 w_{spc}
Kidneys flow	3.26 $w_{kidneys}$	Fat flow	0.027 w_{fat}	Muscle flow	0.037 w_{muscle}
Skin flow	0.12 w_{skin}	Gut flow	0.8 w_{gut}	Crc flow	0.048 w_{crc}

Crc: carcass; SPC: spleen.
[1] Tissue weight is in kilograms, and tissue blood flow is in ml/min.
[2] Here flow represents the blood flow. Nominal tissue blood flow is calculated by taking into account the tissue weight.
W_{tissue} is the nominal weight of the tissue.

respectively) compared to the weight of 0.47 kg taken by Higgins. For the kidneys, guts, and liver, the peak concentrations do not seem to differ as much as their times of occurrence. This can be explained by the assumption of a flow-limited or diffusion-limited models: in Wada's case, the drug seems to take longer to reach its optimum level because of the diffusion phenomenon, which was ignored in the model by Higgins.

4.9 Model-Based Predictive Control Design Using the New Dynamic Model

For many years workers have long characterized anesthetic drugs, as well as analgesics, by pharmacokinetic parameters and have used these values within computer programs to administer these drugs according to target concentration infusion (TCI) [31] or computer-assisted infusion (CAI) concentration [32] schemes rather than simple manual infusions for maintenance of anesthesia and pain relief. All these computerized drug delivery schemes are open-loop systems which assume that the models upon which their control law is based are good representatives of average-population dynamics, but they proved very popular with anesthetists as they do not require any measurement, hence the absence of any feedback. However, in the case of fentanyl, for example, research work conducted by Jacobs and Williams [33] argued that the concentration at the "effect compartment" is the proper variable to target rather the concentration at the "central compartment." In a recent work that we conducted on propofol, we went a step further by postulating that for adequate maintenance of anesthesia, feedback control is more effective even in terms of drug consumption. Hence, in the case of fentanyl we propose to extend the pharmacokinetic model of the brain (site of action of anesthetics and analgesics) to include pharmacodynamics and derive a an adaptive control law using nonlinear generalized predictive control (NLGPC).

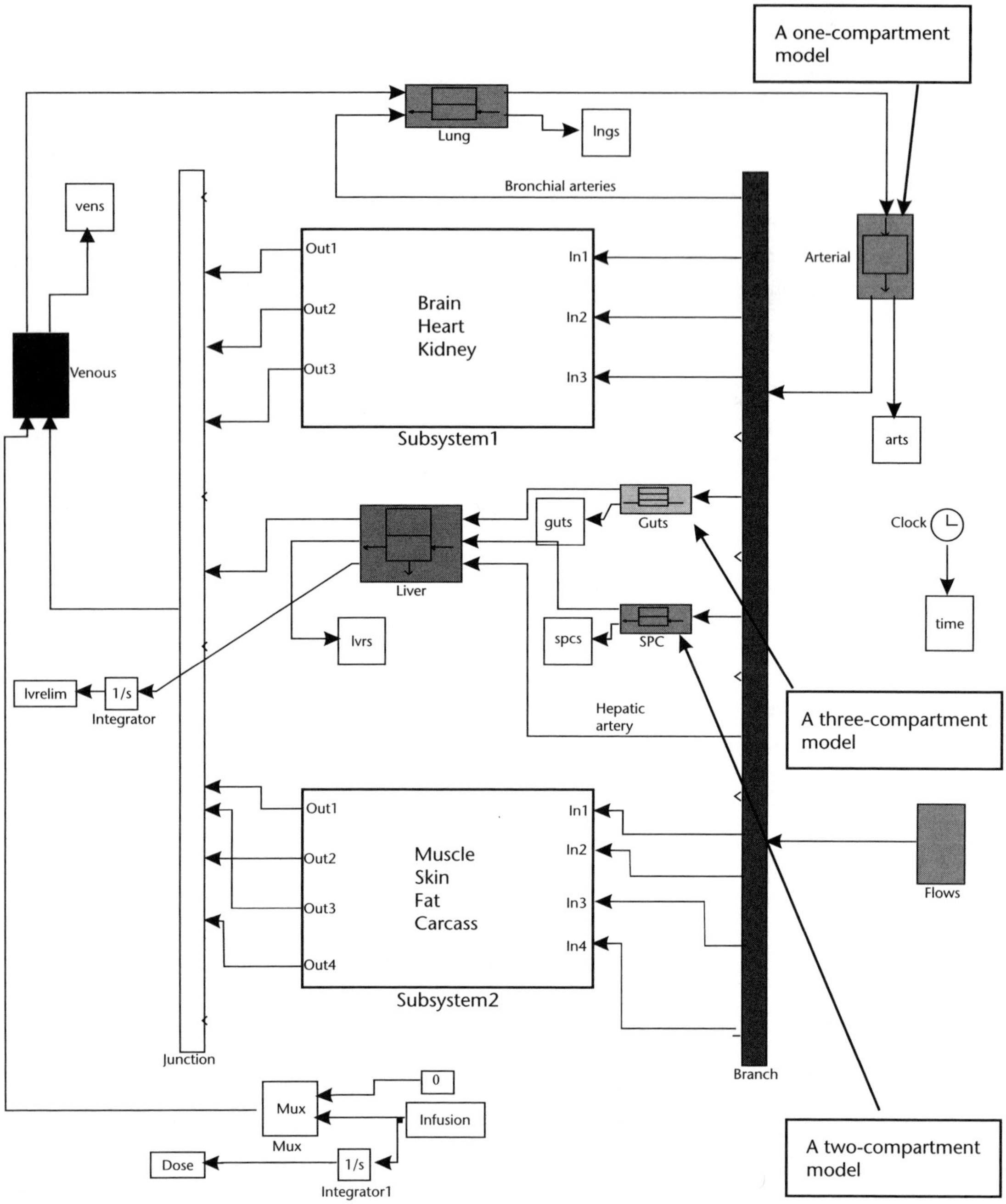

Figure 4.21 MATLAB-SIMULINK representation of the physiological model for fentanyl in humans as described by Wada et al. (*After:* [22].)

In pharmacokinetic modeling, the relationship between the drug dose and drug concentration is analyzed. In pharmacodynamic modeling, the relationship between the drug concentration in a certain site and the drug effect is analyzed. The drug concentration in the sample brain of the Mapleson-Higgins model for fentanyl is used to bridge the gap between the two branches. In the following, only one method of rep-

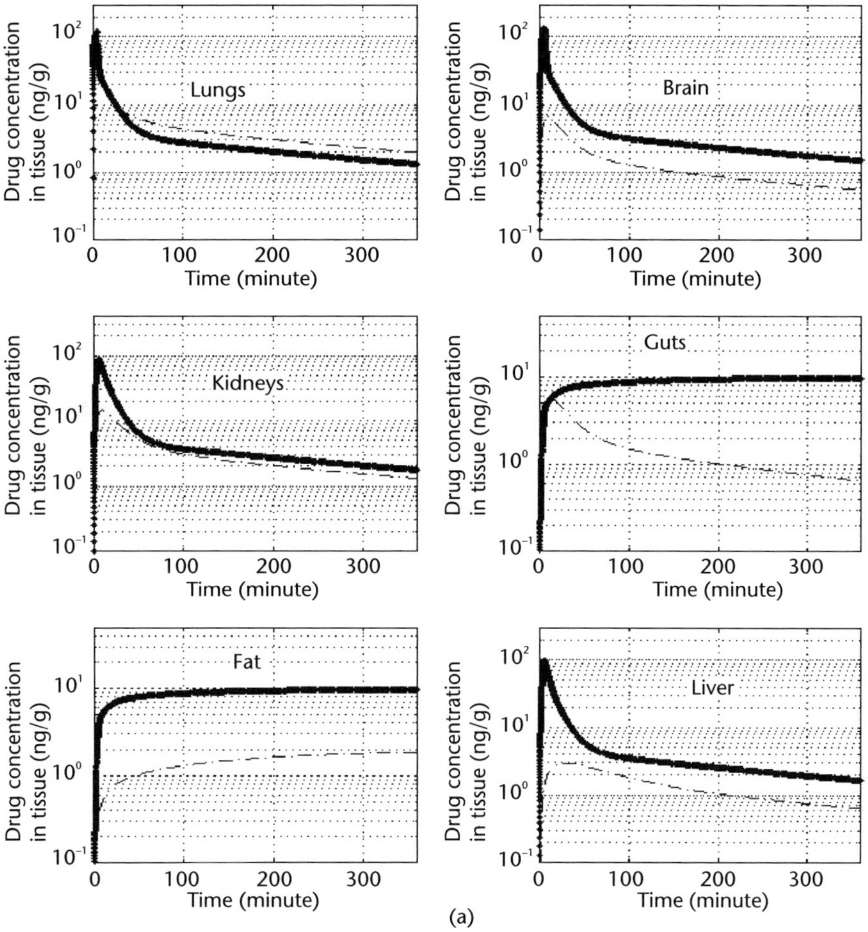

(a)

Figure 4.22 Plot concentrations for the various tissues in humans as simulated by (1) dash-dot (Wada's model), and (2) dot (Higgins' algebraic model): (a) full time scale; (b) reduced time scale.

resenting such a relationship is reviewed, which is based on the concept of effect compartment. Other methods of representing such a relationship have not been considered here due to the difficulty in obtaining the relevant data for quantification.

If the concentration of the sample brain can be written as C_b, then the drug effect can be defined using either of the following two quantities:

$$\text{Effect}_1(C_b) = \frac{C_b^{\gamma}}{EC_{50}^{\gamma} + C_b^{\gamma}} \tag{4.67}$$

and

$$\text{Effect}_2(C_b) = E_0 - \frac{C_b^{\gamma}}{EC_{50}^{\gamma} + C_b^{\gamma}} E_{\max} \tag{4.68}$$

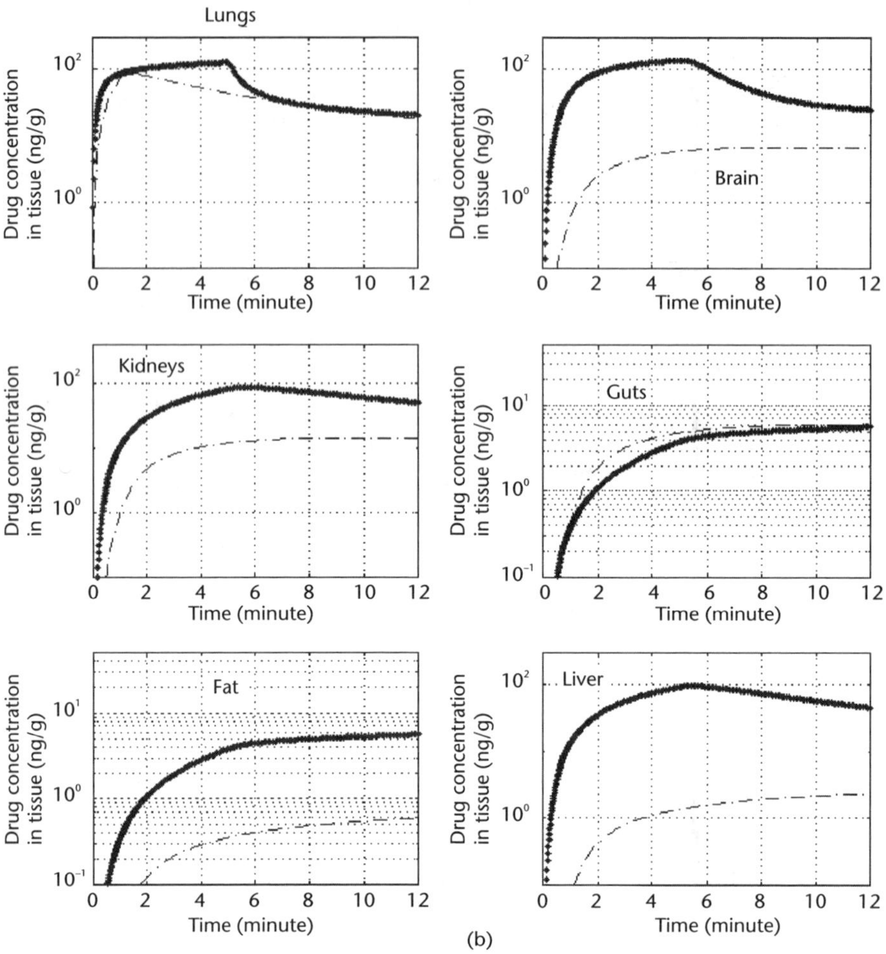

Figure 4.22 (continued.)

where EC_{50} is the effect site concentration at 50% of drug effect. Nominal values of EC_{50} are reported as 6.9, 7.8, and 8.1 ng/ml. The value of γ is often known as the steepness (also known as the slope factor) of the Hill equation.

Definition of (4.68) is mainly the frequency response concept, where the drug effect is also known as the spectral edge. The constant E_0 is defined as E_0 and the frequency at the maximum drug effect is often given by E_{max} = 15 Hz [33].

Such pharmacodynamic modeling with respect to the Brain can be very crucial in establishing the correct levels of fentanyl administered to obtain a certain depth of anesthesia/analgesia, instead of relying solely on the drug concentration in blood. Figure 4.23 shows the SIMULINK representation of the overall model with the brain submodel being extended to include such a pharmacodynamic study.

4.9.1 Nonlinear Generalized Predictive Control

Recall from Chapter 2 the locally linearized discrete model in the backward shift operator z^{-1}:

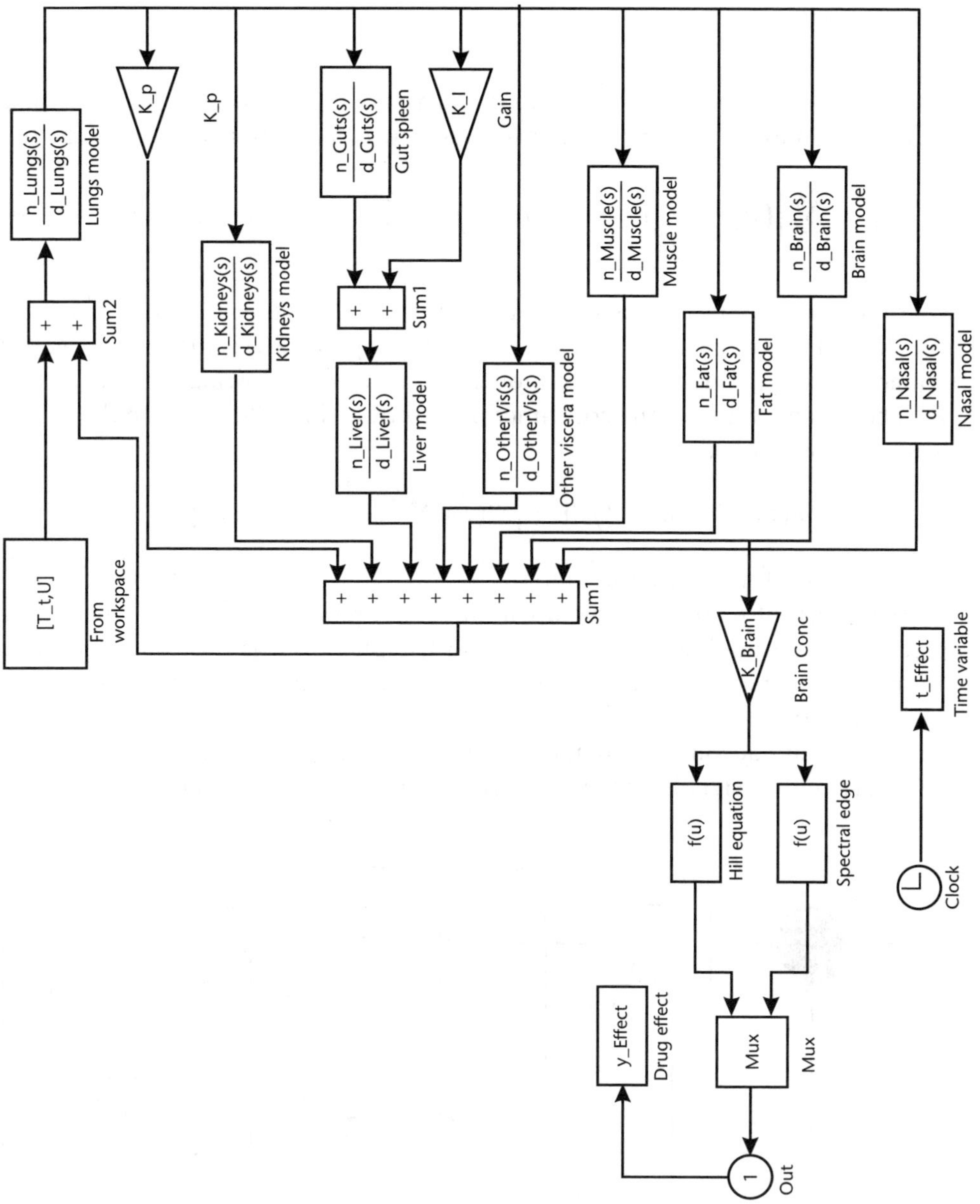

Figure 4.23 SIMULINK representation of the dose-effect relationship.

$$A(z^{-1})\Delta y(t) = B(z^{-1})\Delta u(t-1) + C(z^{-1})\xi(t) \tag{4.69}$$

where:

$A(z^{-1}) = 1 + a_1z^{-1} + a_2z^{-2} + \dots + a_nz^{-n}$
$B(z^{-1}) = b_1 + b_2z^{-1} + b_3z^{-2} + \dots + b_mz^{-m+1}$
$C(z^{-1}) = c_0 + c_1z^{-1} + c_2z^{-2} + \dots + c_pz^{-p}$
$\xi(t)$ is an uncorrelated random sequence.
$\Delta = 1 - z^{-1}$

where $u(t)$ represents the control input and $y(t)$ is the measured variable. The controller computes the vector of controls using optimization of a function of the form

$$J_{GPC} = \sum_{j=N_1}^{N_2}\left[\left(P(z^{-1})\hat{y}(t+j) - \omega(t+j)\right)^2\right] + \sum_{j=i}^{NU}\left[\lambda(j)\left(\Delta u(t+j-1)\right)^2\right] \tag{4.70}$$

where N_1 is the minimum costing (output) horizon, N_2 is the maximum costing horizon, NU is the control horizon, ω is the future set point, $\lambda(j)$ is the control weighting sequence, and $P(z^{-1})$ is the inverse model in the model-following context.

The minimization of the cost function described in (4.70) leads to the following projected control increment:

$$\Delta u(t) = \bar{g}^T(\omega - \Psi),\ \Psi = \left[\Psi(t+N_1), \dots, \Psi(t+N_2)\right] \tag{4.71}$$

(See Chapter 3 for all other definitions.)

In the case of a Wiener-Structure nonlinearity representation (i.e., linear element in series with a nonlinear block—see also Chapter 2), it is possible to modify the controller architecture to absorb such a nonlinearity as shown in Figure 4.24. The controller works as follows.

Taking into account the diagram in the above figure and using GPC's strategy, the future output signals of the linear part can be written as follows:

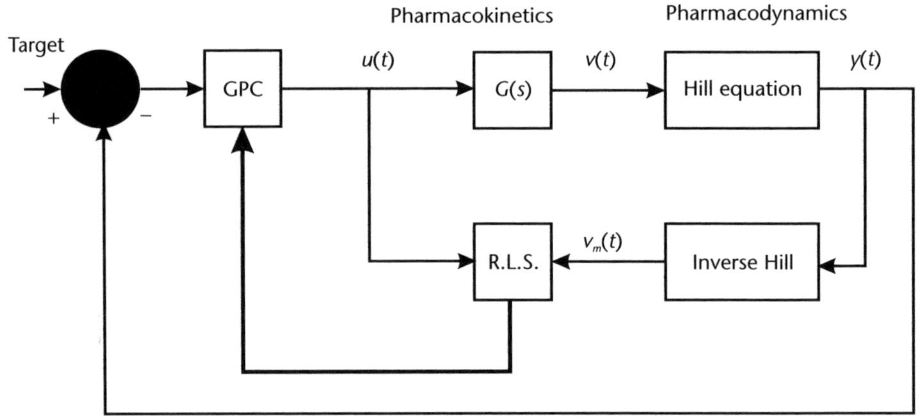

Figure 4.24 A schematic diagram depicting the structure of NLGPC.

$$\begin{cases} \hat{v}_m(t+1) = g_0 \Delta u(t) + f(t+1)_m(t+1) \\ (t+2) = g_1 \Delta u(t) + g_0 \Delta u(t+1) + f(t+2) \\ \vdots \\ \hat{v}_m(t+N_2) = g_{(N_2-1)} \Delta u(t) + \ldots + g_{(N_2-NU)} \Delta u(t+N_2-NU) + f(t+N_2) \end{cases} \tag{4.72}$$

where g_j, $j = 0, 2, \ldots, N_2 - 1$ are elements of the dynamic matrix G_d, and

$$\begin{cases} f(t+j) = G_j \Delta u^f(t-1) + F_j \hat{v}_m^f(t) \\ v_m(t) = D_m \left(\dfrac{y}{1-y} \right)^{1/\alpha_m} \end{cases} \tag{4.73}$$

G_j and F_j being polynomials in z^{-1} and α_m and D_m represent the assumed model parameters of the Hill (Sigmoid) equation. Hence, the nonlinear signal $y(t)$ can be written as follows:

$$y(t) = \frac{\hat{v}_m(t)^{\alpha_m}}{\left(\hat{v}_m(t)^{\alpha_m} + D_m^{\alpha_m}\right)} \tag{4.74}$$

Using the following properties of the backward shift operator z:

$$\begin{cases} z^j \phi(t) = \phi(t+j) \\ z^j \left[\phi(t) \cdot \Psi(t)\right] = \phi(t+j) \cdot \Psi(t+j) \end{cases} \tag{4.75}$$

One can write the predicted nonlinear signals $\hat{y}(t+j)$ as follows:

$$\begin{cases} \hat{y}(t+1) = \dfrac{\hat{v}_m(t+1)^{\alpha_m}}{\left(\hat{v}_m(t+1)^{\alpha_m} + D_m^{\alpha_m}\right)} \\ \hat{y}(t+2) = \dfrac{\hat{v}_m(t+2)^{\alpha_m}}{\left(\hat{v}_m(t+2)^{\alpha_m} + D_m^{\alpha_m}\right)} \\ \vdots \\ \hat{y}(t+N_2) = \dfrac{\hat{v}_m(t+N_2)^{\alpha_m}}{\left(\hat{v}_m(t+N_2)^{\alpha_m} + D_m^{\alpha_m}\right)} \end{cases} \tag{4.76}$$

The signals $\hat{v}(t+j)$, $j = 1, 2, \ldots, N_2$ are the ones predicted by the linear GPC controller according to (4.71).

The cost function (4.70) to be optimized remains the same except that it is no longer differentiable, and a solution using numerical optimization techniques such as the modified simplex method of Nelder-Mead or genetic algorithms can be used. We refer to this control architecture as nonlinear generalized predictive control.

4.9.2 Simulation Results

Simulated control experiments considered the model of Figure 4.23. A series of normalized EEG targets of 80%, 95%, then 90% were specified to NLGPC, which assumed the following tuning parameters: $N_1 = 1$; $N_2 = 10$; $NU = 1$; $\lambda = 0$; $P(z^{-1}) = 1$; $T(z^{-1}) = (1 - 0.8z^{-1})^2$. The RLS algorithm assumed a first-order linear discrete model with one $\hat{a}$ and two $\hat{b}$'s. A sampling interval of 1 minute was assumed throughout.

The result of a typical run is shown in Figure 4.25 where it can be seen that the administration of fentanyl is carefully monitored. Such a feedback control scheme can also be used as a basis for an open-loop TCI system in the operating theater, which in our opinion will be more effective than the existing target infusion systems that take into account only an average pharmacokinetic model. It is believed that this latter is in no way representative of what happens to the drug once inside the human body.

A graphical user interface (GUI) for the overall fentanyl model was produced using MATLAB-SIMULINK, which includes modeling, simulation, model reduction analyses, and control. Figure 4.26(a–c) relates to the GUI, which was built to navigate elegantly through the various structures of this complex model structure. It is also worth noting that this complex physiological modeling study is generic in nature and has been shown to also work successfully with other drugs, such as pethidine [28].

4.10 Conclusions

This chapter focused on an interesting facet of modeling drug behavior via an approach that, I am sure any reader would agree, can deliver a lot, but at a price. This approach, referred to as physiological modeling, should do exactly what its name implies—that is, mimic as closely as possible the physiological phenomena that happen immediately after a drug is administered into the body. Indeed, this chapter started by setting the scene of what physiological modeling can and cannot do. The good news relates to the accurate predictions these structures can provide, and the bad news is that such models are not easy to quantify due to their too "white" architectures. The study then went on to review some of the classical mod-

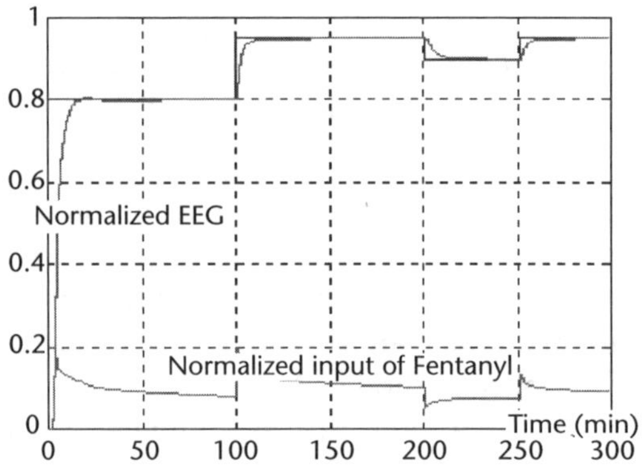

Figure 4.25 Generalized predictive control of anesthesia via EEG measurements.

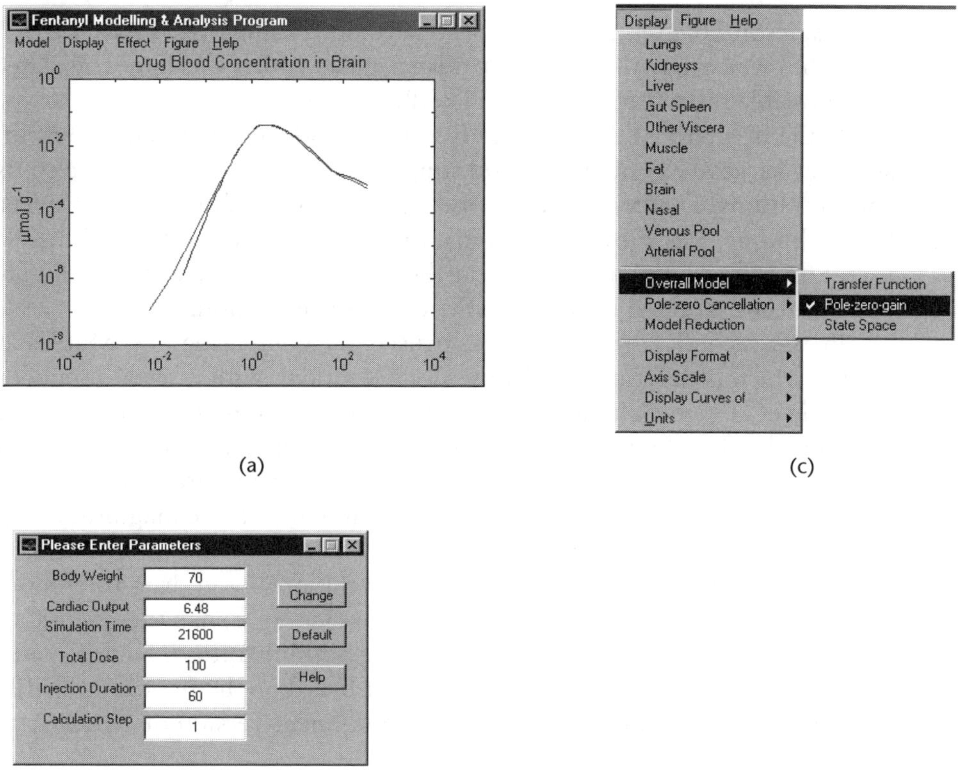

Figure 4.26 (a–c) GUI interface relating to the new representation of the Mapleson-Higgins model for fentanyl.

els, which were proposed as early as 1919 [10], including the Krogh cylinder model, Kety's model [19], the models by Lutz [12] and Mapleson [24]. The most important message from this brief review was that analytical equations describing the movement of blood (drug) across the very often complicated highway of tissues and capillaries do exist, but vital specific data for their corresponding numerical solutions can represent a significant bottleneck. What should sound like an old cliché now is that a compromise between complexity and significance must be reached.[5] In other words, one should, at this stage, look for a structure that is close to this concept of physiological modeling with a possibility of quantifying such model structure. In my search for a model that is able to fulfill such an often difficult balance to strike, I found the Mapleson model structure to be indeed suitable. In order to demonstrate the idea of integrating a systems-engineering approach to such recirculatory model representations, this current study focused on the work that Higgins conducted as part of his M.Sc. degree at Glasgow University in 1994 [27], which investigated the modeling of fentanyl, an analgesic drug used routinely during surgery. Hence, a dynamic modeling approach for this model, which was referred to throughout as

5. Should the reader feel at this stage that this statement is tantamount to capitulation, then this feeling is legitimate.

the Mapleson-Higgins model, has been developed using MATLAB-SIMULINK. The algorithm was originally written in Visual-Basic which was later translated into C^{++} with possible integration with MATLAB.

In the first part of the study, a drug-flow model using an input-output relationship was successfully obtained, and supplemented by models relating to tissue drug concentrations. Furthermore, a model reduction, which uses the balanced realization technique, was shown to lead to a fourth-order overall model. This will be very useful for control design purposes at the brain level, if the effect of the drug in this tissue is to be targeted rather than the drug concentration.

In the second part, the study focused on comparing the Mapleson-Higgins model, which is described by a flow-limited model, with a second model proposed by Wada et al., which assumes that fentanyl pharmacokinetics must include organs described by flow-limited models and others described by diffusion-limited models. A simulation study showed differences in concentration predictions more apparent in the early disposition of drug disposition, and also in the magnitude of peak concentrations and time of occurrence.

The last part of the study showed how the proposed drug-flow model can be used not only for prediction of drug behavior (purely from a therapeutic viewpoint) but for designing "optima" recipes for drug administration. Such a control design has been carried within the realm of a model-based predictive control (MBPC) strategy (see Chapters 3 and 5), particularly a nonlinear version of NLGPC.

References

[1] Wise, M. E., "Negative Power Functions of Time in Pharmacokinetics and Their Applications," *Journal of Pharmacokinetics and Biopharmaceutics*, Vol. 13, 1985, pp. 309–346.

[2] Jacquez, J. A., *Compartmental Analysis in Biology and Medicine*, Amsterdam: Elsevier, 1972.

[3] Godfrey, K., *Compartmental Models and Their Application*, New York: Academic Press, 1983.

[4] Hull, C. J., *Pharmacokinetics for Anesthesia*, Oxford, U.K.: Butterworth Heinemann, 1991.

[5] Jarvis, D. A., "Physiological Pharmacokinetic Models: A Review of Their Principles and Development," *Anesthetic Pharmacology Review*, Vol. 2, 1994, pp. 214–230.

[6] Tucker, G. T., "Pharmacokinetics and Pharmacodynamics: Evolution of Current Concepts," *Anesthetic Pharmacology Review*, Vol. 2, No. 3, 1994, pp. 177–187.

[7] Bjorkman, S., et al., "Comparative Physiological Pharmacokinetics of Fentanyl and Alfentanil in Rats and Humans Based on Parametric Single-Tissue Models," *Journal of Pharm. and Bioph.*, Vol. 22, No. 5, 1994, pp. 381–409.

[8] Crone, C., "Capillary Permeability: Techniques and Problems," in C. Crone and N. A., Lassen, (eds.), *Capillary Permeability, Proceedings of the Alfred Benzon Symposium II*, Copenhagen, 1970, pp. 1–31.

[9] Rowland, M., and N. T. Tozer, *Clinical Pharmacokinetics: Concepts and Applications*, Philadelphia, PA: Lea and Febiger Publishers, 1989.

[10] Krogh, A., "The Number and Distribution of Capillaries in Muscles with Calculations of the Oxygen Pressure Head Necessary for Supplying the Tissue," *Journal of Physiology*, Vol. 52, 1919, pp. 409–415.

[11] Renkin, E. M., "Exchangeability of Tissue Potassium in Skeletal Muscle," *American Journal of Physiology*, Vol. 197, 1959, pp. 1211–1215.

[12] Lutz, R. J., R. L. Dedrick, and D. S. Zaharko, "Physiological Pharmacokinetics: An In Vivo Approach to Membrane Transport," *Pharmacology and Therapeutics*, Vol. 11, 1980, pp. 559–592.

[13] Lopez-Curto, J. A., et al., "Microvascular Anatomy of the Adult Canine Tibial Diaphysis," *Journal of Bone Jt. Surg.*, Vol. 62, 1980, pp. 1362–1369.

[14] Grunewald, W., "The Influence of the Three-Dimensional Capillary Pattern on the Intercapillary Oxygen Diffusion: A New Composed Model for Comparison of Calculated and Measured Oxygen Distribution," in M. Kessler, (ed.), *Oxygen Supply*, Baltimore, MD: Waverly Press, 1973, pp. 5–17.

[15] Frasher, W. G., and H. Wayland, "A Repeating Modular Organization of Microcirculation of Cat Mesentry," *Microvasc. Res.*, Vol. 4, 1972, pp. 62–76.

[16] Rose, C. P., C. A., Goresky, and G. G. Bach, "The Capillary and Sarcolemmal Barriers in the Heart: An Exploration of Labeled Water Permeability," *Circulation Research*, Vol. 41, 1977, pp. 534–545.

[17] Bassingthwaighte, J. B., and C. A. Goresky, "Modeling in the Analysis of Solute and Water Exchange in the Microvasculature," in E. M. Renkin and C. C. Michel, (eds.), *Handbook of Physiology*, Chapter 13, Bethesda, MD: American Physiology Society, 1984, pp. 549–626.

[18] Crone, C., "The Permeability of Capillaries in Various Organs as Determined by Use of the 'Indicator Diffusion' Method," *Acta Physiologica Scandinavica,* Vol. 58, 1963, pp. 292–305.

[19] Goresky, C. A., W. H. Ziegler, and G. G. Bach, "Capillary Exchange Modeling," *Circulation Research*, Vol. 27, 1970, pp. 739–764.

[20] Kety, S. S., "Exchange of Inert Gas at Lungs and Tissues," *Pharmacological Reviews*, Vol. 3, 1951, pp. 1–41.

[21] Zuntz, N., "Zur Pathogenese and therapie de durch rashe Luftdruckanderungen erzeugten Kraukheiten Fortschr," *Med*, Vol. 15, 1987, pp. 632–639.

[22] Wada, D. R., D. R. Stanski, and W. F. Ebling, "A PC-Based Graphical Simulator for Physiological Pharmacokinetic Models," *Computer Methods and Programs in Medicine*, Vol. 46, 1995, pp. 559–592.

[23] Ebling, W. F., E. N. Lee, and D. R. Stanski, "Understanding Pharmacokinetics and Pharmacodynamics Through Computer Simulation: I. The Comparative Clinical Profiles of Fentanyl and Alfentanil," *Anesthesiology*, Vol. 72, 1990, pp. 650–658.

[24] Mapleson, W. W., "Circulation-Time Models of the Uptake of Inhaled Anaesthetics and Data for Quantifying Them," *British Journal of Anaesthesia*, Vol. 45, 1973, pp. 319–334.

[25] Davis, N. R., and W. W. Mapleson, "A Physiological Model for the Distribution of Injected Agents with Special Reference to Pethedine," *British Journal of Anesthesia*, Vol. 70, 1993, pp. 248–258.

[26] Davis, N. R., "Pharmacokinetics of Injected Analgesics," Ph.D. Dissertation, University of Wales College of Medicine, 1986.

[27] Higgins, M. J., "Clinical and Theoretical Studies with Opioid Analgesic Fentanyl," M.Sc. Thesis, University of Glasgow, 1990.

[28] Mahfouf, M., D. A. Linkens, and D. Xue, "Model Reduction for Complex Physiologically-Based Drug Models," *Proceedings of the Annual Conference of the United Kingdom Simulation Society (UKSIM'99)*, Nottingham Trent University, 1999, pp. 182–188.

[29] Moore, B., "Principal Component Analysis in Linear Systems: Controllability, Observability, and Model Reduction," *IEEE Trans. on Automatic Control*, Vol. AC-26, 1981, pp. 17–32.

[30] Sherwood, L., *Human Physiology: From Cells to Systems*, 3rd ed., Belmont, CA: Wadsworth Publishing Company, 1997, p. 307.

[31] White, M., and G. N. C. Kenny, "Intravenous Propofol Anesthesia Using a Computerized Infusion System," *Anesthesia*, Vol. 45, 1990, pp. 204–209.

[32] Schuttler, J., et al., "Total Intravenous Anesthesia with Propofol and Alfentanil by Computer Assisted Infusion," *Anesthesia (Supplement)*, Vol. 43, 1988, pp. 3–7.

[33] Jacobs, J. R., and E. A. Williams, "Algorithm to Control Effect Compartment Drug Concentrations in Pharmacokinetic Model-Driven Drug Delivery," *IEE Trans. on Biomedical Engineering*, Vol. 40, No. 10, 1993, pp. 993–999.

CHAPTER 5

A Hybrid System's Approach to Modeling and Control of Unconsciousness

This chapter aims at exploiting the power of the model architectures introduced in Chapter 4 (the recirculatory physiological models) to design monitoring and closed-loop control strategies for drug administration during surgery. Indeed, the unconstrained and constrained versions of predictive control are evaluated for online administration of anesthetic drugs during surgery. First, a patient simulator is developed using a physiological model of the patient, and the necessary control software is validated via a series of extensive simulation studies. The validated system is then transferred into the operating theater for a series of clinical evaluation trials. The trials were performed with little involvement of the design engineers, and good regulation of the blood pressure was achieved using fixed-parameter versions of the algorithm. In addition, the model architecture is extended to include a supervisory layer in the form of a Mamdani-type fuzzy logic based system. Under normal operating conditions, predictive control provides regulation around target blood pressure values and the switch over to the fuzzy based control system occurs only when a fault is detected.

5.1 Introduction

It is clear from the study that was described in Chapter 3 that the measure of anesthetic depth during surgical anesthesia has always represented a tough challenge, and the experience of the anesthetist is required to control the patient's anesthetic state. Controlling anesthesia means that the patient is maintained at a suitable level of sedation in order to allow the surgeon to proceed with the operation without causing *awareness in the patient*. Hence, a number of drugs have to be administered carefully to achieve an appropriate depth of anesthesia level without compromising the patient's health. There have been reports of incomplete general [1] by patients who were pharmacologically paralyzed while under anesthesia, such as the following testimony by a patient who underwent surgery that required anesthesia [2]:

> The feeling of helplessness was terrifying. I tried to let the staff know I was conscious but I could not move even a finger nor eyelid. It was like being held in a vise and gradually I realized that I was in a situation from which there was no way out. I began to feel that breathing was impossible and I just resigned myself to dying.

Nowadays, safer drugs are used with patients, better monitoring machines help the anesthetist make an informed decision during surgery, and state-of-the-art delivery equipment has helped the anesthetist maintain a more stable anesthetic depth. Such clinical measures, however, are still open to the anesthetist's scrutiny, as patients react differently to different drugs, and as a result, different patients at the same anesthetic depth can present different clinical measures.

Several researchers have argued that if automatic control of anesthetic depth relies on its accurate measurement, then more efforts must be directed towards devising ways of measuring DOA directly. Previous research work by Schwilden and coworkers [3] used quantitative EEG analysis in humans to give an indication of the anesthetic state. However, the interpretation of the tracings proved difficult and a subjective task. The information proved unreliable even when interpreted by experienced staff, since the characteristic patterns were often disturbed by factors such as anoxia and surgical stimulations. As will be seen in Chapter 6, auditory evoked potentials (AEP) have been shown to be good indicators of anesthetic depth [4, 5], they are said to have a graded response with changing anesthetic depth, they demonstrate little interpatient variability, and they measure the underlying anesthetic depth as opposed to the anesthetic concentration in the blood. However, the extraction, analysis, and the online use of such signals in the operating theater is not a straightforward task.

Alternatively, as stated in Chapter 3, a direct measurement of unconsciousness must involve one or more cardiovascular measurements such as blood pressure, which still represents a credible tool in noncritical situations. More recently, a multitasked closed-loop control system consisting of two controllers was presented by Gentilini and coworkers [6]. In this work, the authors controlled mean arterial pressure and hypnosis through the Bi-Spectral Index (BIS), a measure of the EEG, via isoflurane administration.

In the study reported in this chapter, two interesting themes will be explored, one relating to modeling and the other concerning control. As far as the former is concerned, and in contrast to the compartmental models [7] already seen in Chapter 3, this study will look at physiological modeling, which despite their relative complexity, tend to be nearer to the "actual" physiological processes [8]. It must be stressed here that although compartmental modeling is a form of physiological modeling in itself, a recirculatory physiological model, albeit using the compartmental modeling idea, provides a better approximation and a more transparent view of drug interaction(s) with the body.

Hence, this study will outline a closed-loop (recirculatory) model of humans in relation to the anesthetic isoflurane (see Chapter 3 for more information on this drug). From a control viewpoint, this chapter will explore the theme of *optimal* control via the adoption of constraints (i.e., constrained control). Indeed, in this particular study, all known physical constraints will be included in the objective (cost) function prior to the optimization operation,[1] in this case the control sequence can be guaranteed to be optimal. There is no doubt that constrained control, similarly to any design that claims to tackle one particular problem, will endemically open up to other complex issues relating, for instance, to "feasibility" of the solution(s) [9], but

1. As opposed to an ad hoc optimization, whereby the constraints are added after the control sequence has been calculated by "clipping" the variables given the minimum and maximum values.

as will be shown in the later sections, possible solutions to deal with these issues can also be suggested. Perhaps the most interesting aspects of this study are the integration of several approaches to solve the problem of feedback control of a *very subjective* variable (i.e., unconsciousness), leading to a *hybrid* control architecture, but also successful transfer of a closed-loop constrained control strategy directly into the operating theater to be used during surgery operations in a series of 10 clinical trials which all required anesthesia control.

5.2 The Mean Arterial Pressure Physiological Model

The model, associated with the anesthetic drug isoflurane, which is used in this study, is represented in Figure 5.1. It consists of two parts: one part for the uptake and distribution of drug in the lungs, and the other part for the circulation of the blood flows. It can be shown that the overall nonlinear model describing such pharmacokinetics (what the patient's body does to the drug) as well as pharmacodynamics (what the drug does to the patient's body) is as follows:

$$\begin{cases} \dot{p}_i = k_i g_{i,0} CO_0 (1 + a_1 p_1 + a_2 p_2 + a_A p_A)(p_A - p_i) \dfrac{1 + b_i p_i}{\sum_{j=1}^{9} g_{j,0}(1 + b_j p_j)} \\ \dot{p}_L = k_L \begin{Bmatrix} \lambda_b (1 - l_s) CO_0 (1 + a_1 p_1 + a_2 p_2 + a_A p_A)(p_V - p_L) + \\ q_{Air}(p_{Air} - p_L) \end{Bmatrix} \\ \dot{p}_A = k_A CO_0 (1 + a_1 p_1 + a_2 p_2 + a_A p_A)\left[p_V l_s + p_L (1 - l_s) - p_A\right] \\ \dot{p}_V = k_V CO_0 (1 + a_1 p_1 + a_2 p_2 + a_A p_A) \left[\dfrac{\sum_{i=1}^{9} g_{i,0}(1 + b_i p_i) p_i}{\sum_{j=1}^{9} g_{j,0}(1 + b_j p_j)} - p_V \right] \end{cases} \tag{5.1}$$

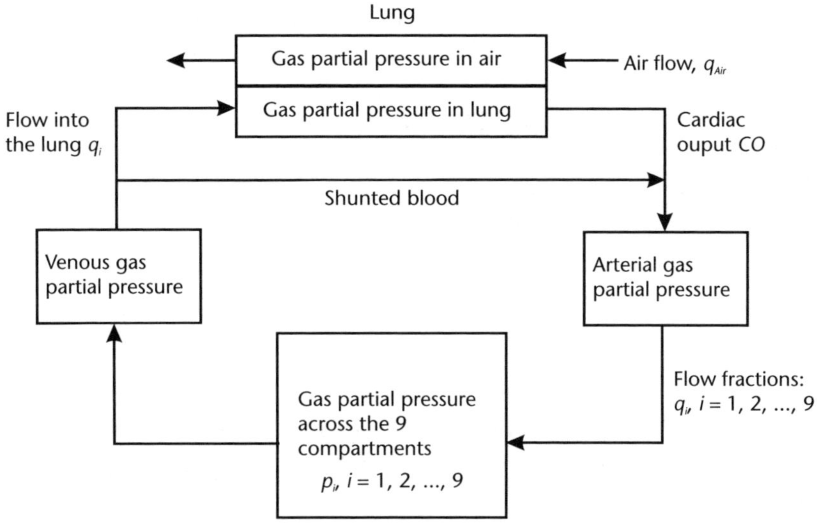

Figure 5.1 The patient physiological model relating to inhalational anesthesia. (*After:* [10].)

where $i = 1, \ldots, 9$ (number of compartments).

The state vector $p(t)$ describes the partial pressure of the anesthetic gas in every compartment, the input being the concentration of the anesthetic gas in the inspired air (p_{Air}), V refers to venous, A refers to artery, and L refers to lungs, $g_{j,0}$, b_i, k_i, CO_0, and λ_i are all terms that can be inferred from the partial pressures or are constants that are either patient or drug dependent [10]. The MAP is given by the following equation:

$$MAP = CO_0 \frac{1 + a_1 p_1 + a_2 p_2 + a_3 p_A}{\sum_{j=1}^{9} g_{j,0}\left(1 + b_j p_j\right)} \tag{5.2}$$

where CO_0 is the total cardiac output prior to any anesthetic being given.

Because giving 100% O_2 can cause the patient to have lung problems, a mixture of 70% N_2O and 30% O_2 is preferred during anesthesia. N_2O, having a mild anesthetic effect, acts as a carrier for isoflurane and lowers the drug equilibrium time. Hence, its effect was modeled by increasing the effective air-flow q_{Air} in (5.1), to take into account the partial pressures in relation to this gas such that for the drug isoflurane:

$$\begin{aligned} q_{Air_{new}} &= q_{Air}\left(1 + K_{q_{Air}}\right) \\ K_{q_{Air}} &= 0.235 \end{aligned} \tag{5.3}$$

It is worth noting that these parameters are subject to variations as high as 4:1 ratio (as are many parameters associated with biomedical systems), an observation which suggests that fixed controllers, such as PID networks, would find it difficult to cope with such variations of dynamics and gains without being retuned.

Moreover, a recent technique that consists of delivering the anesthetic in liquid form, which is transformed into gas almost instantaneously at room temperature, was adopted. This has the advantage of avoiding having to drive a vaporizer with all its software complexity. In order to reflect such a modification, a model that describes the dynamics associated with the vaporization process was elicited through an experimental study using the following first-order differential equation:

$$p_{iso\_gas} = -k_{1g}\, q_{Air} P_{iso\_gas} + k_{2g}\, q_{iso\_liq} \tag{5.4}$$

where p_{iso_gas}, p_{iso_liq} are the concentrations of the anesthetic in gas and liquid forms, respectively, and k_{1g}, k_{2g} are constants that we have determined experimentally as being

$$\begin{aligned} k_{1g} &= 0.23 \\ k_{2g} &= 3.04 \end{aligned} \tag{5.5}$$

The model described by (5.1) to (5.5) forms the basis for a closed-loop control strategy design using constrained model-based predictive control as outlined in the next section.

5.3 Constrained Model-Based Predictive Control Using the Quadratic Programming Approach

The long-range predictive controller, particularly GPC (which was introduced in Chapter 3), is extended in this study to include the following constraints.

As in Chapter 3, consider the following locally-linearized discrete model in the backward shift operator z^{-1}:

$$A(z^{-1})\Delta y(t) = B(z^{-1})\Delta u(t-1) + C(z^{-1})\xi(t) \quad (5.6)$$

where:

$A(z^{-1}) = 1 + a_1 z^{-1} + a_2 z^{-2} + \ldots + a_{n_a} z^{-n}$
$B(z^{-1}) = b_1 + b_2 z^{-1} + b_3 z^{-2} + \ldots + b_{n_b} z^{-m+1}$
$C(z^{-1}) = c_0 + c_1 z^{-1} + c_2 z^{-2} + \ldots + c_p z^{-p}$
$\xi(t)$ is an uncorrelated random sequence
$\Delta = 1 - z^{-1}$
$u(t)$ represents the control input and $y(t)$ is the measured variable

The controller computes the vector of controls using optimization of a function of the form

$$J_{GPC} = \sum_{j=N_1}^{N_2}\left[\left(P(z^{-1})\hat{y}(t+j) - \omega(t+j)\right)^2\right] + \sum_{j=1}^{NU}\left[\lambda(j)\left(\Delta u(t+j-1)\right)^2\right] \quad (5.7)$$

where N_1 is the minimum costing (output) horizon, N_2 is the maximum costing horizon, NU is the control horizon, ω is the future set point, $\lambda(j)$ is the control weighting sequence, and $P(z^{-1})$ is the inverse model in the model-following context with $P(1) = 1$. Furthermore, the $C(z^{-1})$ polynomial in (5.6) is replaced by a fixed polynomial $T(z^{-1})$ known as the observer polynomial for the predictions $P(z^{-1})\hat{y}(t+j)$. This filter enables to offset the effect of the Δ operator as a highpass filter on the input-output data.

When the control horizon NU (which reflects the number of degrees of freedom for the controller) is greater than 1, the solution of (5.7) in the *unconstrained* case (physical and terminal constraints not included prior to optimization) differs from that in the *constrained* case (physical and terminal constraints included before optimization takes place). Hence, one way of solving (5.7) in the constrained case is to consider the following Least Squares Inequality (LSI) problem [11]:

$$\text{Minimize } \|A\tilde{u} - b\| \text{ subject to } H\tilde{u} > h \quad (5.8)$$

where $\tilde{u}$ is the NU solution vector, $\boldsymbol{H}$ is the static/dynamic constraints information matrix, and $\boldsymbol{h}$ is a vector containing the lower and upper limits of the constraints. In the case of (5.7), we have

$$A = \begin{bmatrix} G_d \\ \lambda^{1/2} \end{bmatrix}; \; b = \begin{bmatrix} \omega - P \cdot \hat{y} \\ 0 \end{bmatrix} \quad (5.9)$$

G_d is the dynamic (step-response) matrix of the form

$$G_d^T = \begin{bmatrix} g_0 g_1 \cdots g(N_2 - 1) \\ 0 \quad g_0 \cdots g(N_2 - 2) \\ \vdots \\ 0 \quad 0 \ldots g(N_2 - NU) \end{bmatrix} \tag{5.10}$$

where g_i are the step-response coefficients [12]. $\boldsymbol{H}$ and $\boldsymbol{b}$ will depend on the types of constraints which are considered (i.e., input rate constraints, input magnitude constraints, and output magnitude constraints). If all three types of constraints are considered, then we write the conditions as follows:

$$\begin{cases} \delta_{\min} \le \Delta u(t + j - 1) \le \delta_{\max} \\ \beta_{\min} \le u \le \beta_{\max} \\ \gamma_{\min} \le \Phi(t + j) \le \gamma_{\max} \end{cases} \tag{5.11}$$

For example, for $N_2 = 3$ and $NU = 2$, $\boldsymbol{H}$ and $\boldsymbol{b}$ in (5.8) can be written as follows:

$$H = \begin{bmatrix} 1 & -1 & 0 & 0 & 1 & -1 & 1 & -1 & g_0 & -g_0 & g_1 & -g_1 & g_2 & -g_2 \\ 0 & 0 & 1 & -1 & 0 & 0 & 1 & -1 & 0 & 0 & g_0 & -g_0 & g_1 & -g_1 \end{bmatrix}^T$$

$$b = \begin{bmatrix} \delta_{\min} \\ -\delta_{\max} \\ \delta_{\min} \\ -\delta_{\max} \\ \beta_{\min} - u(t-1) \\ u(t-1) - \beta_{\max} \\ \beta_{\min} - u(t-1) \\ u(t-1) - \beta_{\max} \\ \gamma_{\min} - \Phi(t+1) \\ \Phi(t+1) - \gamma_{\max} \\ \gamma_{\min} - \Phi(t+2) \\ \Phi(t+2) - \gamma_{\max} \\ \gamma_{\min} - \Phi(t+3) \\ \Phi(t+3) - \gamma_{\max} \end{bmatrix} \tag{5.12}$$

where $\delta_{\min}$, $\delta_{\max}$, $\beta_{\min}$, $\beta_{\max}$, $\gamma_{\min}$, $\gamma_{\max}$, and $\Phi(t + j)$ are the minimum and maximum allowed control increments, absolute control moves, the outputs, and the free predicted outputs, respectively. This quadratic programming (QP) problem can be solved using the method proposed by Lawson and Hanson [11]. Also, when using both input and output constraints simultaneously, *infeasibility* problems may be encountered [9] (when the optimizer cannot satisfy all constraints at once). Several methods can be used to circumvent such a problem, but the one we used in this instance is the hierarchical removal of output constraints starting from the bottom predictions until the optimizer is capable of returning a *feasible* solution [13]. In the case of such incompatibility, the following hierarchical algorithm is adopted:

1. Perform the GPC calculations: Calculate the G_d matrix.
2. Perform SVD on A of (5.9).
3. Use the least distance programming (LDP) method to find a solution.
4. Test for Incompatibility.
5. If "Constraints Compatible," then apply the control solution and **GO TO 1**.
6. If "Constraints Incompatible," then test if this is the first incompatibility within the sample.
7. If "Not the First Incompatibility Encountered," then remove one horizon constraint at a time, starting from the bottom predictions, then **GO TO 3**.
8. If "First Incompatibility Encountered," then test first the output constraint corresponding to the k predictions, k being the assumed model time delay, then **GO TO 3**.

This hierarchical removal of constraints was found to be more effective than *constraints relaxation*, which we believe defeats the objective of setting up constraints in the first place.

5.4 A Review of Faults Associated with the Anesthesia Control System

With reference to Figure 5.2, which represents the anesthesia control system, the faults associated with the system can be divided into three types: sensor failures, actuator failures, and algorithmic failures:

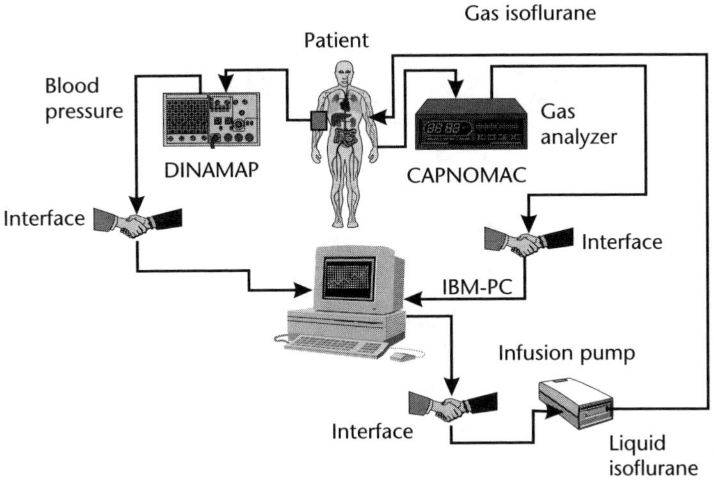

Figure 5.2 Diagram representing the closed-loop control system as used in the operating theater to monitor anesthesia via blood pressure measurements.

5.4.1 Sensor Failures

Sensor failures are mainly due to the DINAMAP instrument, which provides blood pressure measurements. Diathermy (which produces and electrical interference) as well as various surgical stimuli can also cause false readings (corresponding to a blood pressure of 0 mmHg or a sudden increase in blood pressure of more than 5 mmHg, respectively).

5.4.2 Actuator Failures

The actuator, being typified in this case by a syringe pump and which is responsible for administering isoflurane through an infusion line, can sometimes fail to fulfill this task due to a breakdown or to a communication error. The result is a slow and continuous rise in the recorded MAP level.

5.4.3 Algorithmic Failures

Another type of failure that can occur in such circumstances is a failure in the actual algorithm responsible for the control strategy where two individual tasks are involved: the recursive parameter estimation and the control solution calculation.

A Fault in the Estimator

Although faults associated with a recursive parameter estimator can be numerous, those associated with the covariance matrix are the most usual ones. The latter should remain positive definite and its trace (the sum of its diagonal elements) should decrease as the parameter estimates converge to sensible values. Hence, a quick check on this trace value should prove whether the estimator has incorrectly learned the process.

5.5 The Hierarchical Supervisory Level: Structure and Algorithm

The overall anesthesia control system represented in Figure 5.2 has been given an intelligent and autonomous structure by superimposing a supervisory layer that will have the task of monitoring and assessing its performance at every sample (see Figure 5.3).

5.5.1 Detection

Since the structure of the adaptive constrained GPC includes inherently an identification mechanism, it was decided to adopt the process model-based method for detection. The model-based method uses the least number of sensors to monitor the operating state of the system.

Consider a healthy process with the following linear model representation and prediction (residual):

$$A\left(z^{-1}\right)\tilde{y}(t) = B\left(z^{-1}\right)u(t) + x(t) \tag{5.13}$$

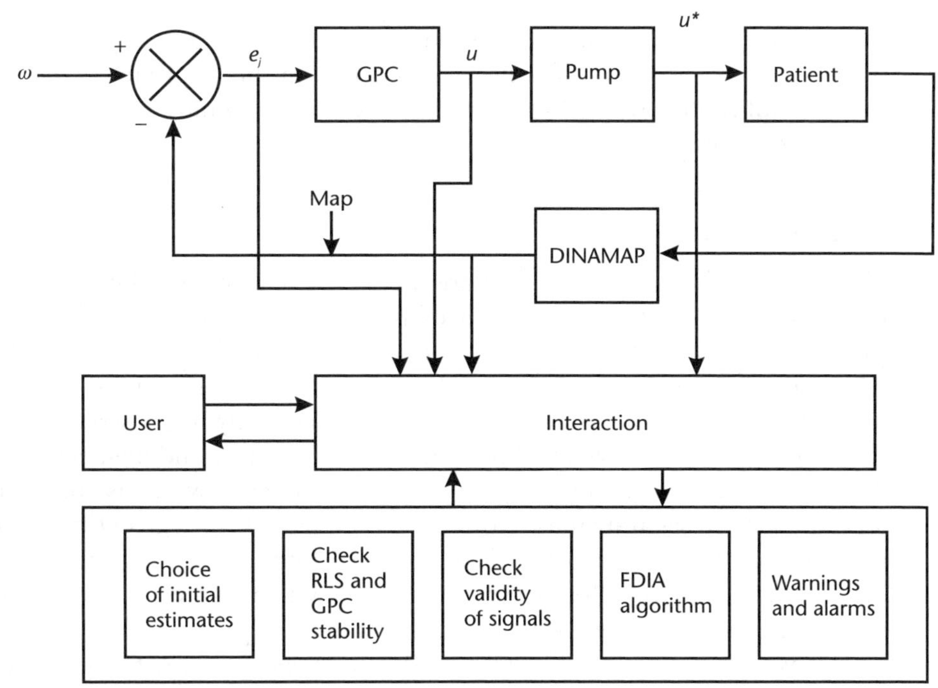

Figure 5.3 The anesthesia control system including the hierarchical supervisory level.

and

$$|\tilde{\varepsilon}| = \left|\tilde{y} - \hat{\theta} \cdot \Phi^T\right| \tag{5.14}$$

where $\tilde{y}$ is the observed output, $\hat{\theta}$ is the estimates, and Φ is the measurement vector that includes the input-output information.

For a healthy system, there is no significant model mismatch, leading therefore to a negligible residual. Any significant residual level emanates from persistently exciting noise, sudden disturbances, and so forth. The effect of these uncertainties causes the residual to deviate from the zero baseline; however, it is limited within some predefined threshold entirely correlated with the so-called *fault* in the system.

Now, consider a faulty model and its residual as follows:

$$\left\lfloor A(z^{-1}) + \delta A(z^{-1}) \right\rfloor \tilde{y}(t) = \left\lfloor B(z^{-1}) + \delta B(z^{-1}) \right\rfloor u(t) \tag{5.15}$$

and

$$|\tilde{\varepsilon}| = \tilde{y} + \delta\tilde{y} - \Phi^T\left(\hat{\theta} + \delta\hat{\theta}\right) \tag{5.16}$$

where δA and δB represent the discrepancies due to the changes in the system structure, $\delta\tilde{y}$ may be due to failure caused by a bias of the sensor or disturbances acting on the system, and $\delta\tilde{\theta}$ denotes a fault in the system structure and in the state variable itself. Hence, when a fault occurs, the value of $|\tilde{\varepsilon}|$ in (5.16) will be greater than that of (5.14); this forms the basis for the fault detection mechanism used throughout (i.e., the residual value is checked against a preset threshold drawn from the *normal* state of operation). The main difficulty associated with this approach is the selection of the threshold value, which has to be chosen so as to minimize the rate of *false alarms* and *fault misses*.

5.5.2 Isolation (Diagnosis)

In order to isolate a fault, checks are carried out on the absolute signals and their trends. For the absolute signal, a check is made on the evolution of the error signal over a window length of k samples. Here again threshold values have to be defined, which set tolerance bands. Two signal categories can be defined: low and high, and a check on the absolute signal trend would confirm or deny the existence of a particular fault. Here also, thresholds will be specified, and the low and high categories will be defined. Because the assumption of a purely deterministic system is not a realistic one, it is necessary to separate noise from the testing signals through extra filtering, which is performed by an additional filter provided by the LRPI algorithm (see Section 5.5.3), and by adopting the idea of voting, as shown in Figure 5.4, in which the particular fault occurs m samples out of k samples in the a priori chosen window ($m < k$).

Flags are provided for |*Signal Error*| and |*Signal Error Rate*|, which are set to 0 or 1 depending on whether they are smaller or bigger than the preset thresholds, respectively. Table 5.1 is a signature table that includes the early possible signs leading to a fault. The rules used for deciding on a fault are simple:

IF (detected sign = signature) **THEN**
Fault type = sign #

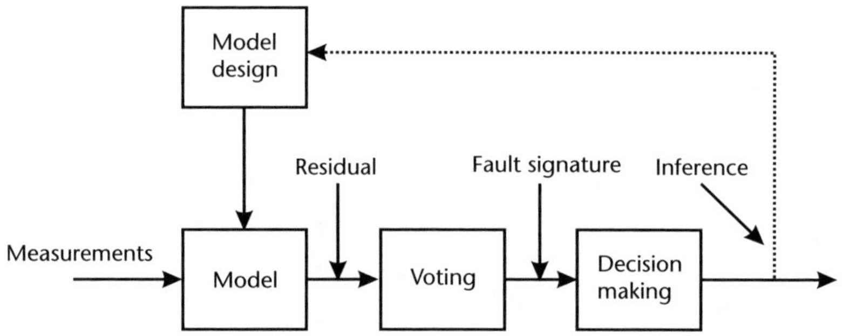

Figure 5.4 Stages of model-based fault detection and diagnosis.

Table 5.1 Binary Codes for Early Signs Prior to Diagnosis: Fault Signatures

Sign	|*Error Signal*|	|*Rate*|	|*Trace*|
1	0	0	0
2	1	0	0
3	1	1	0
4	0	1	0
5	0	0	1

The nature of some possible signs is outlined here:

- Sign 1: Sign of a slow drift. This could be an actuator fault, but it needs k samples to be confirmed. Also, one needs to check that *Error* ≥ 0 and *Error Rate* ≥ 0, and *input* $\neq 0$ for at least $k - 1$ samples.
- Sign 2: Sign of heavy stimulus or the late stage of an actuator fault. One needs k samples to confirm.
- Sign 3: *Error* and *Error Rate* are beyond their respective thresholds. This is definitely the result of stimulus; k samples are needed to confirm whether the fault is recoverable or a permanent.
- Sign 4: *Error* is low but *Error Rate* is high. This feature definitely characterizes a sudden recoverable disturbance from a brief stimulus level.
- Sign 5: The estimator is incorrectly learning the process.

5.5.3 Accommodation (Compensation)

Once the source of the fault has been identified, accommodation or compensation must be provided.

Case of a Sudden Disturbance

A brief stimulus level causes this disturbance. Hence, the use of the observer polynomial $T(z^{-1})$ coupled with the LRPI algorithm based filter $L(z^{-1})$ should suffice to reject it. The filter $L(z^{-1})$ is obtained by extending the long-range predictive idea to identification and by using a series of convolution and deconvolution operations [14]. In this case all input and output signals will be filtered using a filter of the following form:

$$G_F = \frac{\Delta L(z^{-1})}{T(z^{-1})} \tag{5.17}$$

It is worth noting that such filtering only applies to the estimation procedure and not to the control procedure.

Case of a More Permanent Disturbance

For a stimulus level, which lasts for more than 3 samples, the idea of switching to another form of control, such as fixed GPC control or fuzzy logic control, was implemented. The latter alternative scheme is based on a simple Mamdani-type fuzzy PI controller [15] using no more than 25 rules, which were obtained using an optimization algorithm based on symbiotic evolution [16] (see Table 5.2). The definitions for the five fuzzy labels for the inputs and output spaces are given in Figure 5.5. This form of "jacketing" is relatively safer as it does not make use of any online parameter estimation scheme, which, while disturbances are still acting on the system, is not safe to operate as it will undoubtedly lead to biased estimates. Figure 5.6 depicts the overall supervisory control structure as simulated using MATLAB-SIMULINK.

Table 5.2 Fuzzy Control Rule Base

Change-in-Error/ Error	*NB*	*N*	*ZE*	*P*	*PB*
NB	NB	NB	N	N	ZE
N	NB	N	N	ZE	P
ZE	N	N	ZE	P	P
P	N	ZE	P	P	PB
PB	ZE	P	P	PB	PB

NB: negative big; N: negative; ZE: zero; P: positive; PB: positive big; Error = Target – Output.

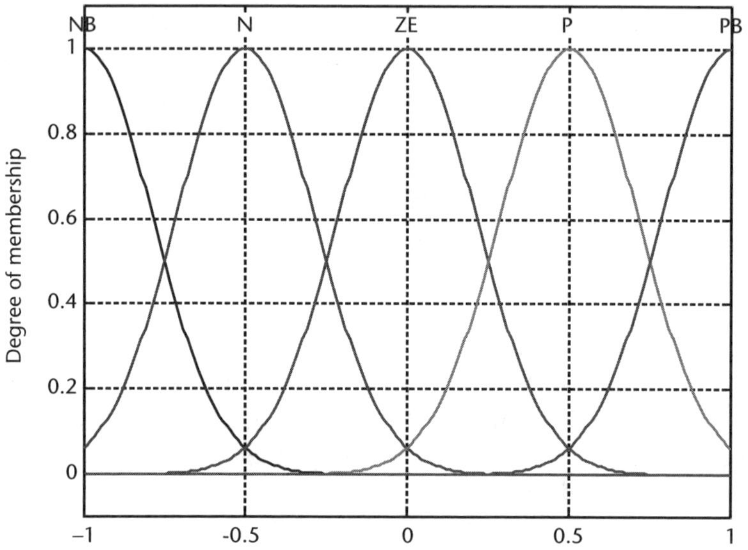

Figure 5.5 Membership functions used in the fuzzy control system.

Case of an Actuator Fault

This form of a fault is not easy to deal with directly; therefore, automatic control should be stopped immediately and manual control via direct administration of the drug by the anesthetist using bolus doses is required in this case until the fault is repaired.

Case of a Problem in the Covariance Trace or the Control Algorithm Itself

When the covariance trace of the parameter estimator is too low, then covariance resetting is one possible solution to restore the above estimator's integrity.

Safe Operation of the Model Adaptation Mechanism

When the system is subjected to a fault, its various states will adopt a behavior that will not reflect its true dynamics, gain magnitude, or its sign. Therefore, keeping the estimation operational will lead to biased parameters, which in turn will lead to unstable control. In light of this we have devised the *jacketing strategy* of Figure 5.7, which allows to switch the adaptation on or off depending on where the value of the controlled variable lies. When adaptation is off, control is either switched to fixed GPC or fuzzy control.

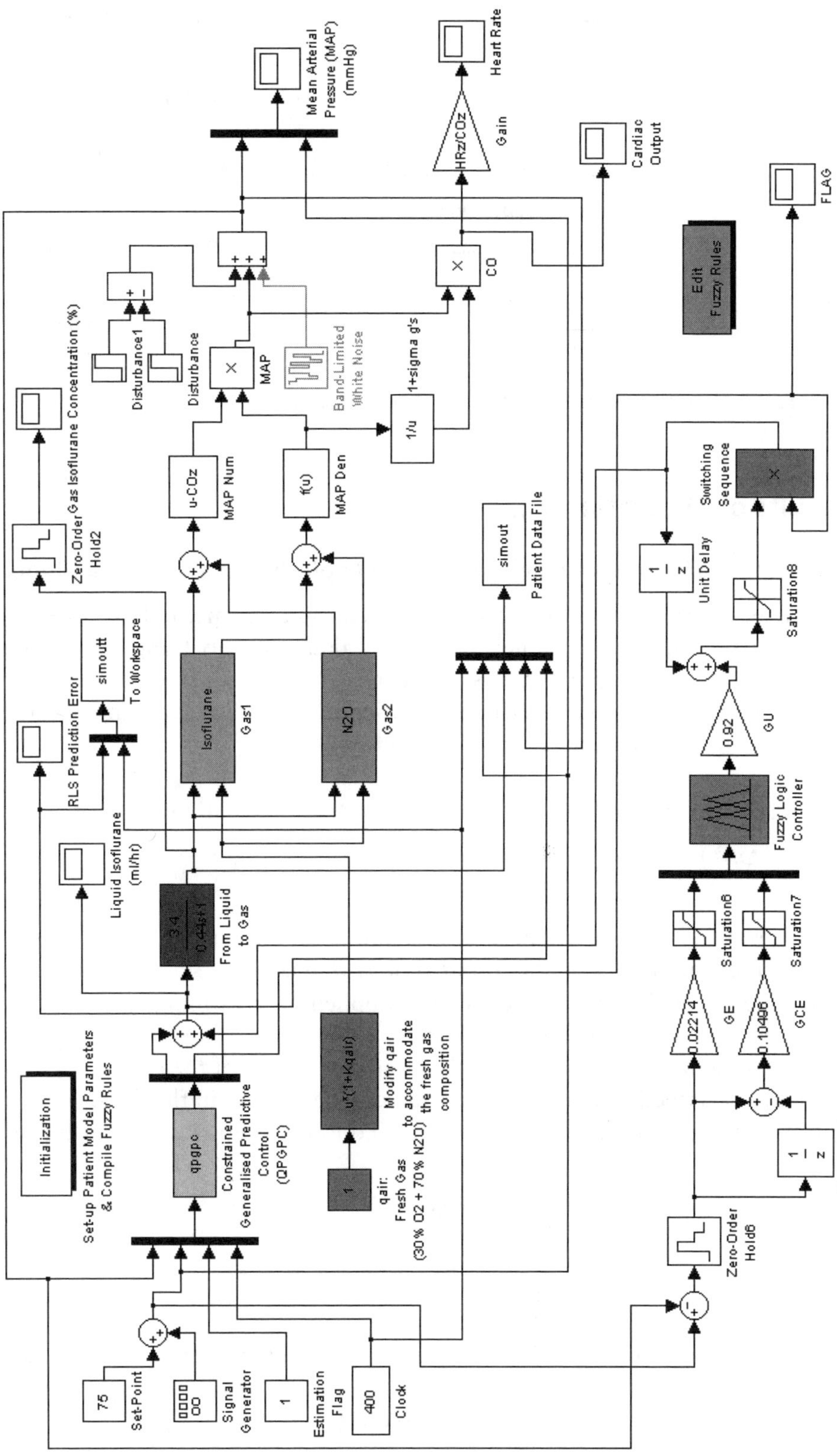

Figure 5.6 MATLAB-SIMULINK representation of the supervisory control scheme, which includes GPC and fuzzy logic control.

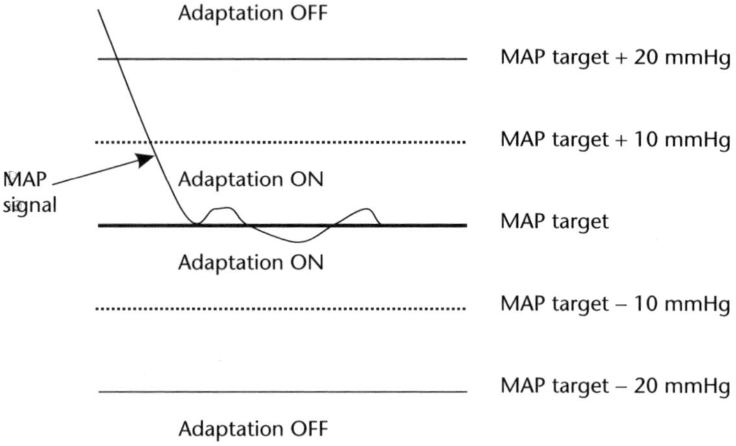

Figure 5.7 Diagram representing the MAP limits for jacketing.

5.6 Results of Simulation Experiments

5.6.1 Identification of Linear and Fuzzy Logic–Based Anesthesia Models

Consider a SISO system that can be modeled using the method proposed by Takagi and Sugeno, and Kang, namely, a fuzzy Takagi-Sugeno Kang (TSK) model [17]. Assuming that the input space is partitioned using p fuzzy partitions and that the system can be represented by fuzzy implications (one in each fuzzy subspace), we can write the fuzzy implication "i" as follows:

$$\begin{aligned} &L^i : IF\ y(t)\ is\ B^i\ THEN\ \Delta y_m^{(i)}(t+1) = -a_1^i \Delta y(t) - \cdots - a_j^i \Delta y(t - n_a + 1) \\ &+ b_1^i \Delta u(t) + \cdots + b_k^i \Delta u(t - n_b + 1) \end{aligned} \tag{5.18}$$

where $y(t)$ and $u(t)$ are the process and controller outputs at time t, $y_m(t + 1)$ is the one-step-ahead model prediction at time t, B^i is a fuzzy set representing the fuzzy subspace in which implication L^i can be applied for reasoning, and $\Delta = 1 - z^{-1}$, with z^{-1} being the backward shift operator.

Such model representation in the consequent part of the above implication was found to be effective against offsets that can be present in the data.

The model parameters can be expressed in the following matrix form:

$$\Theta = \begin{bmatrix} a_1^1 \ldots a_{n_a}^1 & b_1^1 \ldots b_{n_b}^1 \\ \vdots & \vdots \\ a_1^P \ldots a_{n_a}^P & b_1^P \ldots b_{n_b}^P \end{bmatrix} \tag{5.19}$$

The overall fuzzy model output (in incremental form) can be written as follows:

$$\Delta y_m(t+1) = \Theta' \Phi(t) \tag{5.20}$$

where

$$\Phi(t) = \begin{bmatrix} -\Delta y(t), -\Delta y(t-1), \ldots, -\Delta y(t-n_a+1), \Delta u(t), \\ \Delta u(t-1), \ldots, \Delta u(t-n_b+1) \end{bmatrix}^T \tag{5.21}$$

Φ' represents a vector of the β_i – weighted parameters of Θ such that

$$\Theta' \\ \Theta = \begin{bmatrix} a_1' \ a_2' \ldots a_{n_a}' & b_1' \ b_2' \ldots b_{n_b}' \end{bmatrix} \tag{5.22}$$

$$a_j' = \sum_{i=1}^{p} \beta_i \cdot a_j^{(i)} \quad j = 1, \ldots, n_a \tag{5.23}$$

$$b_k' = \sum_{i=1}^{p} \beta_i \cdot b_k^{(i)} \quad k = 1, \ldots, n_b \tag{5.24}$$

and

$$\beta_i = \frac{B^i[y(t)]}{\sum_{i=1}^{p} B^i[y(t)]} \tag{5.25}$$

$B^i[y(t)]$ is the grade of membership of $y(t)$ in B^i, and β is a vector of the weights assigned to each of the p implications at each sampling instant.

For data generation we applied the input profile (liquid isoflurane) shown in Figure 5.8. A total of 400 data points were used for identification.

For off-line model identification, the following second-order model structure for the ith fuzzy partition was used:

$$y_m(t+1) = -\hat{a}_1^i y(t) - \hat{a}_2^i y(t-1) + \hat{b}_1^i u(t) + \hat{b}_2^i u(t-1) + \hat{b}_3^i u(t-2) + \hat{d} \tag{5.26}$$

where $\hat{d}$ is the offset term due to the initial blood pressure value (MAP baseline value).

As already mentioned in Chapter 3, previous studies [18] and the author's research group's own studies suggested that the delay associated with blood pressure dynamics is usually of the order of 25 to 40 seconds, which given our sampling time of 1 minute, represents a fractional value. Hence, the best way to reflect such a delay in the modeling structure is to expand the $B(z^{-1})$ polynomial to include three $\hat{b}$'s rather than two [see (5.26)].

Different ways of fuzzy-partitioning the input space were explored, as shown in Figure 5.9. Tables 5.3 and 5.4 show the model parameters derived from the identification process using the above input/output data. Table 5.3 relates to the model with the offset, whereas Table 5.4 corresponds to the structure without the offset term. Using these tables together with Figures 5.10 and 5.11, it can be observed that the fuzzy TSK–based structures provided more accurate models than the linear model, and such accuracy improved as the number of partitions increased. However, it was also observed that it is not advantageous to increase the number of partitions beyond five, as the accuracy gained as a result will be offset by the heavy

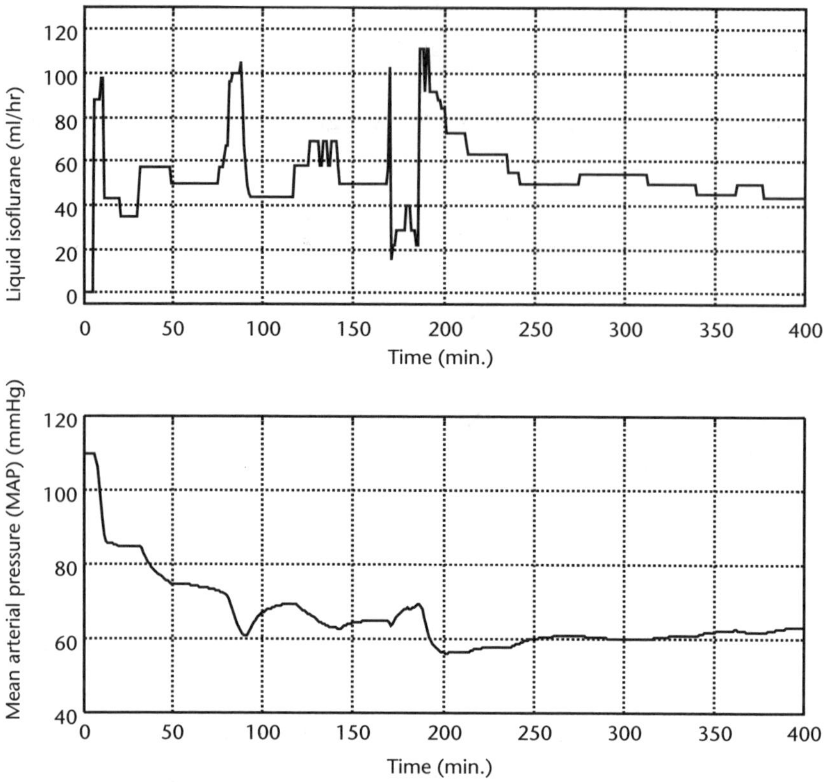

Figure 5.8 The input-output data set used for linear model identification.

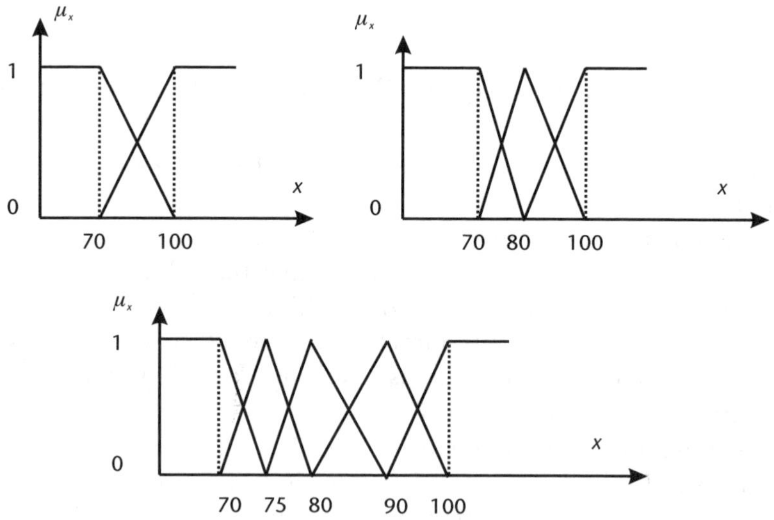

Figure 5.9 Fuzzy partitioning of the input space for deriving the fuzzy TSK–based patient models.

computational burden incurred. Moreover, the inclusion of the offset term in the structure (20) gave a more accurate model, but because the control strategy will be based on the CARIMA model structure this term will not be estimated. This identifi-

Table 5.3 Effect of the Number of Fuzzy Partitions of the Input Space on the Modeling Accuracy of the Patient Dynamics (Offset Term Included)

Model	$\hat{a}_1$	$\hat{a}_2$	$\hat{b}_1$	$\hat{b}_2$	$\hat{b}_3$	$\hat{d}$	*MSE*
Linear	−0.9675	0.062	−2.3842	−1.3737	1.0477	3.8937	0.0854
2-Partition TSK	−1.8197	0.8248	−1.6249	0.3846	1.0901	0.3961	0.0035
	−0.9031	0.0009	−3.7255	−2.3613	0.1014	10.6587	
3-Partition TSK	−1.8964	0.8985	−1.5833	0.4893	1.0917	0.1249	0.0019
	−1.7111	0.7472	−1.7464	0.7106	1.9624	2.1785	
	−0.8308	0.0008	−3.6271	−2.7202	−0.8465	18.61	
5-Partition TSK	−1.8961	0.8977	−1.5689	0.4827	1.0912	0.0911	0.0015
	−1.5631	0.5364	−2.3799	−0.4188	1.1454	−1.1907	
	−1.4362	0.4258	−3.1109	0.6487	2.0801	−0.9406	
	−1.7267	0.5549	−2.2556	−0.9766	0.0010	−13.275	
	−1.0597	0.0008	−3.5245	−1.8939	0.4539	−6.5724	

Table 5.4 Effect of the Number of Fuzzy Partitions of the Input Space on the Modeling Accuracy of the Patient Dynamics (Offset Term Excluded)

Model	$\hat{a}_1$	$\hat{a}_2$	$\hat{b}_1$	$\hat{b}_2$	$\hat{b}_3$	*MSE*
Linear	−1.0090	0.0056	−2.0279	−1.1212	2.4342	0.2543
2-Partition TSK	−1.6836	0.6775	−1.7781	0.1107	1.0169	0.0327
	−1.0011	−0.0013	−3.9111	−2.0753	2.1249	
3-Partition TSK	−1.8985	0.8986	−1.5774	0.4851	1.1031	0.0037
	−1.1983	0.1733	−2.9847	−1.3157	0.3002	
	−1.0000	0.0009	−3.5458	−2.0947	0.4388	
5-Partition TSK	−1.9093	0.9097	−1.5656	0.5030	1.1039	0.0015
	−1.9162	0.9128	−2.5875	0.3655	1.7011	
	−1.8574	0.8756	−2.6855	1.7948	3.4131	
	−1.1911	0.1601	−2.5781	−1.9935	1.0648	
	−0.9999	0.0008	−3.5552	−2.1111	0.1033	

cation structure allowed one also to quantify the patient's time constants which were of the order of $T_{c_1} = 0.20$ minute; $T_{c_2} = 25$ minutes. These values suggest that the system is mainly a first-order system, but due to the lag introduced by the process of transforming liquid isoflurane into gas, a second-order model will be considered throughout.

5.6.2 Closed-Loop Control Experiments

The simulation study considered the continuous nonlinear system (5.1) to (5.3), which was represented in MATLAB-SIMULINK (under fault-free and faulty conditions), using a sampling interval of 1 minute, while the external constrained predictive control module was coded in C. For parameter estimation, a UD-factorization method [19] was used on incremental data using the following linear model structure:

$$y_m(t+1) = -\hat{a}_1 y(t) - \hat{a}_2 y(t-1) + \hat{b}_1 u(t) + \hat{b}_2 u(t-1) + \hat{b}_3 u(t-2) \tag{5.27}$$

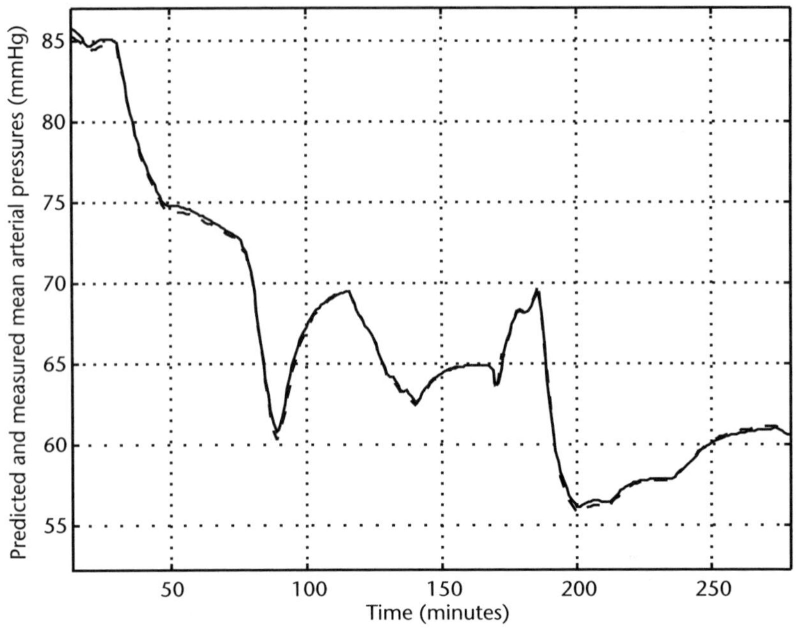

Figure 5.10 Predicted and measured outputs when a linear model is used. Dashed line: predicted; solid line: measured.

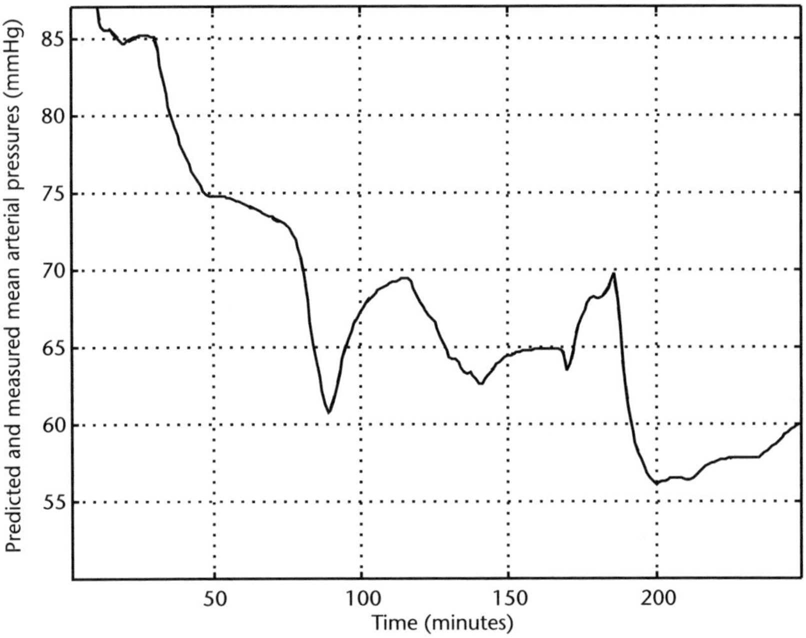

Figure 5.11 Predicted and measured outputs when a two-partition fuzzy TSK model is used.

It is worth noting that the offset was ignored here since GPC considers a CARIMA model structure. Furthermore, as already stated, the rationale behind the use of 3 $\hat{b}$-estimates in (5.27) is to allow any existing fractional delay to be absorbed in the structure.

At time $t = 0$ an initial arterial pressure of $MAP_0 = 110$ mmHg was assumed. The set-point command was 90 mmHg, then 100 mmHg for a 400-minute total simulation time. The GPC algorithm used a combination of tuning factors of (1, 8, 2, 0) for (N_1, N_2, NU, λ) together with a filter polynomial $T(z^{-1}) = (1 - 0.8z^{-1})^2$. Different fuzzy partitions of the input space can be used; we chose triangular shapes for simplicity. The algorithm used the three types of constraints with the following limits:

$$\begin{aligned} &-0.2 \leq \Delta u(t + j - 1) \leq 0.2 \\ &0 \leq u(t + j - 1) \leq 5 \\ &\omega - 5 \leq \Phi(t + j) \leq \omega + 5 \\ &j = 1, \ldots, NU \end{aligned} \tag{5.28}$$

The first experiment considered the unconstrained algorithm. This led to the output response of Figure 5.12, where despite the minimum variance, the control signal displayed a high activity and large excursions in places leading to large drug absorptions by the patient, which even if they last only a few samples can lead to undesirable consequences.

The second experiment considered the constrained algorithm as described in Section 5.3. This led to the output response of Figure 5.13(a), where it can be seen that the controller tracked the set-point changes efficiently but in response to the stimulus level found it hard to track the set point once the disturbance ceased to act on the system. Figure 5.13(b) shows that the controller generated a moderately

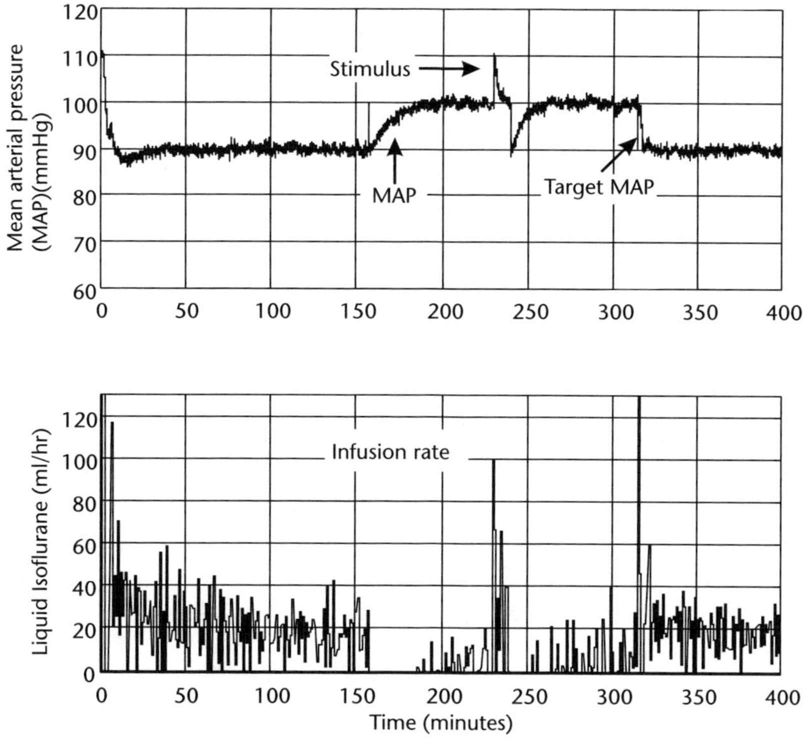

Figure 5.12 Closed-loop control of patient simulated anesthesia using unconstrained GPC with $NU = 2$ when patient body weight = 70 kg.

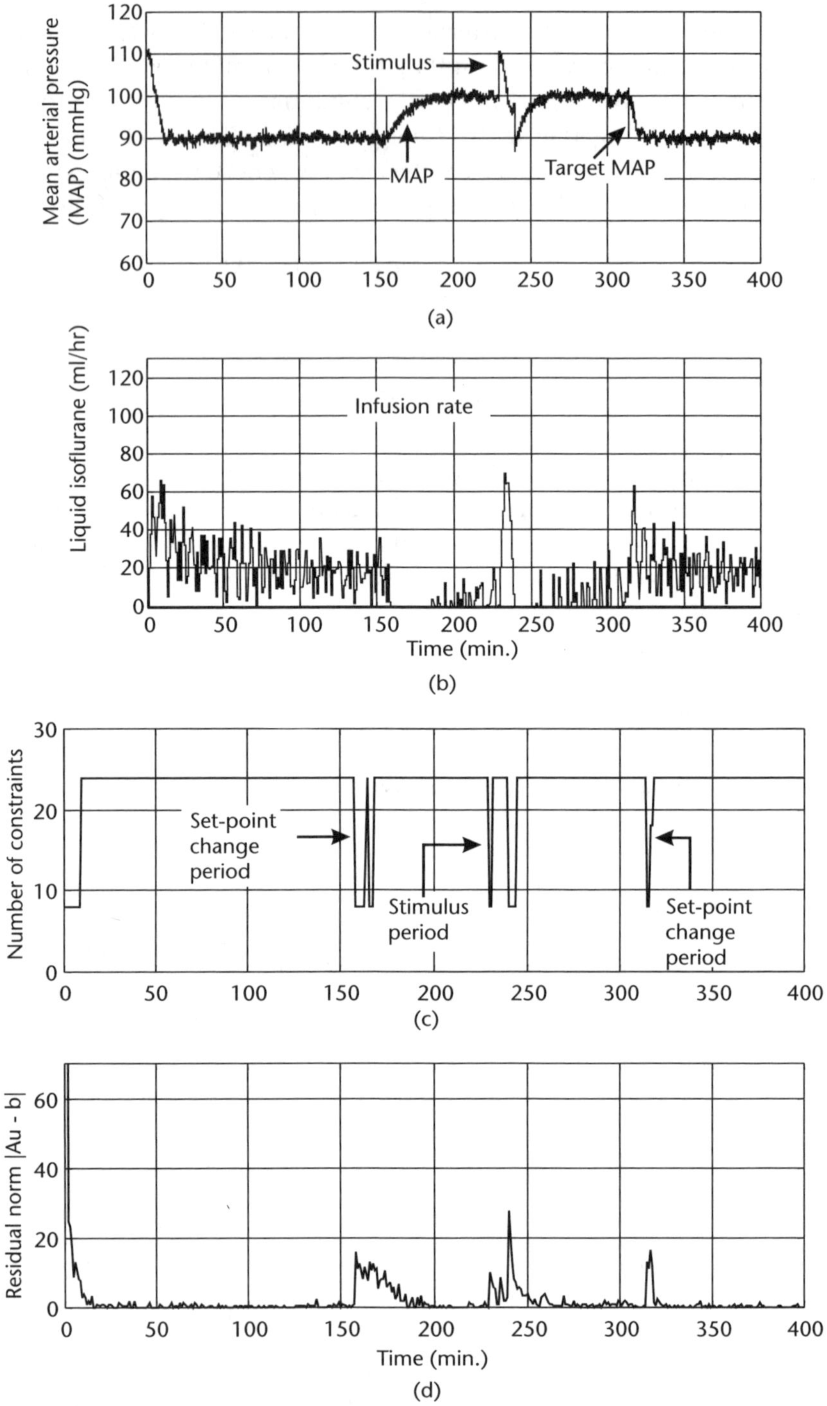

Figure 5.13 (a–d) Closed-loop control of patient simulated anesthesia using constrained GPC with input and output constraints; $NU = 2$ when patient body weight = 70 kg.

active control signal, which normally indicates a reasonable drug consumption. In turn, Figure 5.13(c) shows how the algorithm reverted to an input-constraints algorithm at times when the set point changed and disturbances were occurring by hierarchically removing the constraints conditions it could not satisfy (see Steps 1 through 8 in Section 5.3). Figure 5.13(d) displays the norm-residual of the control solution throughout the run, which remained low (indicating a good control solution), except when the set-point changed and the sudden disturbance occurred.

To reflect the changes in patient physiological parameters, another experiment was conducted in which the weight of the patient was made to change from 70 to 40 kg, leading to a different set of patient dynamics. The closed-loop control experiment using the constrained control algorithm with the same parameters as before led to the output response of Figure 5.14(a, b), where it can be seen that the adaptive capabilities of the algorithm led to good output tracking properties despite the sudden output disturbance. It is worth noting that, given the low patient weight, the drug consumption is consequently less than in the previous cases where the patient was heavier.

The fourth simulation experiment related to the use of a fuzzy TSK CARIMA model structure as described by (5.18) to (5.25). A two-partition fuzzy model of the form described by (5.18) with $i = 2$ was considered. Figure 5.15(a, b) shows the performance of the algorithm, where it can seen that a conservative infusion rate profile was obtained without the large input excursions that were seen in the case of a linear model [see Figure 5.13(a, b)]. This can also be explained by Figure 5.15(c), where it can be seen that there were less violations of constraints between times 150 and 250 minutes. Consequently, Figure 5.15(d) shows a residual norm plot with lower magnitudes.

Finally, the last experiment was conducted in order to test the performance of the supervisory layer, which consists of a fuzzy logic control structure as described in Section 5.5.3. In order to evaluate the level of threshold for the prediction error that will allow one to declare a fault or not, the evolution of the signal that related to the prediction error [see (5.16)] was monitored under fault-free conditions. It was decided on a threshold level for fault detection of $Det_{Thr} = 0.05$. Hence, a stimulus fault was introduced at time $t = 100$ minutes, which lasted for 10 minutes. Figure 5.16 shows the evolution of the model prediction error, which registered a peak when the fault occurred. However, when the FDIA scheme was switched on and the fault detected and diagnosed as a heavy stimulus (see Sign #2 in Section 5.5.3), it was accommodated for by the hybrid structure via the switching of the control from constrained GPC to a simple Mamdani-type fuzzy control system [see Figure 5.17(a, b)]. The same figure also shows that the MAP signal did not drop as low as in the previous case.

This simulation study and others (not reported here) formed the basis for the transfer of the overall closed-loop control system to the operating theater for administration of isoflurane during surgery, as the next section explains.

5.7 Real-Time Closed-Loop Control Experiments in the Operating Theater

The real-time closed-loop control system that was transferred to the operating theater comprises the following (see Figure 5.18):

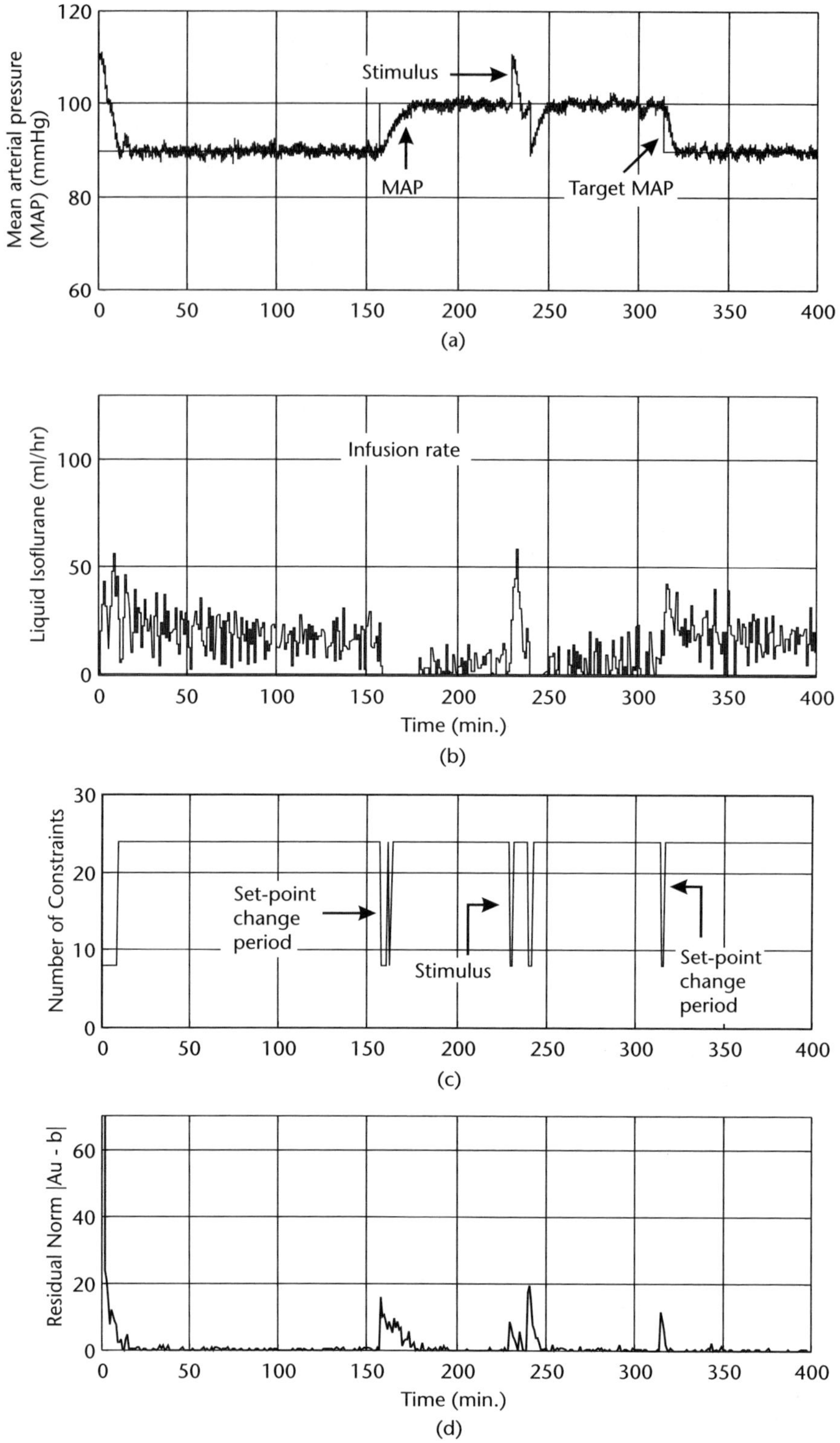

Figure 5.14 (a–d) Closed-loop control of patient simulated anesthesia using constrained GPC with input and output constraints; $NU = 2$ when patient body weight = 40 kg.

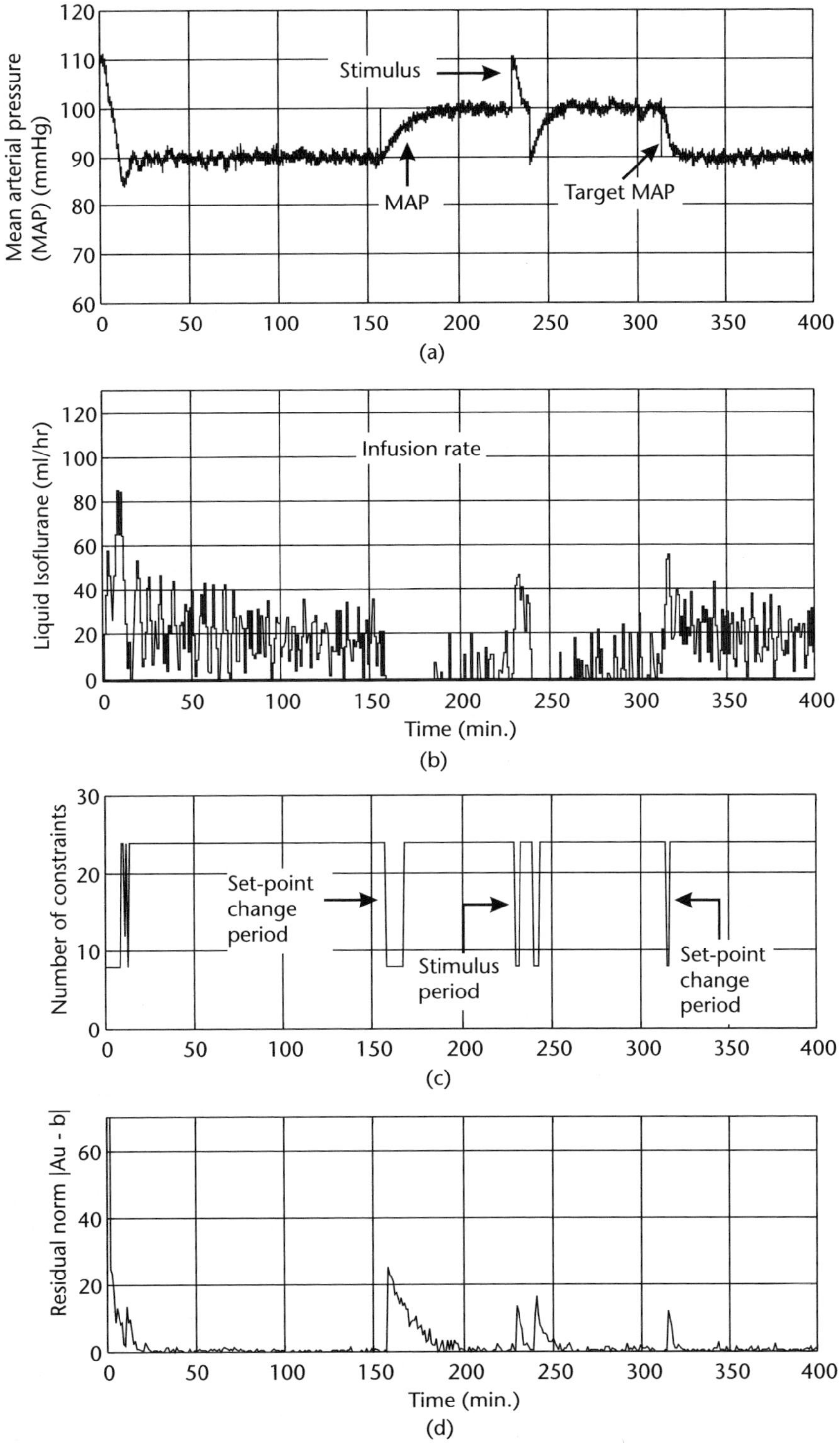

Figure 5.15 (a–d) Closed-loop control of patient simulated anesthesia using constrained GPC with input and output constraints with a two-partition fuzzy TSK model; $NU = 2$ when patient body weight = 70 kg.

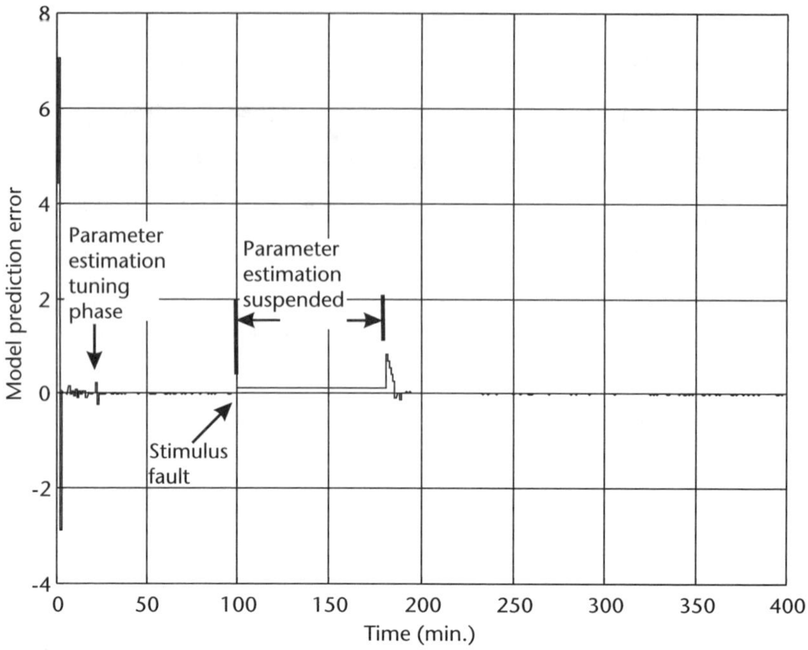

Figure 5.16 Recorded model prediction error during a simulation run under faulty conditions.

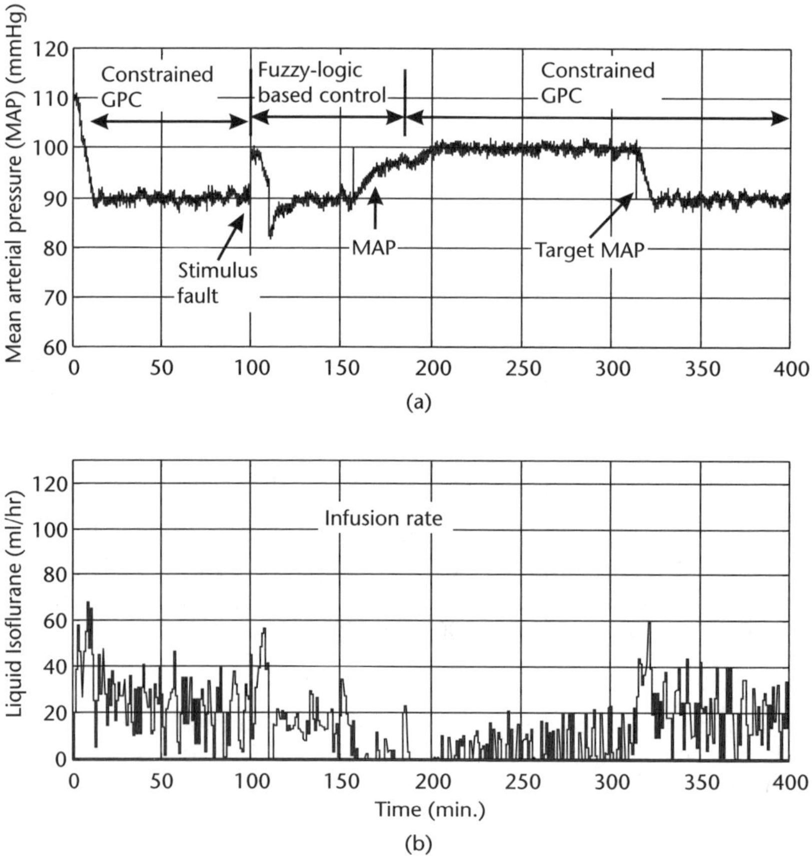

Figure 5.17 (a, b) Supervisory control of anesthesia under faulty conditions.

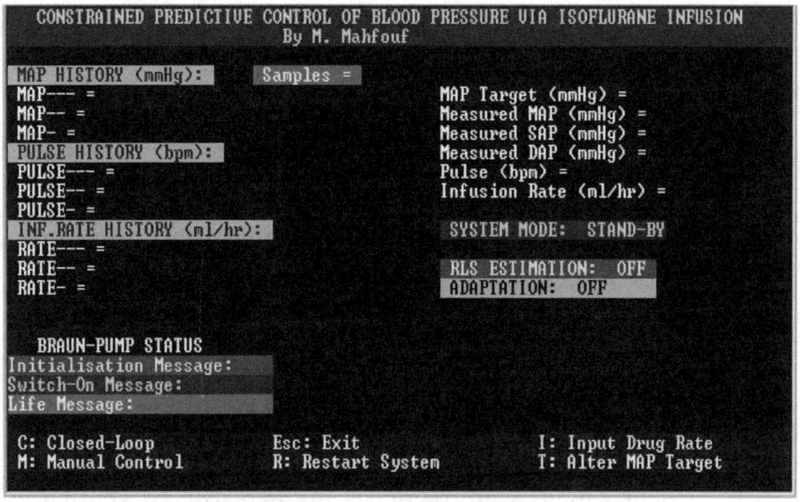

Figure 5.18 A screen shot of the interactive closed-loop control environment that the anesthetist is presented with during the clinical validation of the software.

- An IBM-compatible microcomputer which incorporates the control system;
- A BRAUN PERFUSOR SECURA digital pump driving a disposable syringe containing a liquid solution of isoflurane;
- A DINAMAP instrument for measuring the arterial blood pressure;
- A CAPNOMAC ULTIMA device for measuring the inspired and expired isoflurane concentrations.

The links between the syringe pump, the Capnomac machine, the blood pressure monitor, and the computer are via three RS-232 serial ports.

After local Ethics Committee approval, 10 patients were selected for the experiments, and they all underwent surgery that required anesthesia. The anesthetist has the options of "manual control" or "automatic control," "set-point changes," "fixed control mode (RLS OFF)," or "adaptive control mode (RLS ON)." The same figure also shows that the anesthetist monitors the patient through other variables such as the heart rate (HR), systolic arterial pressure (SAP), and diastolic arterial pressure (DAP).

5.7.1 Clinical Preparation of Patients Before Surgery

Patients were invited to participate in the trials and the all gave written consent on a form approved by the local Ethics Committee. The following patients were excluded from the trials:

- Those with high blood pressure;
- Those taking medication known to adversely interact with anesthetic drugs;
- Those with medical conditions that would be unlikely to give adequate control.

All patients underwent highly stimulating, major body surface surgery that requires general anesthesia. Information relating to the patients is presented in Table 5.5.

A commonly used method of anesthesia was standardized to deal with patients of different weights. The main components were:

- Induction of anesthesia by injecting drugs directly into the veins (including morphine);
- Insertion of a tube into the trachea;
- Inflation of the lungs with isoflurane vapor in 30% oxygen and 70% nitrous oxide (N_2O);
- Muscle paralysis maintained with doses of atracurium.

The patient's lungs were regularly inflated using a ventilator, whose volumes and frequency were initially adjusted to maintain a stable end-tidal carbon dioxide concentration. A standard semi-closed breathing circuit was used, which admits the continuous flow of nitrous oxide and oxygen and isoflurane vapor, uses them to inflate the patients lungs, and then recirculates the gasses through a carbon dioxide absorber before reusing them in the next breathing cycle.

Throughout anesthesia and surgery, all patients were given enough salt solution (Hartmann's) into the veins to maintain the blood volume, allowing for blood losses during surgery. Throughout anesthesia and surgery the inhaled and exhaled gas concentrations and the patients' vital signs, including heart rate, oxygen saturation, systolic and mean arterial pressure were monitored.

When the patients were stable and safe, the isoflurane vapor supply was turned off and the online controlled injection of isoflurane liquid into the breathing circuit was started, and its performance closely monitored. The isoflurane was injected continuously into a transparent part of the breathing circuit 100 cm from the patient; the volumes injected were vaporized immediately, and no drops of isoflurane liquid were ever seen.

Table 5.5 Summary of Patients' Personal Details

Patient	*Sex*	*Age (years)*	*Weight (kg)*	*Duration of Procedure (min)*	*Automatic Control Duration (min)*
1	F	70	60	54	46
2	F	62	45	40	34
3	F	51	55	42	41
4	F	59	45	56	53
5	F	65	65	56	54
6	F	55	63	50	49
7	F	68	55	51	47
8	F	67	65	62	6
9	F	47	85	53	51
10	F	58	68	76	51

At the end of the operations, the control was discontinued, and the patients were allowed to regain consciousness. The patients were transferred to the recovery room for further supervision before transfer to the ward. All patients were visited the day after surgery; none had any recall of the intraoperative period and all were satisfied with their management.

5.7.2 Results and Discussions

All 10 trials were conducted using a sampling-time interval of 1 minute. Control and estimation were performed every minute. Parameter estimation was triggered at time $t = 4$ minutes (to allow the measurement vector to gather enough input-output information for effective and faster convergence) with a covariance matrix and forgetting factor values of $P = 10^3 \cdot I$ and $\rho = 0.975$, respectively. Parameter adaptation, which was only carried out for patients 6, 7, 8, 9, and 10 using the same model structure as (5.27) (estimated parameters fed back to the control law), was initiated at sample time $t = 10$ minutes. The reference MAP level was specified by the anesthetist depending on the surgeon's requirements at the time of the surgery, but in all ten trials this reference ranged from 70 to 95 mmHg. Results corresponding to each patient are presented in two parts. The first part consists of two traces: the upper trace represents the recorded MAP and the reference level together with some events that occurred during the trial; the lower trace shows the variation of the infusion rate of isoflurane in ml/hr. The second part of the results includes the parameter estimates variations and/or the other cardiovascular variables such as the SAP, DAP, and the averaged HR. Because of a lack of space (this is not a selectivity that is based on whether the algorithm worked or failed), only 8 of the 10 experiments will be described here, illustrating lessons learned from all the trials.

Patient 1

For this experiment, the constrained GPC protocol embodied a combination of (1, 8, 2, 0) for (N_1, N_2, NU, λ) and a second-order observer polynomial $T(z^{-1}) = (1 - 0.8z^{-1})$, which was the same for estimation and control. The lower and upper constraints were as follows:

$$-0.1 \le \Delta u(t+j-1) \le 0.1$$
$$0.001 \le u(t+j-1) \le 0.2$$
$$j = 1, \ldots, NU$$

Parameter estimates of the linear model were initialized to some values that reflected a second-order model with a gain and two time constants. As shown in Figure 5.19(a), the MAP signal tracked the target MAP level (80 mmHg) efficiently despite the significant level of noise acting on the system. In turn, the control signal of Figure 5.19(b) suggests a moderately active signal despite the high control of $NU = 2$. Figure 5.19(f) displays the evolution of the five-parameter estimates together with the prediction error. The final parameter estimates corresponded to a second-order system with the following time constants:

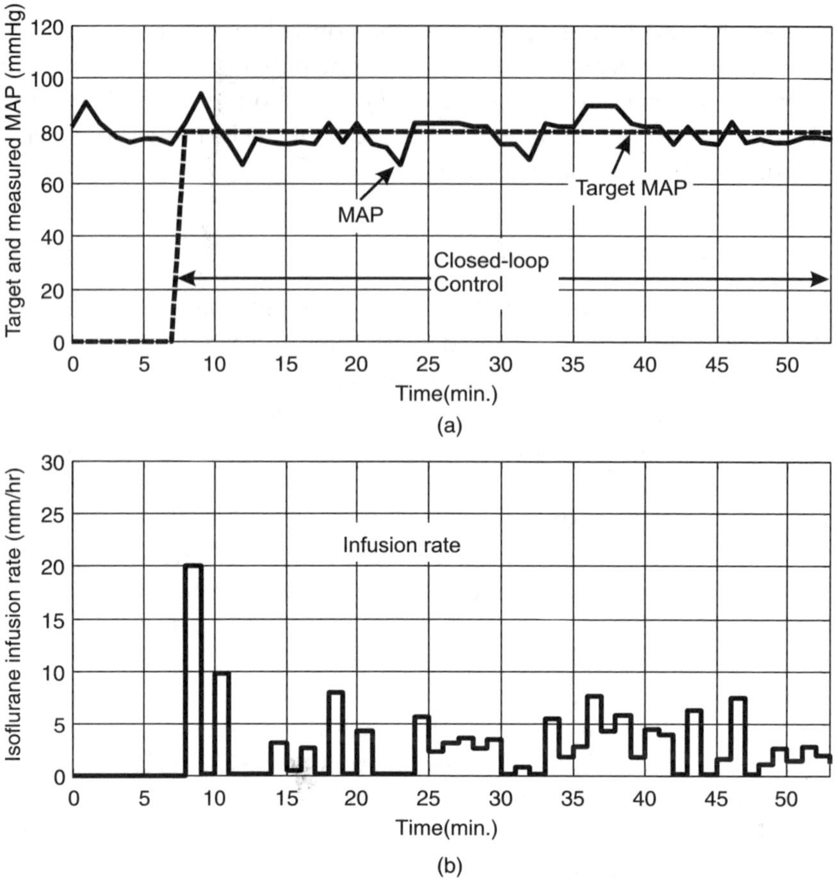

Figure 5.19 Patient 1: (a) recorded MAP during surgery under feedback control; (b) drug infusion rate during surgery; (c) measured systolic arterial pressure (SAP); (d) measured diastolic arterial pressure (DAP); (e) measured average heart rate (HR); and (f) system parameter estimates and prediction error for model of (5.27).

$$\hat{T}_{c_1} = 16.35 \text{ minutes}; \hat{T}_{c_2} = 0.66 \text{ minute}$$

At the end of this operation the patient recovered very well.

Patient 3

The same parameters as in Patient 1 were assumed here except that the following limits for the constraints on the control increments were set:

$$-0.2 \le \Delta u(t + j - 1) \le 0.1$$
$$0.001 \le u(t + j - 1) \le 0.2$$
$$j = 1, \ldots, NU$$

The result of the run is shown in Figure 5.20(a), where it can be seen that the patient displayed some resistance to the drug since it took approximately 15 minutes to bring the MAP signal down to the target MAP level (90 mmHg). Despite such

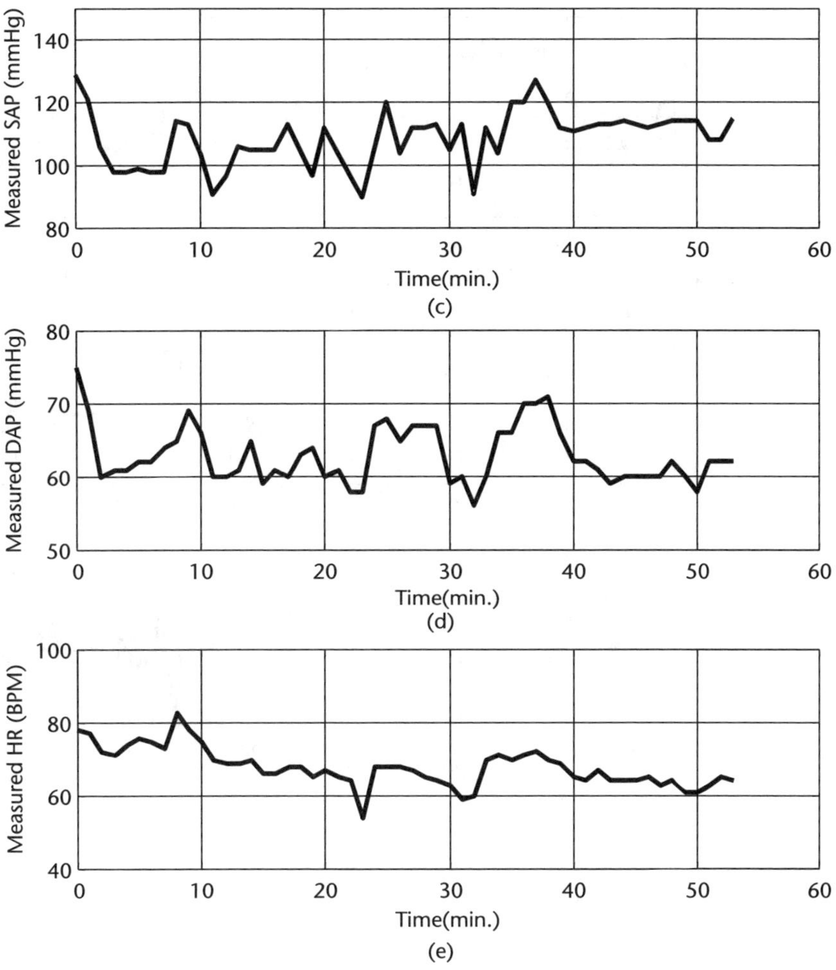

Figure 5.19 (continued.)

resistance, the infusion rate of Figure 5.20(b) was moderately active although the MAP signal did not display much stochastic activity when compared to the signal of Figure 5.20(a). The final parameter estimates of Figure 5.20(c) [see also (5.27)] corresponded to a second-order system with the following time-constants:

$$\hat{T}_{c_1} = 14.58 \text{ minutes}; \hat{T}_{c_2} = 2.35 \text{ minutes}$$

Patient 5

For this experiment the same parameters as in Patient 1 were assumed except that the following limits for the constraints on the control increments were set:

$$-0.1 \le \Delta u(t + j - 1) \le 0.1$$
$$0.001 \le u(t + j - 1) \le 0.2$$
$$j = 1, \ldots, NU$$

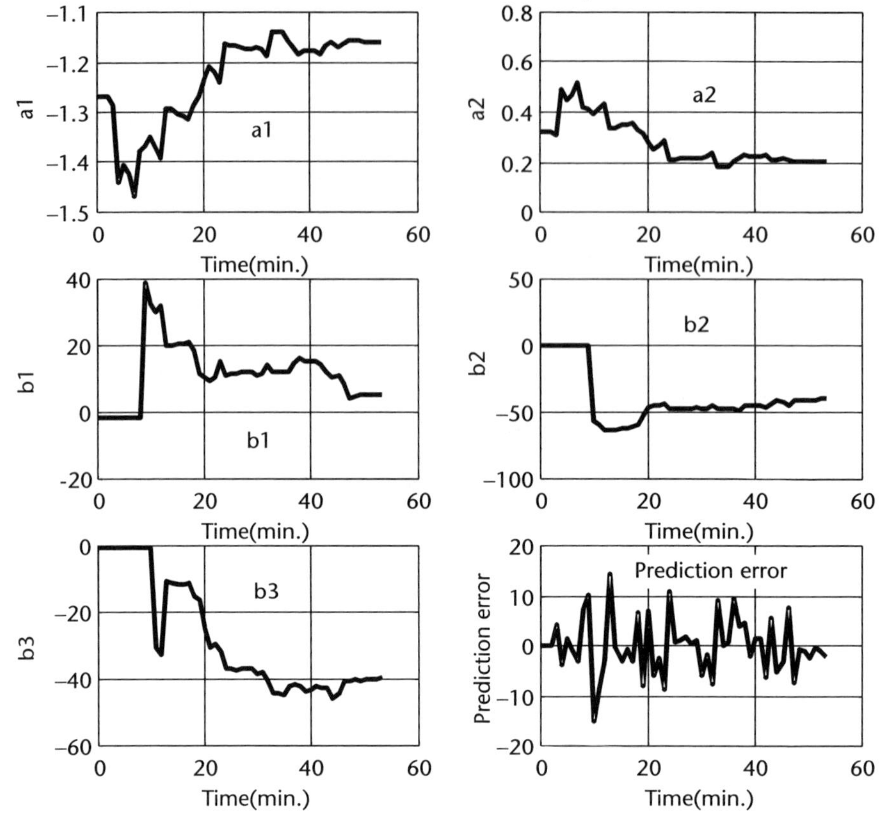

(f)

Figure 5.19 (continued.)

The rationale behind such a modification was that the anesthetist anticipated that he was dealing with a resistive patient and specified that the constraints bounds be set accordingly. Also, to add more robustness to the GPC design, the second-order observer polynomial used for estimation $T(z^{-1})$ was altered to have a unity gain such that

$$T_{estimation}\left(z^{-1}\right) = 25\left(1 - 0.8z^{-1}\right)^2$$

$$T_{control}\left(z^{-1}\right) = \left(1 - 0.8z^{-1}\right)^2$$

Figure 5.21(a) shows how the patient was resistive to the drug, as it took approximately 20 minutes to see a real drop in the blood pressure signal (target level 85 mmHg). The model prediction error of Figure 5.21(c) was low indicating good parameter estimates, which reflected the following time constants at the end of the run:

$$\hat{T}_{c_1} = 14.82 \text{ minutes}; \hat{T}_{c_2} = 1.21 \text{ minutes}$$

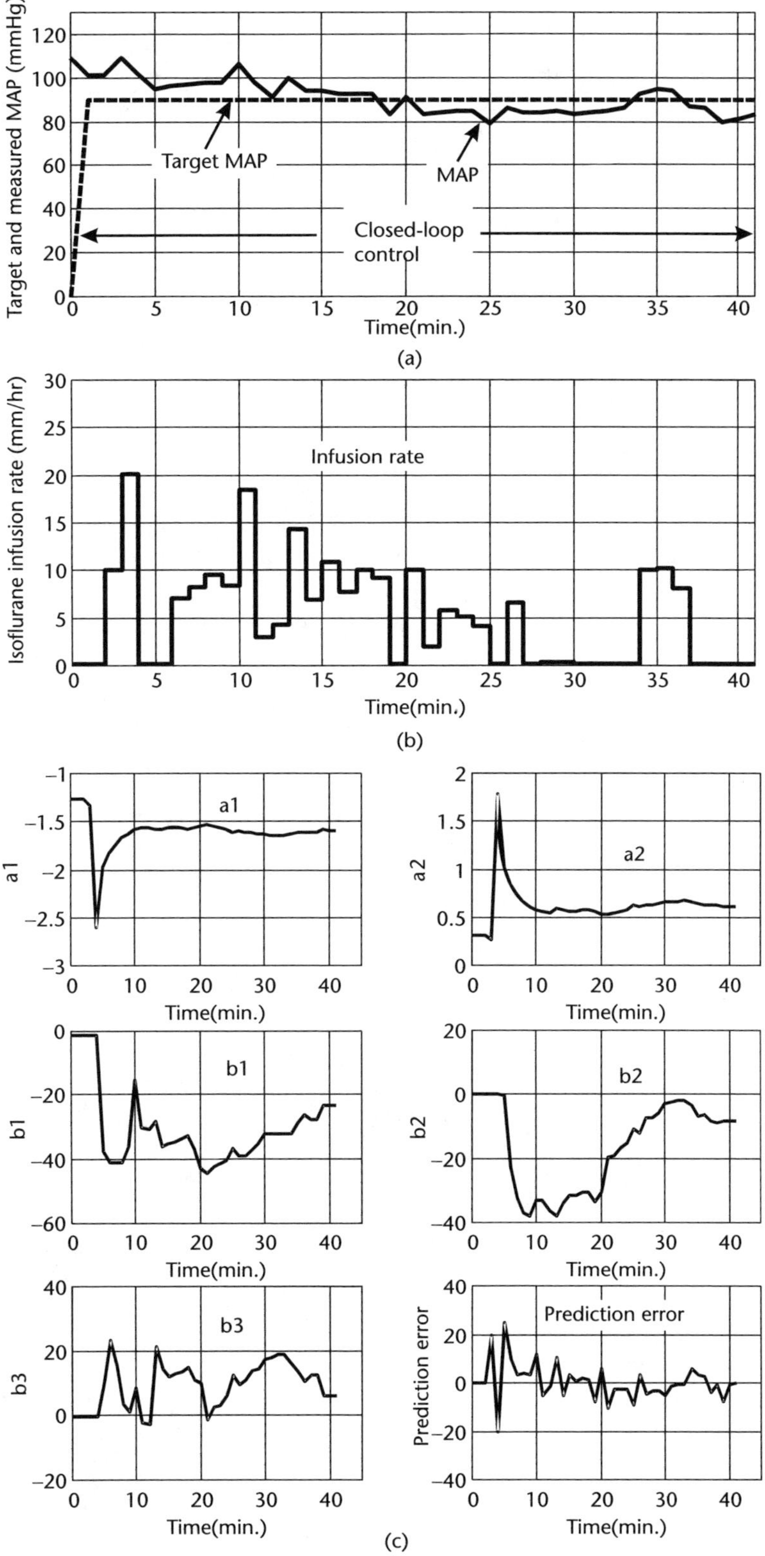

Figure 5.20 Patient 3: (a) recorded MAP during surgery under feedback control; (b) drug infusion rate during surgery; and (c) system parameter estimates and prediction error for model of (5.27).

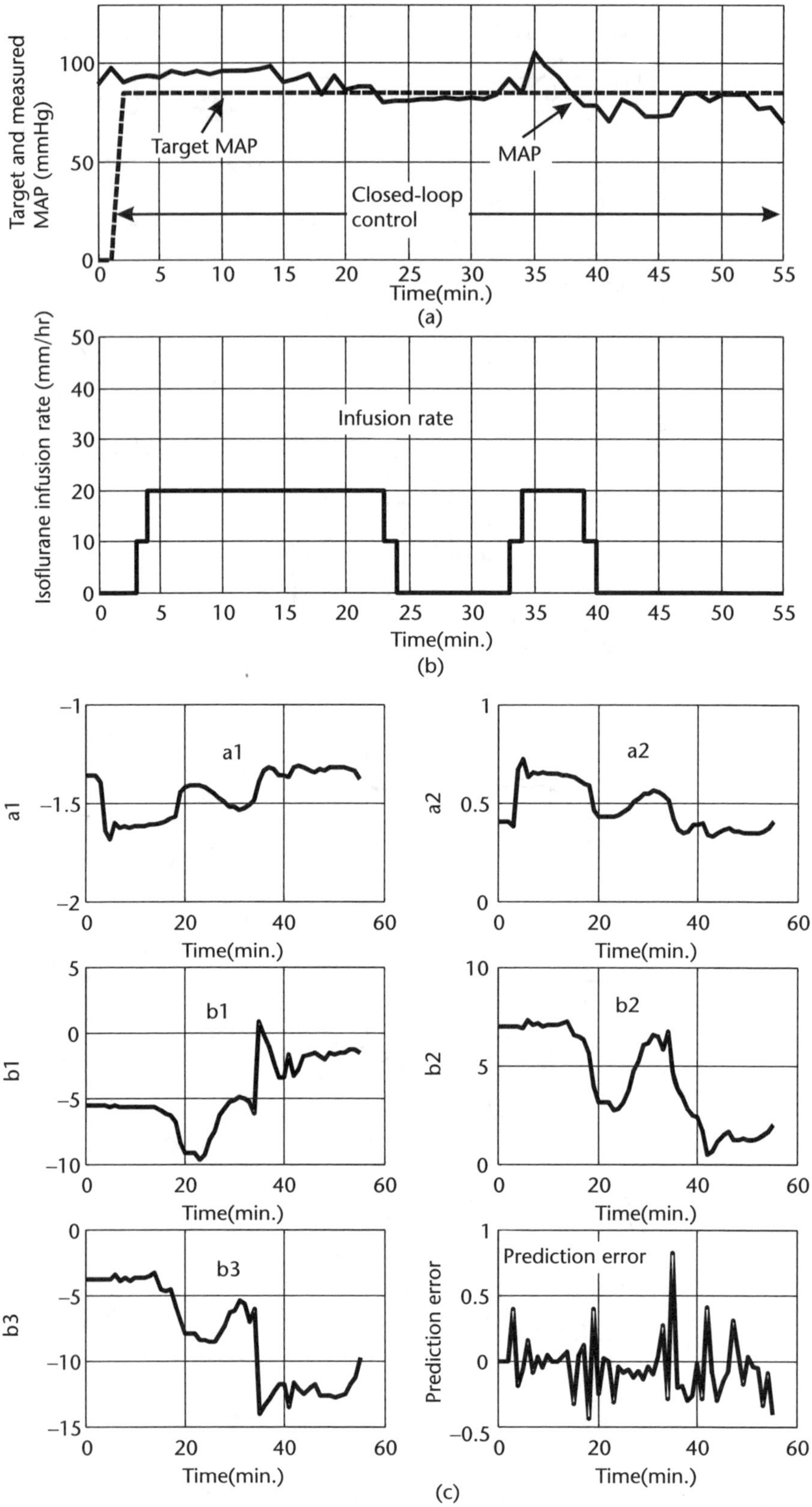

Figure 5.21 Patient 5: (a) recorded MAP during surgery under feedback control; (b) drug infusion rate during surgery; and (c) system parameter estimates and prediction error.

Patient 6

For this experiment the same parameters as in Patient 1 were assumed except that the following limits for the constraints on the control increments were set:

$$-0.2 \le \Delta u(t+j-1) \le 0.1$$
$$0.001 \le u(t+j-1) \le 0.3$$
$$j = 1, \ldots, NU$$

This time, adaptation was switched on at time t = 10 minutes. Figure 5.22 shows the steady level of arterial pressure obtained in this case (target level 70 mmHg). The final parameter estimates obtained reflected the following time constants at the end of the run:

$$\hat{T}_{c_1} = 131.60 \text{ minutes}; \; \hat{T}_{c_2} = 1.15 \text{ minutes}$$

The large time constant $\hat{T}_{c_1}$ could either be due to the fact it *actually* reflects the patient dynamics (which is not unusual as biomedical physiological phenomena are slow in nature) or it could be due to the fact that not enough persistent excitation was present to allow such dynamics to converge to the patient true dynamics. Having said this, according to the principle of *certainty equivalence*, the self-tuning properties still held.

Patient 7

For this experiment the same parameters as in Patient 6 were assumed with the adaptation again being switched on at t = 10 minutes. Figure 5.23 shows how the target levels of MAP of 75 mmHg, then 80 mmHg were tracked efficiently throughout the run. This was judged by the anesthetist to be the best run so far. The final parameter estimates obtained reflected the following time constants at the end of the run:

$$\hat{T}_{c_1} = 13.11 \text{ minutes}; \; \hat{T}_{c_2} = 1.43 \text{ minutes}$$

Patient 8

For this patient, the adaptive version of the constrained algorithm was also applied, but because of the weight of this patient (85 kg), the controller did not succeed in maintaining a "minimum" variance around the MAP target of 90 mmHg (when compared to some of the previous performances) (see Figure 5.24). This is also reflected in the infusion rate, which was not steady. The final parameter estimates obtained reflected the following time constants at the end of the run:

$$\hat{T}_{c_1} = 17.22 \text{ minutes}; \; \hat{T}_{c_2} = 1.62 \text{ minutes}$$

Patient 9

For this experiment, a target MAP was set at 95 mmHg and the adaptation-jacketing procedure according of Figure 5.7 was adopted. As can be seen from

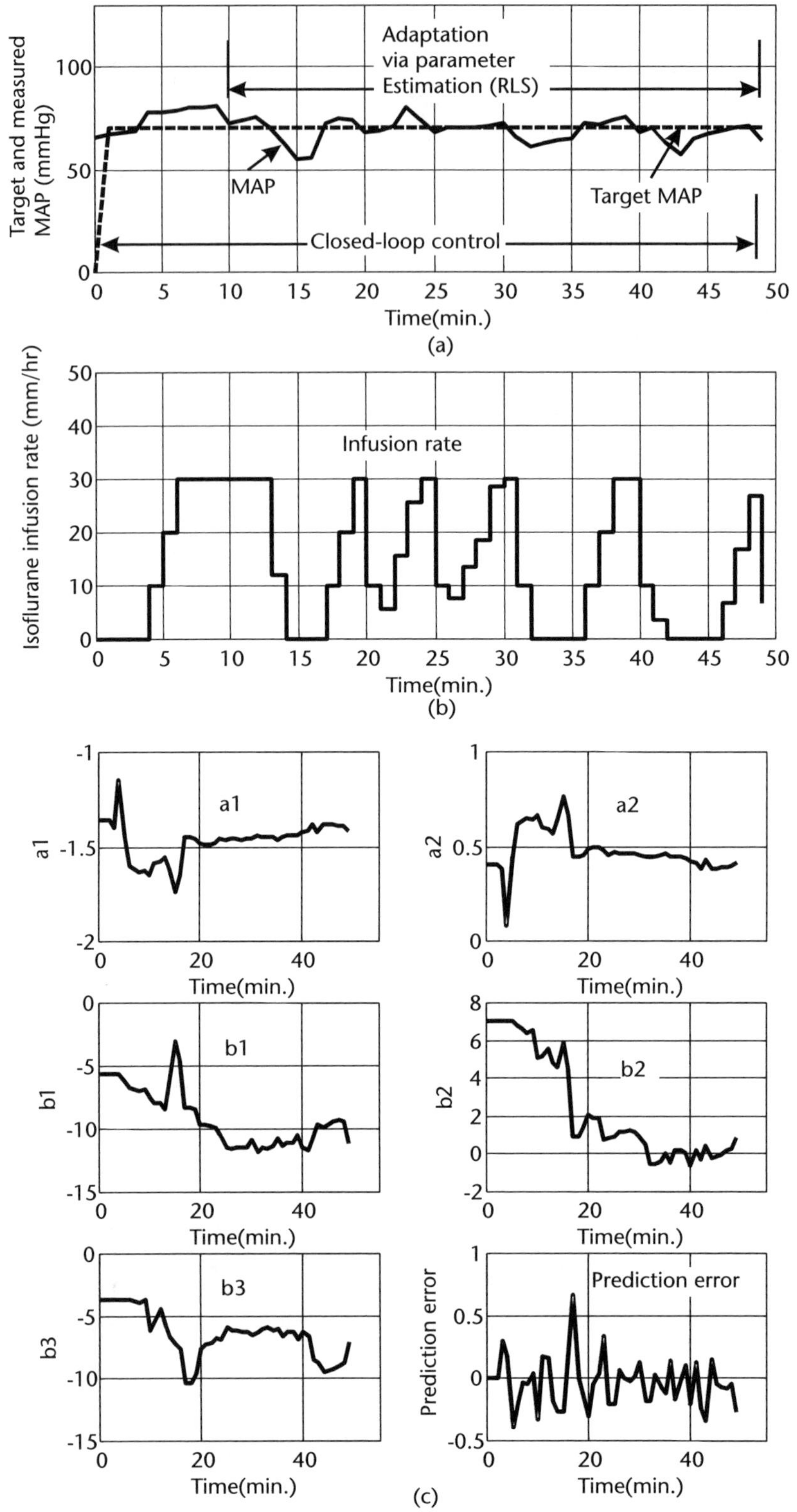

Figure 5.22 Patient 6: (a) recorded MAP during surgery under feedback control; (b) drug infusion rate during surgery; and (c) system parameter estimates and prediction error of model of (5.27).

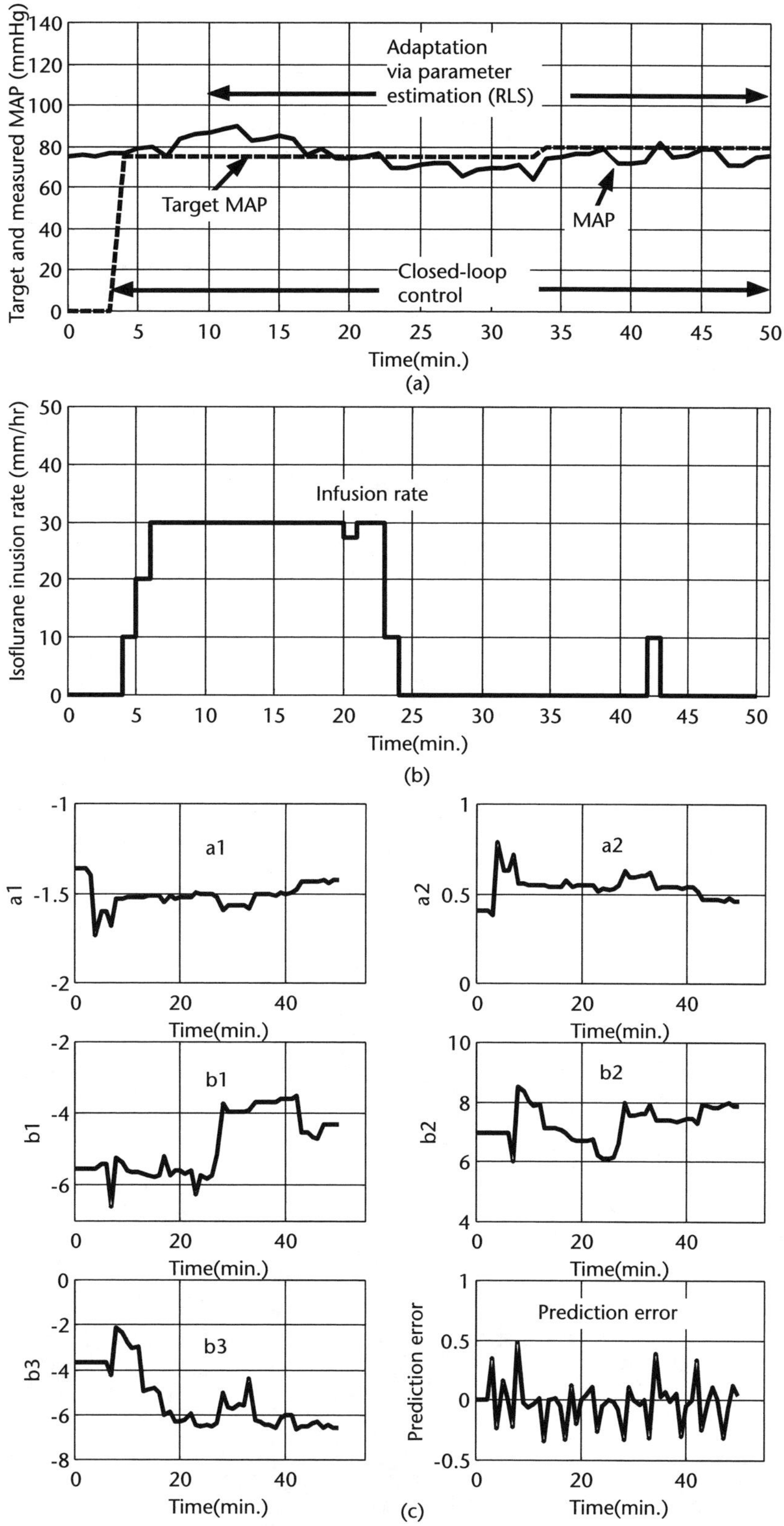

Figure 5.23 Patient 7: (a) recorded MAP during surgery under feedback control; (b) drug infusion rate; and (c) system parameter estimates and prediction error for model of (5.27).

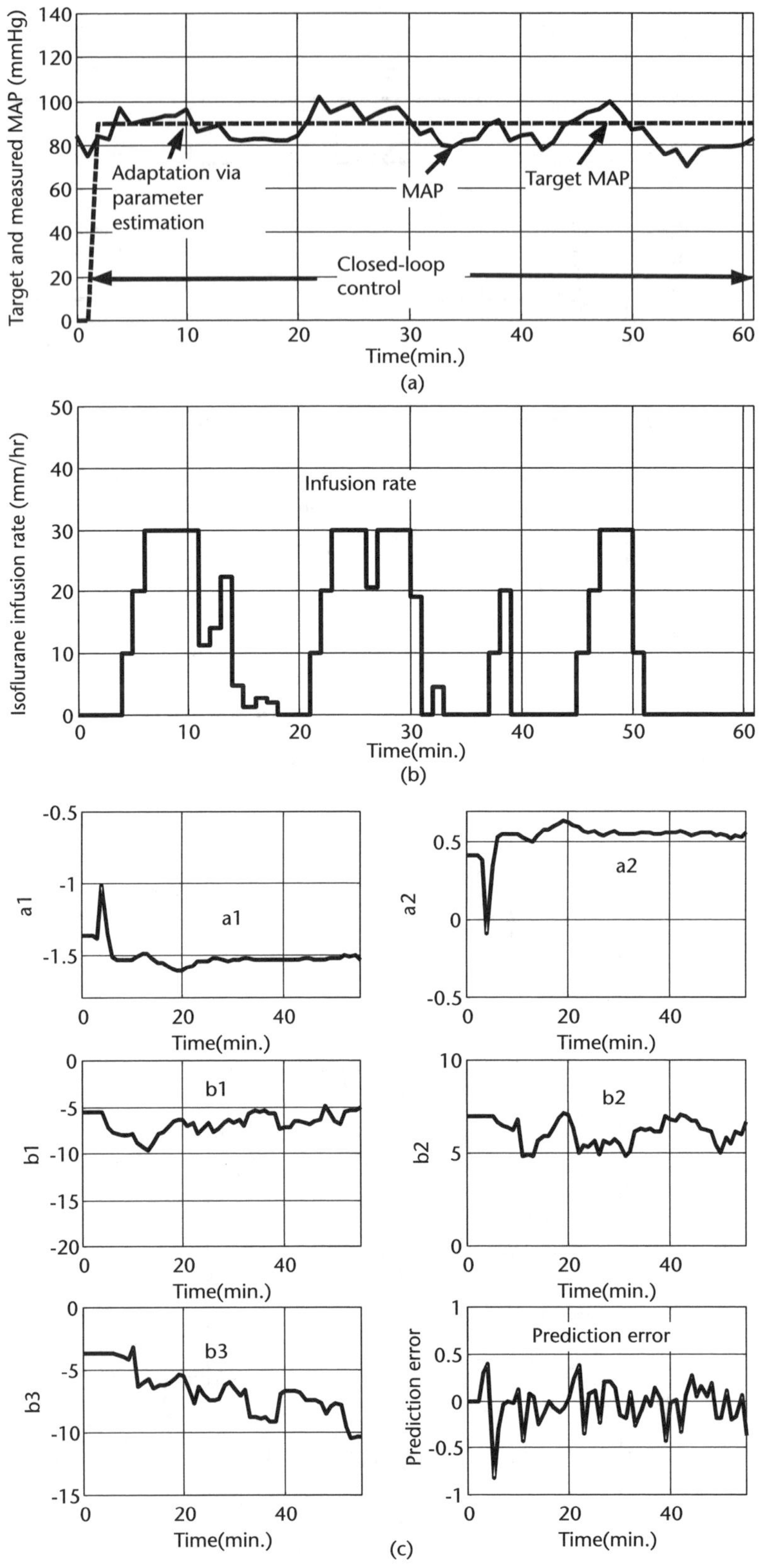

Figure 5.24 Patient 8: (a) recorded MAP during surgery under feedback control; (b) drug infusion rate; and (c) system parameter estimates and prediction error for model of (5.27).

Figure 5.25(a), the adaptation was switched off when the MAP signal violated the range of target + 20 mmHg, but it was switched back on when the MAP signal entered the safe zone. Despite the noise level, which was particularly high in this case, the algorithm maintained good control and the parameter estimates assumed the following final values:

$$\hat{T}_{c_1} = 14.68 \text{ minutes}; \hat{T}_{c_2} = 0.79 \text{ minute}$$

Patient 10

The final experiment was the longest in the group, lasting 76 minutes as it included automatic and manual control modes. Also for this experiment, the adaptation-jacketing procedure according to Figure 5.7 was adopted in order to protect the RLS estimator from spurious measurements. Hence, as seen in Figure 5.26, closed-loop control lasted until time t = 54 minutes when manual control was resumed with an averaged steady level drug infusion of approximately 14 ml/hr, which was deemed necessary by the anesthetist. During automatic control, the MAP signal tracked the target MAP better than when manual control was operational, and it can also be seen that the jacketing procedure worked well for the RLS estimator was immediately switched off at time t = 16 minutes when the MAP signal registered a big jump from its steady level.

At the end of the automatic control period the parameter estimates led to the following time constants:

$$\hat{T}_{c_1} = 124.57 \text{ minutes}; \hat{T}_{c_2} = 0.30 \text{ minute}$$

5.8 Analyses of the Data

To analyze the data, three indices were used: the mean value, the standard deviation, and the root-mean square deviation—the last two indices being commonly used to give an indication of the spread of a set of values around the mean value and the target value, respectively. The indices are defined by the following expressions:

$$\overline{X} = \frac{1}{N}\sum_{i=1}^{N} X_i \tag{5.29}$$

$$SD = \sqrt{\frac{1}{N-1}\sum_{i=1}^{N}\left(X_i - \overline{X}\right)^2} \tag{5.30}$$

where X_i, $\overline{X}$, and N are the current measurement, the mean value, and the total number of points considered, respectively.

$$RMSD = \sqrt{\frac{1}{N-1}\sum_{i=1}^{N}\left(X_i - X_{Trgt}\right)^2} \tag{5.31}$$

Here, X_{Trgt} is the reference MAP level. Table 5.6 summarizes these values for each of the 10 patients in the trials.

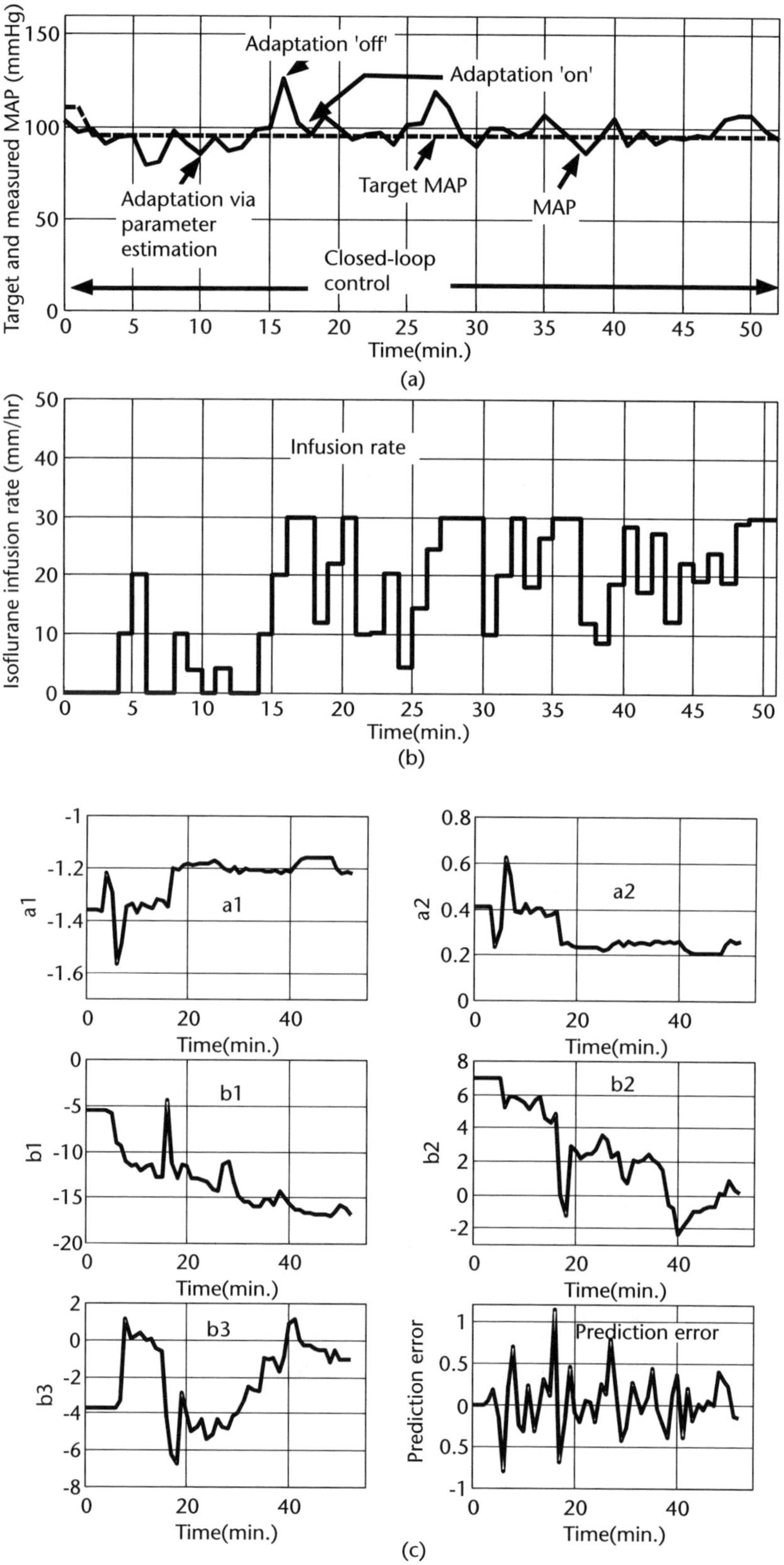

Figure 5.25 Patient 9: (a) recorded MAP during surgery under feedback control; (b) drug infusion rate; and (c) system parameter estimates and prediction error for model of (5.27).

As shown in Table 5.6, the mean values of MAP suggest that the MAP level obtained with the 10 patients was very satisfactory. It is not easy to quantify such performances in terms of overshoots, as the environment under which the controller was working can be described as very dirty in terms of artifacts originating from the mains, surrounding equipment, and so forth. The best performances so far obtained were those of Patients 1, 3, 4, 5, 6, 7, and 10, which produced mean MAP levels close to the respective targets; that is, 79.33 (SD 5.66), 90.85 (SD 7.56), 78.36 (SD 11.54), 86.54 (SD 8.11), 70.02 (SD 6.17), 76.73 (SD 7.03), and 81.04 (SD 8.07), respectively, and this despite the fact that in some cases constrained GPC was operating in fixed mode (nonadaptive) and was switched on during steady-state conditions. Despite the interpatient and intrapatient parameter variability, the controller's performance did not deteriorate because of the inclusion of the filter polynomial $T(z^{-1})$, which acts as an observer for the predictions and compensates for any unmodeled dynamics, especially at higher frequencies [20].

Table 5.6 also shows that all SD and RMSD values are relatively low for all patients; this suggests that good control around the target values was achieved. Indeed, in anesthesia changes of blood pressure of ±10 mmHg are common without any major repercussions on the patient's anesthetic state.

As far as infusion rate variations for the eight patients are concerned, an analysis of the mean isoflurane drug consumption per minute per kilogram body weight was performed. Table 5.7 summarizes such evaluation. As shown in the table, the highest dose of isoflurane was recorded with Patient 7, which was not surprising since she proved rather resistive to the drug, evidenced by a plateau of 20 ml/hr which lasted almost 20 minutes and yet was not enough to bring the blood pressure down to its target (see Figure 5.23). The lowest drug doses were those of Patients 1 and 2, although the input signal was less active in the case of Patient 1 than Patient 2 (SD values of 0.99 and 2.19, respectively). The large values of SD suggest that all control signals were relatively active, this being due to the high level of noise present in the blood pressure signal as explained previously. A summary of all the results for Patients 1 through 10 is given in Table 5.8, where the mean and SD indices are included for the duration of the control period, the mean consumption, mean RMSD, and SD.

The value of 8.23 mmHg for the mean of standard deviations indicates that, generally, a steady level of MAP was obtained. Moreover, the fact that this value was so close to the mean of the RMSD of MAP (8.45 mmHg) also implies that the MAP level was relatively close to the target; it should be noted that the reference MAP level was not the same for all trials, and perhaps more justice would be done to the control algorithm if such statistical analyses were conducted only if the reference MAP values were the same for all the 10 trials. Finally, it is worth noting that the mean dose of isoflurane of 2.67 μl/kg/min was lower than that which would be obtained routinely using conventional manual control. This can, in fact, be verified by the trial for Patient 10, where the anesthetist gave a steady level of isoflurane of 14 ml/hr for a period of only 19 minutes, equivalent to an average consumption of 2.83 μl/kg/min, which was higher than the average consumption level obtained for the 10 patients (the automatic control mode resulted in an average dose of 3.74 μl/kg/min in three times this period). Moreover, for this patient the anesthetist could only achieve a mean level of MAP of 73.52 mmHg (SD 5.62 and a target MAP of 80

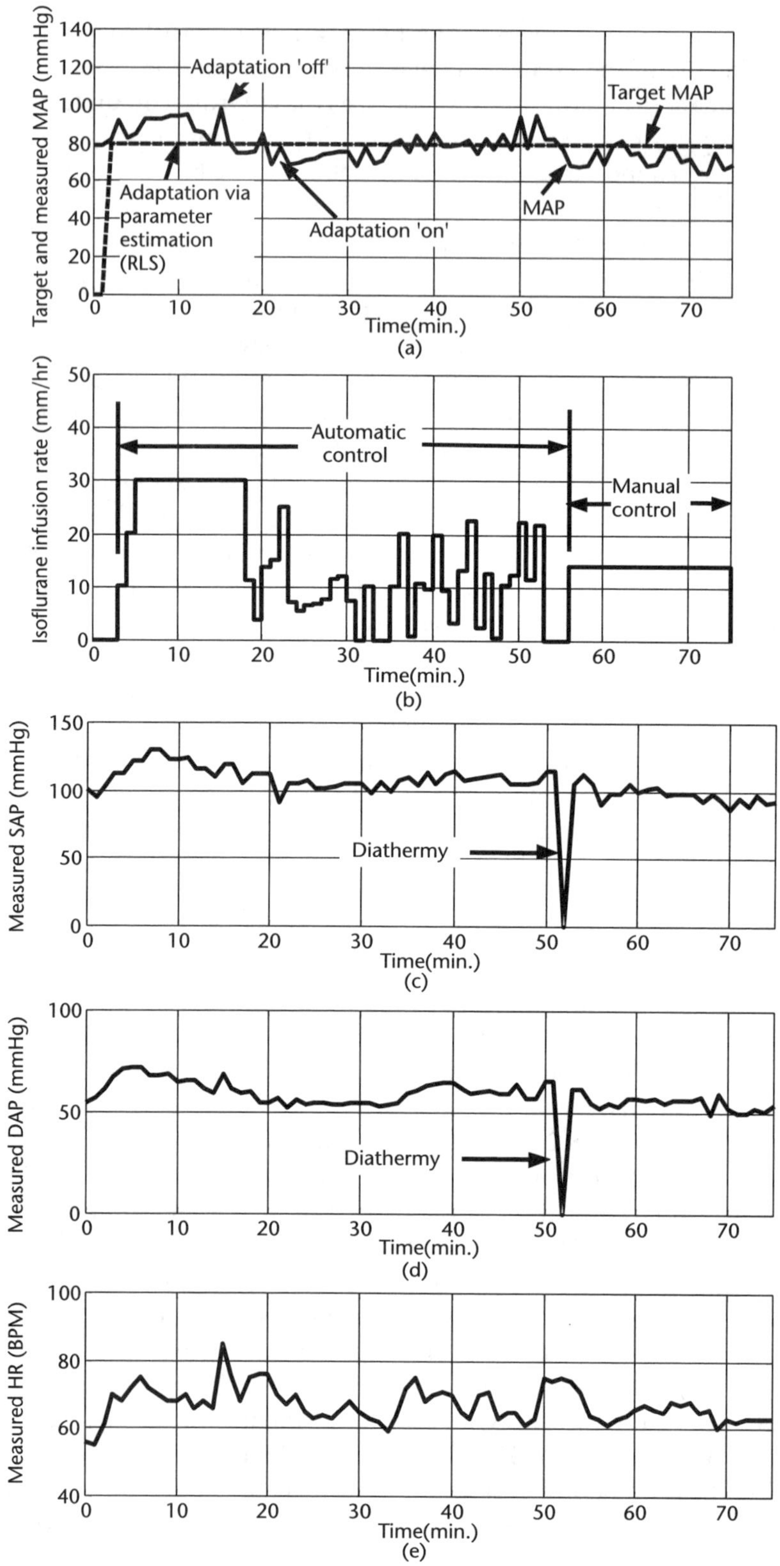

Figure 5.26 Patient 10: (a) recorded MAP during surgery under feedback control; (b) drug infusion rate; (c) Measured systolic arterial pressure; (d) measured diastolic arterial pressure; (e) measured average heart rate; and (f) system parameter estimates and prediction error for model of (5.27).

Table 5.6 Summary of Patients' Controlled Output Performances

Patient	*Total Points*	*Reference MAP Level (mmHg)*	*Mean MAP (mmHg)*	*SD (mmHg)*	*RMSD (mmHg)*
1	46	80	79.33	5.66	5.70
2	29[a]	90	94.28	12.81	13.53
3	41	90	90.85	7.56	7.61
4	50[a]	80	78.36	11.54	11.66
5	54	85	86.54	8.11	8.25
6	49	70	70.02	6.17	6.17
7	30[a,b]	75	76.73	7.03	7.24
8	60	90	87.05	6.96	7.57
9	51[a]	95	97.27	8.34	8.65
10	51[c]	80	81.04	8.07	8.14

a The anesthetist altered the reference MAP level either after starting closed-loop control or during the trial.

b For this patient, the data analysis was carried out for the first set-point phase only (i.e., Map target = 75 mmHg—the first 30 data points only).

c This trial consisted of a combination of automatic control and manual controls.

Table 5.7 Summary of Patients' Input Drug Consumptions

Patient	*Total Points*	*Mean Dose (μl/kg/min)*	*SD (μl/kg/min)*
1	46	0.82	0.99
2	29[a]	0.79	2.19
3	41	1.57	1.64
4	50[a]	1.87	3.32
5	54	2.48	2.48
6	49	3.67	3.17
7	30[a,b]	5.54	4.26
8	60	2.80	3.21
9	51[a]	3.44	2.09
10	51[c]	3.74	2.59

a The anesthetist altered the reference MAP level either after starting closed-loop control or during the trial.

b For this patient, the data analysis was carried out for the first set-point phase only (i.e., Map target = 75 mmHg—the first 30 data points only).

c This trial consisted of a combination of automatic control and manual controls.

Table 5.8 Summary of Overall Patients' Performances

Parameter	*Mean*	*SD*	*Range*
Control duration (min)	48.6	7.23	34–60
Dose (μl/kg/min)	2.67	1.48	0.79–5.54
Mean of MAP (mmHg)	85.14	8.53	70.02–97.27
RMSD of MAP (mmHg)	8.45	2.40	5.70–13.53
SD of MAP (mmHg)	8.23	2.27	5.66–12.81

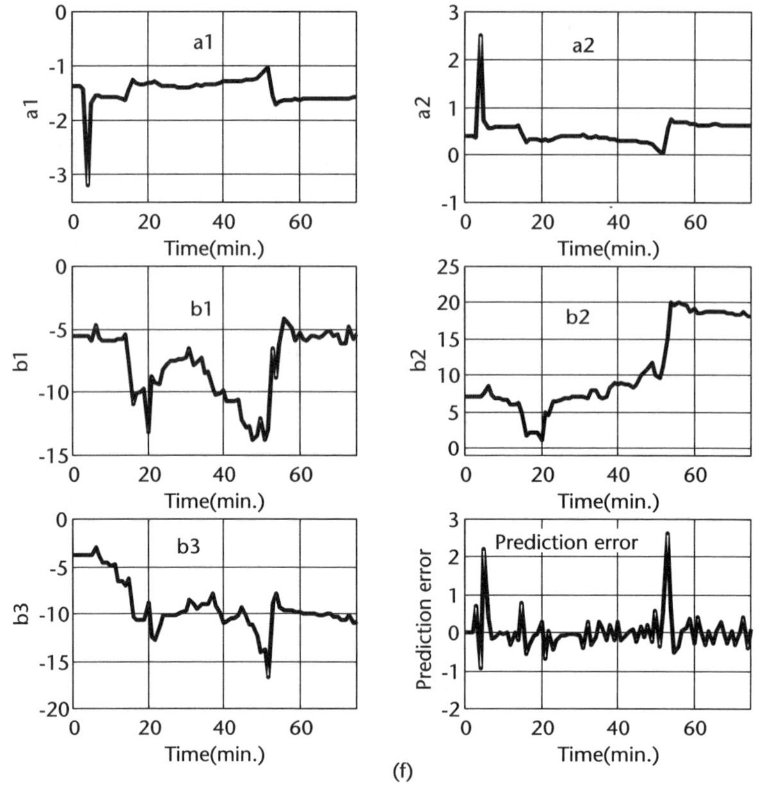

Figure 5.26 (continued.)

mmHg) during this period of manual control (see Figure 5.26), hence justifying the need for automatic control for a steady level of MAP and an adequate dose of administered drug.

5.9 Conclusions

The application of constrained GPC, which uses the QP approach to online control of anesthesia via MAP measurements, has been successful from several viewpoints: (1) achieving good anesthesia by keeping a steady level of mean arterial pressure, which in the eyes of many anesthetists represents a good indicator of DOA (in other words, the patients recovered quicker, they were not aware during the operation, and consequently went home earlier); (2) administration of a low dosage of isoflurane when compared to manual control via a gas vaporizer based on the personal assessment of the anesthetist; and (3) relieving the anesthetist of tedium during normal periods of operations and enabling him or her to concentrate on the higher level supervisory control aspects for which no form of automation could be envisaged.

The algorithm operated successfully in adaptive/nonadaptive modes and proved robust with respect to noise, such as stimuli and heavy electrical interference from surgical diathermy and patient parameter variability. In our opinion, this is due to the use of the observer polynomial $T(z^{-1})$, which we stated on several occasions must represent an integral design part in the GPC design. There are various new aspects in the control design that need to be clinically validated such as the inclusion of output constraints with the hierarchical removal of constraints as described in Section 5.3.

Due to its simple and friendly design (see Figure 5.8), the overall computer control system proved very easy to manage, with all the clinical trials being performed by the anesthetist at the Glasgow Western Infirmary and without the presence of an engineer. He was particularly amenable to the idea of constraints and control moves and how this relates to optimizing drug consumption, and hence the impact this might have as far as safety and economic factors are concerned.

From a design viewpoint, extensions to the algorithm have already considered the inclusion a supervisory higher level that monitors the actions of the lower level (constrained level) and provides jacketing when severe faults are detected. Such a level takes the form of a simple Mamdani-type fuzzy controller with 25 optimal fuzzy rules obtained using an optimization algorithm. The closed system, together with the patient simulator, will prove to be a very valuable asset for training junior doctors and especially nurses, who, it is projected, will be given greater autonomy in the very near future in so far as patient care is concerned.

References

[1] Jessop, J., and J. Jones, "Conscious Awareness During General Anaesthesia: What Are We Attempting to Monitor?" *British Journal of Anaesthesia*, Vol. 66, No. 6, 1990, pp. 635–637.

[2] Kulli, J., and C. Koch, "Does Anesthesia Cause Loss of Unconsciousness?" *Tins*, Vol. 14, No. 1, 1989, pp. 6–10.

[3] Schwilden, H., H. Stoeckel, J. and Schuttler, "Closed-Loop Feedback Control of Propofol Anaesthesia by Quantitative EEG Analysis in Humans, *British Journal of Anaesthesia*, Vol. 62, 1989, pp. 292–296.

[4] Kenny, G. N. C., et al., "Closed-Loop Control of Anesthesia," *Anesthesiology*, Vol. 77, No. 3A, 1992, p. A328.

[5] Linkens, D. A., M. F. Abbod, and J. K. Backory, "Fuzzy Logic Control of Depth of Anesthesia Using Auditory Evoked Responses," *IEE Colloquium on Fuzzy Logic Controllers in Practice*, IEE Savoy Place, London, 1996, pp. 4/(1–6).

[6] Gentilini, A., et al., "Multitasked Closed-Loop Control in Anesthesia," *IEEE Engineering in Medicine and Biology*, Vol. 20, No. 1, 2001, pp. 39–53.

[7] Jacquez, J. A., *Compartmental Analysis in Biology and Medicine*, Amsterdam, the Netherlands: Elsevier, 1972.

[8] Jarvis, D. A., "Physiological Pharmacokinetic Models: A Review of Their Principles and Development," *Anesthetic Pharmacology Review*, Vol. 2, 1994, pp. 214–230.

[9] Scokaert, P. O. M., "Constrained Predictive Control," D.Phil. Dissertation, University of Oxford, U.K., 1994.

[10] Derighetti, M., "Feedback Control in Anesthesia," Ph.D. Dissertation, Swiss Federal Institute of Technology, Zurich, 1999.

[11] Lawson, C. L., and R. J. Hanson, *Solving Least-Squares Problems*, Englewood Cliffs, NJ: Prentice Hall, 1974.

[12] Clarke, D. W., C. Mohtadi, and P. S. Tuffs, "Generalized Predictive Control, Part I and II," *Automatica*, Vol. 23, No. 2, 1987, pp. 137–160.

[13] Mahfouf, M., and D. A. Linkens, "Constrained Multivariable Generalized Predictive Control (GPC) for Anesthesia: The Quadratic Programming Approach (QP)," *International Journal of Control*, Vol. 67, No. 4, 1997, pp. 507–527.

[14] Shook, D. S., C. Mohtadi, and S. L. Shah, "Identification for Long-Range Predictive Control," *IEE Proceedings-PtD.*, Vol. 138, No. 19, 1991, pp. 75–84.

[15] Mamdani, E. H., "Application of a Fuzzy Algorithm for the Control of a Simple Dynamic Process," *IEE Proceedings-PtD.*, Vol. 121, 1974, pp. 1585–1588.

[16] Mahfouf, M., and M. Jamei, "Rule-Base Generation Via Symbiotic Evolution for a Mamdani-Type Fuzzy Control System Applied to Car Suspension Design," *Journal of Systems and Control Engineering, IMEchE Proceedings–Part I*, Vol. 218, 2005, pp. 621–635.

[17] Takagi, T., and M. Sugeno, "Fuzzy Identification of Systems and Its Applications to Modeling and Control," *IEEE Trans. of Systems, Man and Cybernetics*, Vol. 15, 1985, pp. 116–132.

[18] Millard, R. K., C. R. Monk, and C. Prys-Roberts, "Self-Tuning Control of Hypotension During ENT Surgery Using a Volatile Anesthetic," *IEE Proceedings-PtD.*, Vol. 125, 1988, pp. 173–178.

[19] Bierman, G. J., *Factorization Methods for Discrete Sequential Estimation*, New York: Academic Press, 1977.

[20] Robinson, B. D., and D. W. Clarke, "Robustnedd Effects of a Prefilter in Generalized Predictive Control," *IEE Proceedings-PtD.*, Vol. 138, 1991, pp. 3–8.

CHAPTER 6

Neural-Fuzzy Modeling and Feedback Control in Anesthesia

This chapter approaches the subject of anesthesia from a totally different angle as far as quantification of unconsciousness is concerned. It includes three main research strands: classification of depth of anesthesia; modeling the patient's vital signs; and control of DOA with simultaneous administration of anesthetic and analgesic drugs. First, a fuzzy relational classifier was developed to classify a set of wavelet-extracted features from the auditory evoked potentials into different levels of DOA. Second, a hybrid patient model using Takagi-Sugeno-Kang fuzzy models was developed. This model relates the heart rate, the systolic arterial pressure, and the auditory evoked potential (AEP) features with the effect concentrations of the anesthetic drug propofol and the analgesic drug remifentanil. The surgical stimulus effect was incorporated into the patient model using Mamdani fuzzy models. Finally, a multivariable fuzzy controller, some parameters of which were optimized using a genetic algorithm, was developed for the simultaneous administration of the two drugs. These three parts were incorporated in a closed-loop simulation system for future use in the operating room.

6.1 Introduction

This chapter should perhaps begin with a "health warning" to those in the health professions, which should read: "If you do not rely on EEG to tell whether a patient is deeply anesthetized or not, please turn to the next chapter." Indeed, because I have been in research in biomedicine for almost 20 years now, I have grown to accept that medical practices very much differ from one hospital to another. For example, propofol is a popular anesthetic in the Sheffield NHS hospitals, whereas isoflurane is the favorite anesthetic at the Glasgow Western Infirmary. This may appear to be a disadvantage, but I, and my Research Group at Sheffield University, always perceive it as an advantage since it allows us to approach anesthesia from often different and interesting perspectives.

Indeed, it also became apparent from the previous chapters that anesthesiologists use a variety of observations, such as blood pressure, heart rate, lacrimation, movement, sweating, and pupil response, to make a judgment on DOA levels. However, the introduction of balanced anesthesia (i.e., the use of three drugs: a muscle relaxant, an analgesic agent and an anesthetic agent) improves the safety of the patient, but the clinical signs are obscured by the effect of the different drugs. Therefore, attention has turned to signals generated from within the central nervous system (CNS) [1, 2]. AEPs, which are the responses in the CNS to an auditory stimulus, have been the focus of much research and have led to some very useful results in monitoring DOA [3–5]. The AEPs show similar graded changes with varying anesthetic concentrations for different agents, as well as appropriate changes to surgical stimuli; it also indicates awareness and light anesthesia [6, 7].

To distinguish between stages (i.e., levels of DOA) that could lead to implicit or explicit memory, with or without pain and an adequate stage of unconsciousness without memory or recall, is not a trivial task [8]. Explicit memory of conscious awareness with pain during anesthesia is different from the situation where there is explicit memory of intraoperative events but no pain (i.e., adequate analgesia but inadequate hypnosis). Conscious awareness with pain perception is the most worrying situation for patients and anesthetists alike. Therefore, it is very important to establish an adequate method for measuring DOA.

Balanced anesthesia has introduced another problem, being that of drug interactions. Anesthetic and analgesic drugs may have different types of interactions: increasing or decreasing the effects of each drug, potentiating the different side effects, or even introducing new side effects [9, 10]. The anesthesiologist needs to be aware of the interactions between the drugs for the safety of the patient. Propofol is a commonly used intravenous anesthetic agent because of its hypnotic properties and almost lack of side effects compared with some inhalational agents [11]. In general, propofol is combined with one of the synthetic opioids to provide analgesia. However, the optimal propofol infusion rate and concentration are significantly affected by the choice of opioid [12, 13]. It is important to analyze the effects of the opioids so as to decrease the amount of drug infused and the recovery time. Remifentanil may be the ideal opioid to use in combination with propofol for continuous intravenous anesthesia [14]. The properties of remifentanil, a short-acting opioid with a very rapid onset and offset of action, lead to more stable operating conditions. Remifentanil potentiates the effects of propofol (i.e., a decrease in the hemodynamic parameters and depression of the CNS), which is reflected in the AEP. The study of the pharmacological interaction of these two drugs is necessary to assure an adequate DOA level and analgesia. This synergistic interaction with propofol can be used to advantage by titrating propofol to lower infusion rates and speeding up recovery. Special care is necessary, however, because if propofol is used to achieve high concentrations, the simultaneous administration of remifentanil will drastically increase propofol side effects. Also, it is of major importance to consider the dynamic interactions of the two drugs. These drug interactions provide practical guidelines for the optimal drug dosing during anesthesia.

In addition, an opioid with a rapid onset of action allows for a rapid response to the stimulus effects. Analysis of the brain signals would improve the quality of analgesia, since there would be more information about the arousal caused by the stimulus on the CNS and subsequent effects on the level of unconsciousness.

In one part of this comprehensive research study, a simulation system replicating a typical patient's behavior during surgical interventions is presented. Data obtained in the operating theatre is used as a description of a routine anesthetic profile in surgical practice. First, a fuzzy relational classifier (FRC) for depth of anesthesia was developed. The FRC uses a set of AEP features to classify DOA into different levels. Second, a hybrid patient model is presented. The patient cardiovascular parameters and the AEP are modeled into TSK fuzzy models [15], using an adaptive network fuzzy inference system (ANFIS). The model for anesthetic drug dynamic interactions can prove to be very useful in understanding the detailed relationship between the concentrations of the two drugs and their vital effects. The patient model also includes a fuzzy stimulus model, describing the effects of the surgical stimulus on the cardiovascular parameters.

The infusion rate of the anesthetic drug is titrated according to the patient's requirements, so as to maintain a certain level of DOA. The patient's clinical signs and/or brain signals are used by the anesthetist to determine the adequate infusion rate. In addition, the anesthetist also establishes the required infusion rate of the analgesic drug, based on the patient's response to surgical stimuli. A closed-loop control system of DOA will help the anesthetist adjust simultaneously the infusion rates of the anesthetic and analgesic drugs. The majority of the researches in the area are mainly concerned with the automatic control of the anesthetic drug, whereas the analgesic is controlled manually by the anesthetist.

In another part of this research study the objective is to design a multivariable control structure for both drugs. The developed patient model will be used to develop a control algorithm relating to the administration of both drugs. The study of the interactions between the anesthetic propofol and the analgesic remifentanil helps to determine the ideal combination of infusion rates. This study will also represent a practical guide for the anesthetist, which would help him or her learn how to adjust the amount of drug infused and hence improve the patient's comfort. First, the patient model will be tested using a series of open-loop simulations with different infusion profiles. Second, the closed-loop structure will be presented, showing the links between the patient model, the DOA classifier (i.e., the FRC presented in another part of this study), and the controller. Third, a multivariable fuzzy controller, developed with the anesthetist's cooperation, is also described. This controller establishes the infusion rates of propofol and remifentanil simultaneously based on the level of DOA, the concentrations of the drugs, and the surgical stimuli. The performance of the controller is tested under different conditions.

6.2 Alternative Assessment Tools of DOA

There is no doubt that the practice of medicine has evolved into what is often referred to as *modern practice*. This is equally true of *modern* anesthetic practice, which uses safer drugs that most of the time tend to mask the "classical" signs of unconsciousness (awareness), such as movement of the eyes, sweating, lacrimation, pupil diameter, respiration heart rate, and blood pressure. Hence, the issue of monitoring anesthetic depth, which is closely related to the technique(s) of measurement, has never been so significantly relevant. Chapter 2 highlighted the issues (limitations) of using (the so-called "golden standard") some of these signs as reliable indicators of unconsciousness, but the next section will discuss in detail the technique behind evoked potentials (EP), particularly auditory evoked potentials, and will review some of the work that has advocated its use as a reliable indicator of DOA.

6.3 Mid-Latency Auditory Evoked Potential

6.3.1 Evoked Potentials

Evoked potentials, or evoked responses, represent modifications of the EEG that are induced by various stimuli, including visual stimuli [visual evoked potential (VEP)], stimulation of peripheral nerves [somatosensory evoked potential (SEP)], and auditory stimuli (AEP). By stimulus, here one means a light flash or a pattern on a screen. As far as the peripheral nerve stimulation is concerned, a small electrical discharge or a needle prick might be considered as a worthy representative.

Once any of the above stimuli is applied to the patient and the EEG recorded via electrodes placed on the patient's scalp, the responses are collected via an interfaced computer and the average response is extracted after appropriate signal amplification[1] and filtering[2] were applied.

The potential application of EPs in DOA monitoring was first highlighted and demonstrated by Uhl and coworkers [16] after they measured VEPs during induction of halothane anesthesia. Since then, the advantages [17, 18] of EPs have been well publicized. Among these advantages one can cite the following:

- The evoked response is an indication of the responsiveness of the central nervous system, whereas EEG reflects the resting level.
- The evoked responses do have an anatomical significance.

Among the multitude of disadvantages that EPs have, one can cite the fact that:

- They can easily be corrupted by artifacts.
- They vary with the intensity of the stimulus as well as with each stimulus application.

Visual Evoked Potential

Although VEPs have been superseded by SEPs and AEPs (see below), these were the first responses to be investigated. Several workers who investigated the flash light evoked response, including Thornton [18], found it to be variable between trials and between subjects, leading to it being labeled as more of qualitative than a quantitative measure. Sebel and his research team [19] also found that increased concentration of nitrous oxide (N_2O) led to increased latency (frequency) of the VEP in more than 50% of the patients.

Somatosensory Evoked Potential

SEPs are produced via stimulation of the sensory system. The advantage of this type of response is that they can be recorded by stimulating almost any nerve trunk at different levels of excitation, however, the same effects can be observed when moving from one anesthetic agent to another. Moreover, when anesthetics are combined with analgesics, SEP tends to reflect changes to the analgesics rather than to the

1. Amplification is necessary because the EP signal has amplitudes of microvolts.
2. Filtering is necessary because EEG is known to include low and high-frequency artifacts.

anesthetic; fentanyl, sufentanyl, and morphine are among the analgesic agents that depress the SEP amplitude significantly [20].

Auditory Evoked Potential

The transition of the electrical activity from cochlea of the ear to the cortex is characterized by the AEP waveform in a series of peaks and troughs. Typically, AEP are responses on the EEG signals to cyclic clicks applied to both ears.

The AEP is represented by three components: the brainstem response (1 to 15 ms), the mid-latency AEP (MLAEP; also known as the early cortical; 15 to 100 ms), and the late latency component (LLAEP; also known as the late cortical; greater than 100 ms). The brainstem response is more suitable when volatile anesthetic agents, such as isoflurane and enflurane are used. The LLAEP is of no clinical relevance in anesthesia for it is easily affected by sleep. However, the MLAEP has been shown to correlate strongly with changes in anesthetic depth. In their 1989 paper, Thornton and Newton [21] suggested four criteria that a signal had to fulfill to be deemed adequate as a *reliable* indicator of DOA:

- Evidence of graded changes with anesthetic concentration;
- Evidence of "similar" changes for different agents;
- Evidence of appropriate changes with surgical events;
- Ability to indicate awareness or "very light" anesthesia.

The same authors concluded that:

- MLAEP shows changes that are dose dependent in a graded fashion with anesthetic concentration.
- Similar graded changes were shown with different anesthetic agents.
- Surgical stimuli have the ability to change the response by reversing the effect of the anesthetic agent.
- Light anesthesia can be identified by a typical waveform (as will be shown in the next section).

6.3.2 Data Acquisition and Feature Extraction

The original pioneering research work on the exploitation of AEP for use as a monitoring tool of DOA was carried out by Thornton and his research team at Northwick Park Hospital [22]. The reader is referred to the first reported research work by James [23], Navaratnarajah [24, 25], and Thornton [22, 26–28] on the effect of anesthetic agents on the brainstem response and MLAEP. All these researchers reported a graded response in the latencies of the brainstem as well as the latencies and amplitudes of the mid-latency components of the AEP during anesthesia.

MLAEP are recorded noninvasively in the operating theater using surface electrodes. The system used in this research study relates to the one developed at the Northwick Park Hospital (see Figure 6.1).

Prior to digitization and extraction of the relevant information, the signal is first analog filtered (an antialiasing bandpass filter with cutoff frequencies of 0.1 to 400

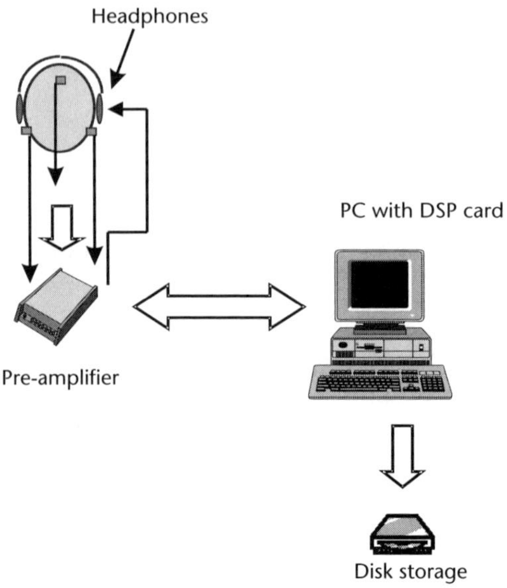

Figure 6.1 The audio evoked response (AER) system.

Hz) in the preamplifier box. The filtered signal is then transmitted to the computer where it is sampled at a rate of 1 kHz. The auditory stimulus was a rarefaction click presented to both ears simultaneously at 75 dB above the average hearing threshold at a rate of 6.1224 Hz. The first 120 ms (corresponding to 121 data points) of data after each stimulus presentation was recorded as the AEP signal.

The AEP has typical amplitudes of a few microvolts in an awake person, while the ongoing EEG has typical amplitudes of tens of microvolts. Where the EEG is an ongoing activity and is seemingly random, the AEP, in contrast, is deterministic: it is time-locked to the stimulus delivered to the ears. The AEP cannot be measured directly from the scalp recordings as they are buried in the ongoing EEG. The signal-to-noise ratio (with the EEG being the noise, unwanted signal, and the AEP being the signal) of typical recordings is less than –40 dB. It is this small signal-to-noise ratio that makes waveform *estimation* and hence signal *classification* a difficult process. Figure 6.2 illustrates the process of *averaging* 188 sweeps[3] of data to form the AER signal of Figures 6.3 and 6.4, showing the MLAEP responses when the patient is awake and anesthetized respectively. The differences in amplitude and latency of the peaks in both figures reflect such a strong correlation between MLAEP and DOA. [Note the coordinates of the peaks (I-V) and (P_a, N_a, P_b, N_b) relating to the brainstem response and the MLAEP, respectively, in both cases.]

Extracting the AEP from the raw EEG signal, which includes stochastic activity, is not an easy task. The following techniques have been investigated:

- Ensemble averaging;
- Digital filtering;
- Artifact rejection;
- Wavelet filtering.

3. Each sweep corresponds to 120 ms worth of data.

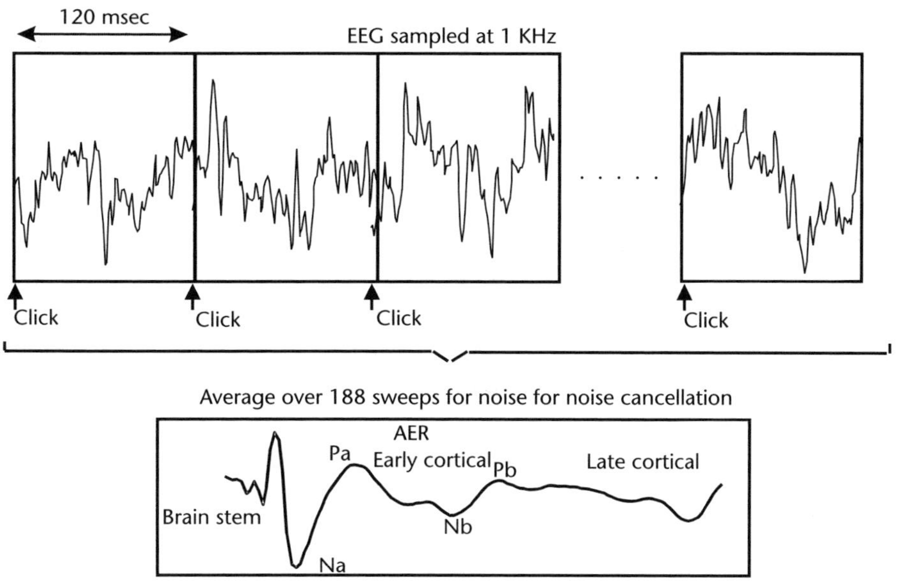

Figure 6.2 The process of collecting auditory evoked responses.

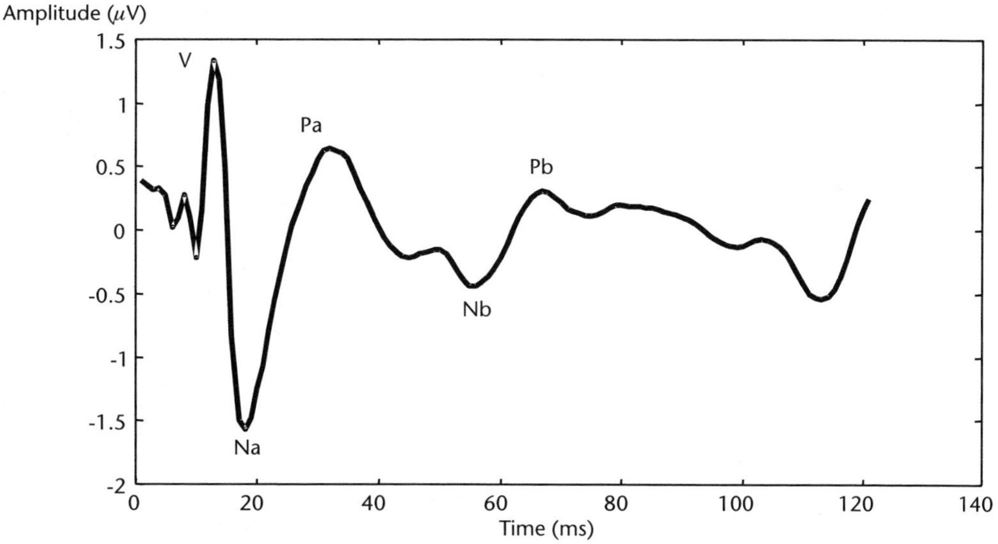

Figure 6.3 The auditory evoked potential for an awake patient; the peaks forming the brainstem response (peaks I-V) are in the first 15 ms of the response. The peaks in the brainstem response are not shown above, as the sampling rate used was too low to make detection of these peaks possible.

While it is not the objective of this chapter to discuss each of these techniques, it will suffice to look at the first and last techniques.

Indeed, perhaps the first technique that was investigated relates to *ensemble averaging* [29]. This technique assumes that the signal (waveform) at the scalp can be expressed as follows:

$$x(t) = s(t) + n(t) \tag{6.1}$$

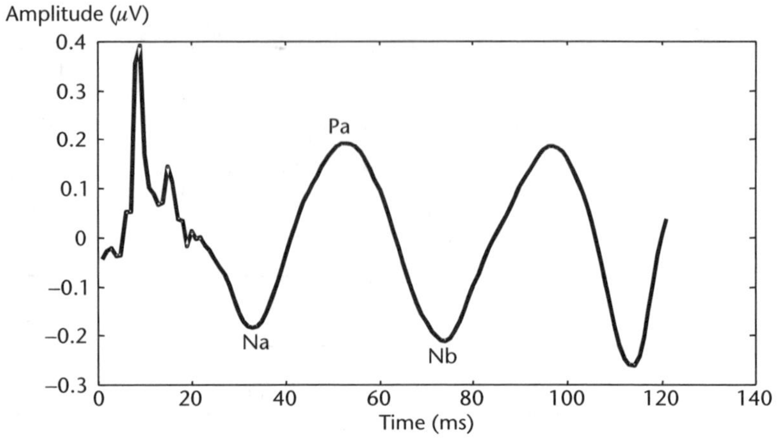

Figure 6.4 The auditory evoked potential for an anesthetized patient.

where $x(t)$ is the measured signal (waveform), $s(t)$ represents the signal associated with the AEP, and $n(t)$ is the signal that is independent of the AEP signal. The ensemble averaging technique was shown to work adequately by decreasing the noise in a directly proportional fashion with respect to the standard deviation of the noise, provided that the signal component is deterministic and that the noise component is stationary.

The digital filtering of AEPs using the wavelet transform (WT) was first introduced by Bertrand [30]. The wavelet concept was deemed useful in this case given that the signals involved seem to contain frequencies that may be varying in time. Hence, the WT technique allows one to extract the relevant frequencies as a function of time; in other words, a two-dimensional representation is possible. The noise that usually contaminates the MLAEP is usually also in the same frequency band of the brainstem response. Using this novel approach, the high frequencies at the beginning of the signal are kept intact but are removed from the MLAEP. Consequently, this technique is very much suitable for finite-duration signals with some nonstationary characteristics. In this research the multiresolution analysis (MRA) technique, allied with WT, was used to extract features of the AEP, which are then used to classify DOA.

The signal is made up of 121 samples (corresponding to the first 121 ms of the evoked response poststimulus), and this is extended to 128 samples so that six levels of decomposition may be achieved using the MRA decomposition algorithm by padding the signal with extra zeroes at the end. The MRA decomposes the AEP signal into several resolutions, which in turn were analyzed for their respective energies using the Daubechies Wavelet [31] of order 6 (giving 12 filter coefficients). This filter order was found to produce acceptable results and was of low enough order for the length of discrete samples (121) to be analyzed. Hence, the energy contained in each detail signal is expressed as follows:

$$D_k = \frac{1}{n_k} \sum_{i=1}^{n_k} \left| S_k(i) \right|^2 \tag{6.2}$$

where:

D_k is the energy contained in detail component k.

n_k is the number of samples in detail component k.

S_k is the detail sequence k.

i is the sample number.

Table 6.1 shows the features that have been analyzed for correlation with DOA as well as the samples making up the components. In this study 10 features were used in the classification process using the FRC. D_{41}, D_{42}, and D_{43} were averaged to produce D_4, and similarly for D_3 and D_5. Several combinations of AEP features were tested. However, the set of features D_1, D_2, D_3, D_{31}, D_{32}, D_{33}, D_4, D_5, D_{51}, and D_{52} gave the best results considering the data available.

6.4 Development of a New Fuzzy Relational Classifier for DOA

6.4.1 Surgical Data

The data used in this study was collected during surgical interventions in the Royal Hallamshire Hospital in Sheffield (United Kingdom). The patients have either an ASA[4] I or II status and gave previous explicit consent to be included in this study. The data collection was restricted to patients who received the same propofol/remifentanil drug profile, and to surgical interventions that did not need other drugs

Table 6.1 Description of the Energy Components Relating to Each Wavelet Detail and Their Corresponding Sample Ranges in the Details Sequence S_k*

Energy Component	*Used Samples*
D_1	9–56
D_2	5–28
D_{31}	3–6
D_{32}	7–10
D_{33}	11–13
D_{41}	2–3
D_{42}	4–5
D_{43}	6–7
D_{51}	2
D_{52}	3

*See (6.2)

4. The American Society of Anesthesiologists' (ASA) physical status classification serves as a guide (although it is imprecise because patients could be placed in different classes by different anesthesiologists) to better communication among anesthesiologists about clinical conditions of patients, in the way to predict their anesthetic/surgical risks—as higher ASA class, as higher the risks. Class I corresponds to a healthy patient with no medical problems, and Class II corresponds to a patient with mild systemic disease.

that could interact with this particular drug profile. The unique properties of remifentanil make it a suitable analgesic drug for use with propofol, and adequate for control in anesthesia.

Only the maintenance[5] phase of anesthesia will be considered, since the interaction of the two drugs is more predominant then, due to the continuous presence of stimuli during this phase. In addition, the initial induction phase has a lot of noise and outside interferences (e.g., intubation, and moving the patient from the preparation room to the operating room). The maintenance phase was considered to start at 1,500 seconds (25 minutes), when the patient was in the operating room and the surgical intervention was about to start. The infusion profiles of the two drugs for one patient are shown in Figure 6.5.

6.4.2 The Classification Algorithm

An FRC trained by fuzzy clustering was applied to features extracted from the AEP, heart rate, and systolic arterial pressure, signals in order to classify DOA. The AEP were recorded in the operating theater and extracted from the EEG trace by filtering and averaging. The appropriate choice of signal features for data compression with acceptable classification performance is of primary concern in the design of automatic DOA monitors. Multiresolution analysis with wavelet transforms was used to decompose the AEP signal into approximations at different scales of resolution. The

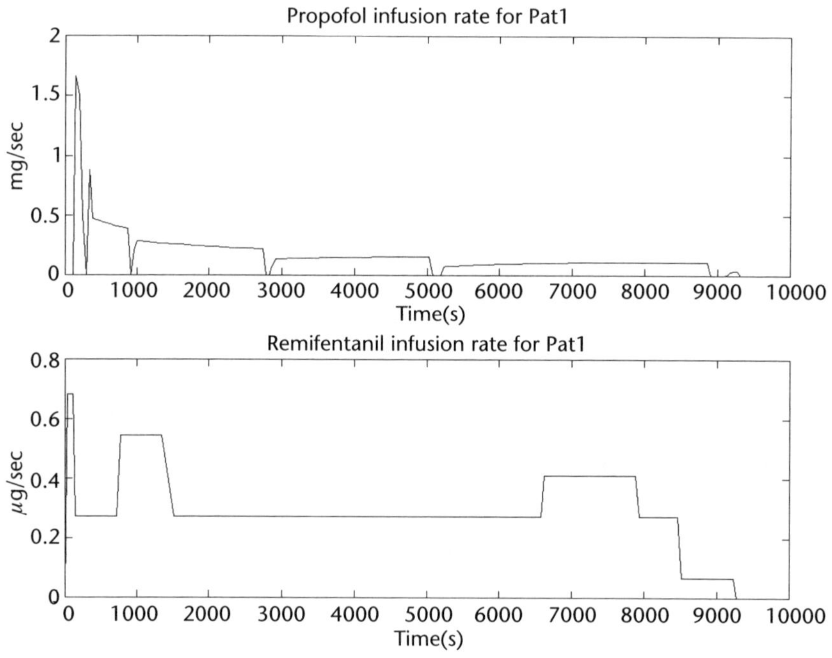

Figure 6.5 Infusion profile for patient "Pat 1."

5. This corresponds to the time when the anesthetist tries to keep a certain level of anesthesia through a continuous drug infusion. It normally follows the induction phase, which consists of administering a large amount of drug to reach a certain level of anesthesia very quickly.

decomposed components were analyzed for their energy content, and these were then used as an indicator of the DOA. The Daubechies wavelet with 12 filter coefficients was used for decomposing the AEP signal. The same approach was used in [1]. The FRC uses a combination of fuzzy clustering and fuzzy relations to establish an inference system. The classifier is trained by unsupervised fuzzy *c*-means clustering, and then a fuzzy relation between the clusters membership functions and the class identifiers is computed. The fuzzy relation specifies the existing relationship between the clusters and the class membership (i.e., the different levels of DOA). The diagram in Figure 6.6 illustrates the training procedure; the basic algorithm of the FRC is presented in [32].

Exploratory Data Analysis

The training data set Z is partitioned into c fuzzy subsets (clusters). The membership function of the data samples in the clusters is described by the fuzzy partition matrix U and each cluster is characterized by its center v_i.

Prior to clustering, several parameters have to be defined in the algorithm: c, m (fuzzy exponent), and ε (termination tolerance). The choice of c can be verified by assessing the validity of the obtained partition. The Xie-Beni index [33] is a commonly used validity criterion and is of the form:

$$\chi(U,V;Z) = \frac{\sum_{i=1}^{c}\sum_{k=1}^{N}\mu_{ik}^{m}\left\|z_k - v_i\right\|^2}{N\left(\min_{i\neq j}\left\{\left\|v_i - v_j\right\|\right\}^2\right)} \tag{6.3}$$

The best partition is the one that minimizes $\chi(U, V; Z)$. The c-means cluster algorithm (see Appendix 6A for more details on fuzzy clustering) was run for different values of c and m, and then several times for each of these settings with different initialization matrices to obtain good parameter settings for this DOA application.

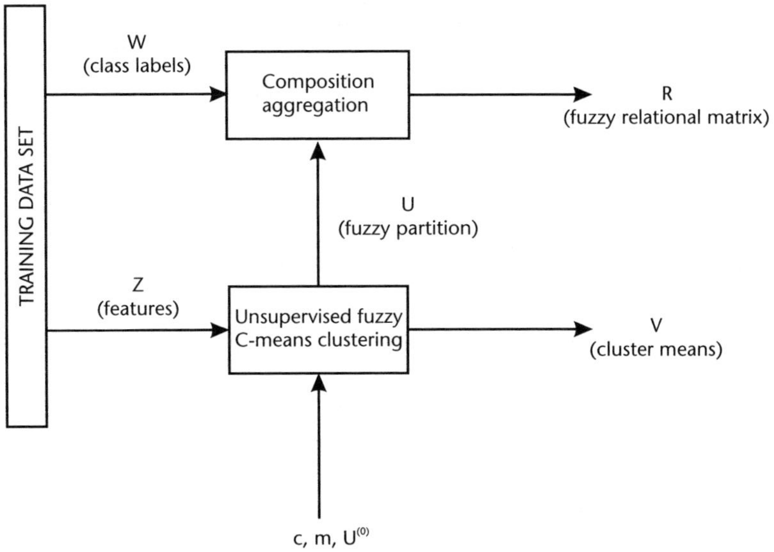

Figure 6.6 A schematic diagram representing the training architecture for the FRC.

Fuzzy Relation

The fuzzy relation encodes the logical relationship between the cluster membership and the class membership. This relation is computed from the information in the fuzzy partition matrix U and in the target vectors containing the membership of the pattern in the classes. The kth target vector is denoted by

$$w_k = \left[w_{1k}, w_{2k}, \ldots, w_{Lk}\right]^T \tag{6.4}$$

where $w_k = [w_{1k}, w_{2k}, \ldots, w_{Lk}]^T$ and $L \in \aleph$ is the number of classes. For the training data, where the classification is exactly known, $w_{Lk} \in \{0,1\}$.

The binary fuzzy relation, R, is a mapping $R{:}[0,1]^c \, x[0,1]^L \rightarrow [0,1]$. It can be represented as a $c \times L$ matrix. The relation R is obtained by aggregating the partial relations R_k, which are computed for the training samples:

$$r_{ij} = \min\left(1, 1 - \mu_{ik} + w_{jk}\right) \quad j = 1, 2, \ldots, L, \quad i = 1, 2, \ldots, \rho \tag{6.5}$$

The aggregation of the relations R_k is computed by means of a fuzzy conjunction operator, implemented element-wise by the minimum function:

$$r_{ij} = \min_{k=1,2,\ldots,N} \left[\left(r_{ij}\right)_k\right] \tag{6.6}$$

Classification of New Patterns

The main objective of any classifier is to determine the class of a new pattern. First, the cluster membership function μ is calculated from the distances to the cluster centers, v_k, as in the fuzzy c-means algorithm. Second, the class membership vector $\boldsymbol{w}$ is computed by a fuzzy relational composition using the Lukasiewicz implication:

$$w_j = \max\left\lfloor \max\left(\mu_i + r_{ij} - a, 0\right) \right\rfloor \quad j = 1, 2, \ldots, L \tag{6.7}$$

where a is considered to be 1 [32]. Finally, defuzzification is applied to obtain a crisp decision, using the maximum method:

$$y = \arg \max_{1 \le j \le L} w_j \tag{6.8}$$

where y is the class index.

Implementation

The FRC is used to classify a set of features into five different DOA levels. The DOA is classified according to the anesthesiologist's expertise as follows:

1. *Awake.* Awake is when the patient response to surgical stimuli is not acceptable anymore.
2. *OK/Light.* OK/Light is when there is a slight patient response (i.e., clinical signs demonstrating that the patient might be gradually "coming around" from unconsciousness.
3. *OK.* OK is when no patient response is observed under surgical stimulus (i.e., the patient is under adequate control).

4. *OK/Deep*. OK/Deep is when there is no patient response, but the patient is slightly deeper than necessary and still stable.
5. *Deep*. Deep is when the patient is clearly in a deep state of DOA, which is not safe for patients and should be an indicator that something is wrong.

The classifier uses two sets of features in the classification process. The first set is constituted from the AEP signal, and the second set is constituted from the cardiovascular parameters. The data from two patients was used to train and test the FRC. The data was labeled by the anesthesiologist based on the observed clinical signs during the operation. The information regarding the AEP (i.e., depression in the CNS) during anesthesia was not made available to the anesthesiologist. The anesthesiologist's opinion was taken as the correct classification since there is no other precise analysis (i.e., gold standard) to evaluate DOA.

Wavelet Extracted AEP Features

The AEPs were recorded in the operating room using surface electrodes. The auditory stimulus was given at a rate of 6.122 Hz. The EEG was sampled at a frequency of 1 kHz after passing through an analog band-pass filter, for a sweep period of 120 ms after each click. The computer produces an average of 188 data sweeps over 30 seconds. A set of 10 features was extracted from the AEP signal using MRA with wavelet transforms—these features reflecting the changes in amplitude and latency associated with DOA [20].

Hemodynamic Parameters

Anesthesiologists have always used the cardiovascular system as an indicator of DOA. Therefore, it is only natural to use these measurements to enhance the overall decision-making process.

Considering that the FRC may not reach a decision when multiple j's satisfy (6.7) or when the vector in (6.6) is all zeros, alternatives to the wavelet AEP features have to be considered. Therefore, the FRC is applied separately to the cardiovascular measurements and trained to obtain a decision based only on these measures. Different combinations of parameters were tested and it was found that HR plus SAP performed the best for classification. The values used are the change of the individual patient baseline of HR and SAP, denoted by ΔHR and ΔSAP. The use of the Δ values allows generalization, since the patient baselines are available before the operation and are considered as an input to the system.

The two FRCs (the first using the wavelet features to classify, the second using ΔHR and ΔSAP) work separately. The main decision belongs to the wavelet features classifier, but if a decision is not reached then the cardiovascular parameters are used to determine DOA.

Results

The training data for the FRC consisted of 2/3 of the available samples, and the remaining 1/3 was used as the testing set (i.e., 276 samples to train and 138 to test). The Xie-Beni index [33] was used to determine the optimal values of c and m for the unsupervised c-means clustering. Table 6.2 shows the classification results of the trained FRC using the wavelet extracted AEP features on the testing data set.

Table 6.2 Performance of the FRC Using the AEP Features on the Testing Data Set

	Correct	*Incorrect*	*Unclassified*
Number of samples	85	53	0
Percentage of samples	61.6%	38.4%	0%

Xie-Beni optimum values of $c = 11$, $m = 2.6$.

The FRC using only the AEP features was able to classify all the samples. As a result, the cardiovascular parameters would not have been used in the classification process. Nevertheless, the FRC was trained and tested with the cardiovascular parameters ΔSAP and ΔHR. This classifier was able to classify only 46.5% of the samples correctly with the Xie-Beni optimum values of $c = 13$ and $m = 2.6$. This result is hardly surprising since the AEPs represent the brain depression of hypnosis. However, a value of 46.5% is a reasonable percentage considering all the complex body organ functions involved in anesthesia.

6.5 Development of a Patient Model

The chosen drug profile plays an important role in anesthesia. The interaction between different drugs also has an influence on the overall efficacy and efficiency of anesthesia.

Propofol and remifentanil have a synergistic relationship. The effect of the combination of these two drugs is greater than that expected based on the concentration-effect relationships of the individual agents. In other words, a smaller amount of both drugs is needed to produce a certain effect compared to when propofol is given as a single agent. The use of remifentanil as the analgesic drug requires more attention than when using other older analgesics. Thus, a patient model is necessary to construct and test a closed-loop control system in anesthesia prior to any transfer of the design to the operating room.

The ΔSAP, ΔHR, and the AEP features were modeled using the drug effect concentrations and the surgical stimuli. Figure 6.7 shows the overall model structure. The drug infusion rates were used as the input to the system, then a pharmacokinetic model was used to determine the plasma concentration of both drugs, independently. An effect compartment transforms the plasma concentrations into effect concentrations and these were used as the input to the ANFIS system to train the fuzzy TSK models for ΔSAP, ΔHR, and the wavelet extracted AEP features. In addition, a stimulus model was developed to establish the effects of the surgical stimulus on the cardiovascular parameters, according to the level of analgesia. This stimulus model is implemented in a Mamdani-type fuzzy model using anesthesiologist's knowledge.

The drug plasma and effect concentrations were modeled independently. Therefore, no pharmacokinetic interactions were considered. This paper is mainly concerned with the pharmacodynamics interaction reflected in the drug effects, since these are very significant in clinical practice.

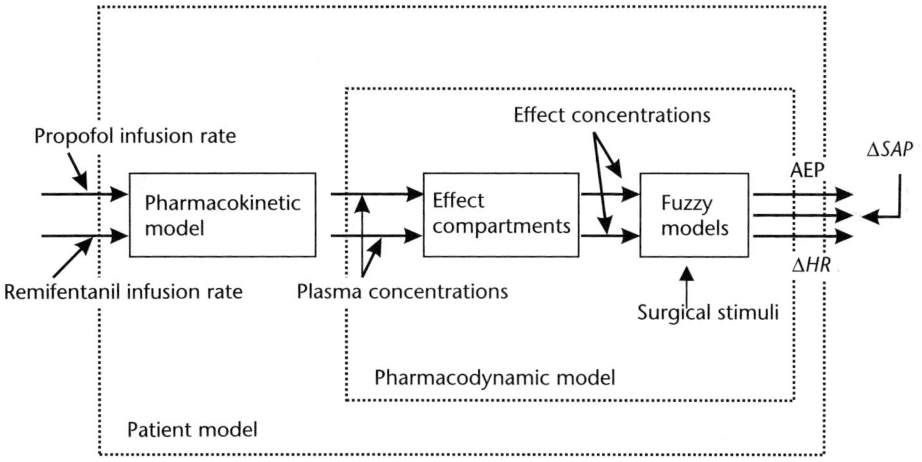

Figure 6.7 Block diagram of the patient model during the maintenance phase.

6.5.1 Pharmacokinetic Models

The pharmacokinetic models of the two drugs were constructed using a three-compartment model. The pharmacokinetic parameters were gathered from the literature and reflect mean population values. For propofol, the parameters from [34], which can also be found in [35], were used, whereas for remifentanil, the parameters from [36] were used.

6.5.2 Pharmacodynamic Models

The pharmacodynamic model comprises two effect compartments, one for each drug, as well as different fuzzy TSK models each describing the relationships between the two drugs concentrations, the cardiovascular parameters, and the AEP features.

Effect Compartment

The effect compartment is a hypothetical compartment describing the delay between the plasma concentration and the effect concentration. The pharmacodynamic parameters k_{e0} used for propofol and remifentanil, are described in [37] and [38], respectively. These papers evaluated the pharmacodynamics of propofol and remifentanil, using the same pharmacokinetic parameters as in the previous section.

Fuzzy Models

A fuzzy inference system was used to model ΔSAP, ΔHR, and the AEP features via fuzzy TSK models. The TSK fuzzy system is constructed from the following rules:

$$\text{IF } x_1 \text{ is } C_1^1 \text{ and } \ldots \text{ and } x_n \text{ is } C_n^1 \text{ THEN } y^1 = c_0^1 + c_1^1 x_1 + \ldots + c_n^1 x_n \quad (6.9)$$

where C_l^1 are fuzzy sets, c_l^1 are constants, and $l = 1, 2, \ldots, m$. That is, the antecedent parts of the rules are the same as in the ordinary fuzzy IF-THEN rules, but the consequent parts are linear combinations of the input variables. Hence, given an input vector $x = (x_1, x_2, \ldots, x_n) \in U \subset R^n$, the output $f(x) \in V \subset R$ of the fuzzy TSK model is computed as the weighted average of the y^{1}'s; that is,

$$f(x) = \frac{\sum_{l=1}^{m} y^1 w}{\sum_{l=1}^{m} w^1} \tag{6.10}$$

$$w^l = \prod_{i=1}^{n} \mu_{C_i^l}(x_i)$$

The ANFIS identifies a set of parameters through a hybrid learning rule combining the back-propagation gradient-descent and the least squares method. This was used to determine the parameters for the fuzzy TSK models [39].

The TSK model uses as inputs the effect concentrations of remifentanil and propofol and as outputs the ΔSAP, or ΔHR, or one of the AEP features. Figure 6.8 shows the propofol and remifentanil effect concentrations obtained using the infusion profiles from Figure 6.5 (maintenance phase) as inputs to the ANFIS. The data from one set of real surgical interventions were divided into training and testing data sets. The following properties were found to provide the best results for the models:

- Subtractive clustering on the training data to generate the initial fuzzy inference system (FIS) structure;
- Gaussian input membership functions;
- Hybrid optimization method.

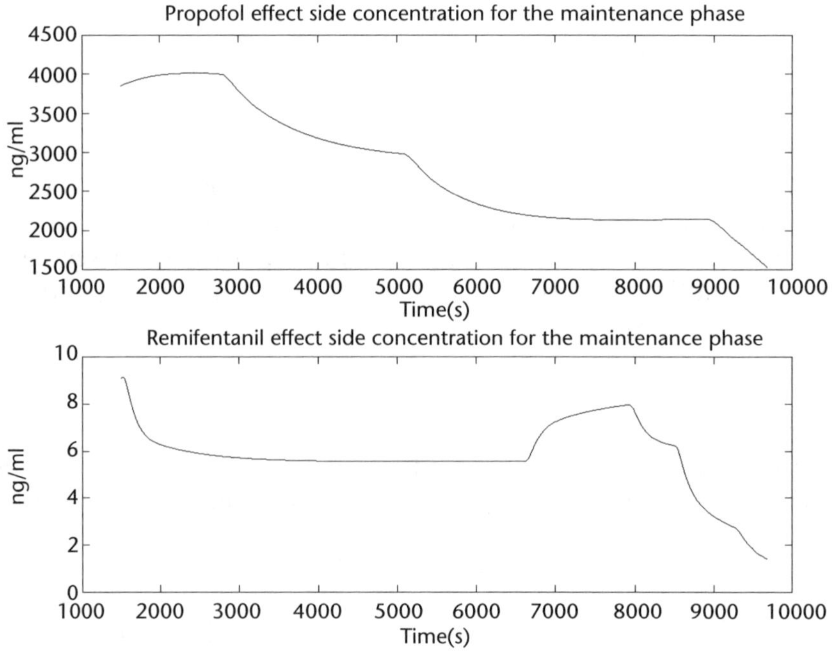

Figure 6.8 Propofol and remifentanil effect concentrations used as inputs to ANFIS (maintenance phase).

The cardiovascular parameters ΔSAP and ΔHR, and the AEP features were modeled separately using the same ANFIS structure. The results of the ΔSAP TSK model on the training and testing data sets are presented in Figure 6.9. The training data set should represent the features of the data that the trained TSK system is intended to model. Therefore, the testing data set consists of four different pieces of the data, making up 1/3 of the total number of samples, the remaining samples being used as the training data set. The mean absolute error on the training and testing data sets were 2.26 and 6.53 mmHg, respectively. The model has a reasonable performance on the training data set with a low error. The results on the testing data set reflect the ΔSAP trend correctly and appear to smooth out the disturbances due to stimuli.

Figure 6.10 shows the results of the ΔHR TSK model on the training and testing data sets. The mean absolute errors on the training and testing data set were 1.12 and 4.54 bpm, respectively. The model reflects the trend in ΔHR and has a reasonable overall performance. Note that the models reflect the changes in effect concentrations and interactions between the two drugs, and therefore, the changes due to outside disturbances are not reflected in the results.

The AEP features are a set of spikes and they do not reflect trends. Figure 6.11 shows the results of the trained TSK model for one of the wavelet extracted AEP features, D_1, which is a low-frequency feature.

The error on the testing data set was lower than the one on the training data set (0.043 versus 0.055), which shows that there was no overtraining of the model. The TSK model does not reflect the spikes in the data, and there is an increase in the D_1 value as a result of the drugs concentration. However, surgical stimuli are affecting

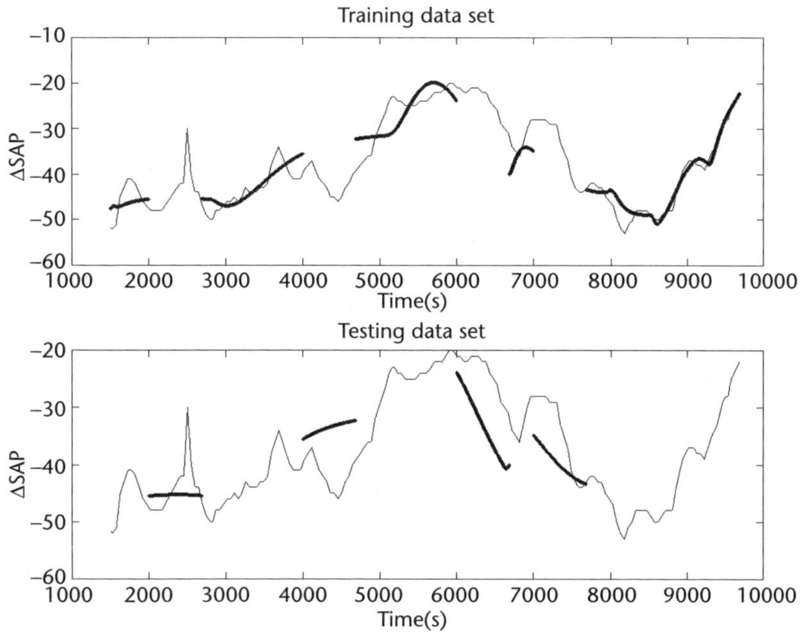

Figure 6.9 Results of the ΔSAP TSK model on the training/checking data sets (strong solid line) versus the actual ΔSAP.

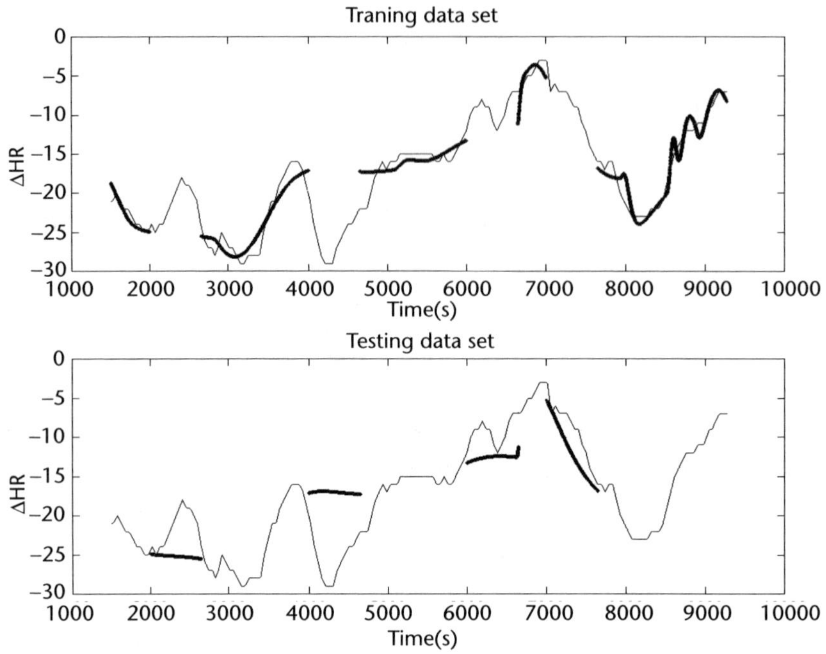

Figure 6.10 Results of the ΔHR TSK model on the training/checking data sets (strong solid lines) versus the actual ΔHR.

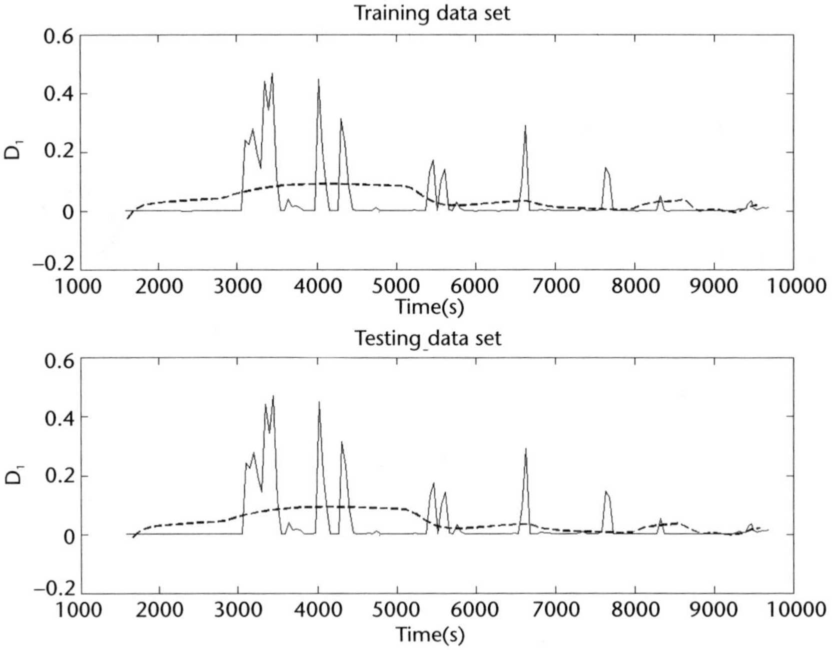

Figure 6.11 Results of the AEP feature D_1 TSK model on the training and testing data sets (dashed line) versus the actual AEP feature D_1.

the values of D_1. The models have different characteristics according to the high and low frequency features. In general, the sharp peaks in the AEP features are the response to surgical stimulus. Full details of the TSK models for all the 10 features are presented in [6].

6.5.3 Surgical Stimuli Model

Surgical stimuli increase the value of SAP and HR. The final values of the cardiovascular parameters are the result of the two drugs and the surgical stimuli. Remifentanil action can be viewed as a reduction in the patient's perception of surgical stimuli, determining the quality of analgesia. First, the stimulus perceived by the patient was modeled, in order to reflect the analgesic action of remifentanil. Finally, the change in the value of SAP and HR was modeled so as to reflect the effect of the perceived stimuli. The overall stimulus model comprises three Mamdani-type fuzzy models based on anesthesiologists' knowledge and represented by IF-THEN rules.

Table 6.3 shows the rule base of the perceived stimulus model for the maintenance phase of anesthesia accounting for surgical stimuli and the remifentanil effect concentration. The surgical stimuli are described as Zero, Low, Medium, and High. The remifentanil effect concentration was labeled by the anesthesiologist as Zero, Low, Medium, High, and Very High, and considered to be between 0 and 10 ng/ml throughout the procedure. The perceived stimulus obtained from the rule base is labeled Zero, Very Low, Low, Medium, High, and Very High.

After considering the effect of the perceived stimulus, the value of SAP is described using the rule base shown in Table 6.4. The values of the different SAP classes are described as follows:

- Baseline SAP > 130 mmHg:
  - *Low*: SAP < 70% of baseline;
  - *Medium*: SAP between 70% and 80% of baseline;
  - *High*: SAP > 80% of baseline.

- Baseline SAP between 120 and 130 mmHg:
  - *Low*: SAP < 75% of baseline;
  - *Medium*: SAP between 75% and 85% of baseline;
  - *High*: SAP > 85% of baseline.

Table 6.3 Perceived Stimulus Fuzzy Rule Base

Surgical Stimulus	*Remifentanil Effect Concentration (ng/ml)*				
	Zero	*Low*	*Medium*	*High*	*Very High*
Zero	Zero	Zero	Zero	Zero	Zero
Low	Low	Very Low	Zero	Zero	Zero
Medium	Medium	Low	Very Low	Zero	Zero
High	High	Medium	Low	Very Low	Zero

Surgical stimuli ∈ [1, 0], remifentanil effect concentration ∈ [0, 10 ng/ml], and perceived stimulus ∈ [1, 0].

Table 6.4 Rule Base for the New Value of SAP After the Stimulus

Perceived stimulus	*Previous Value of SAP*		
	Low	*Medium*	*High*
Zero	Low	Medium	High
Low	Low	Medium	High
Medium	Medium	High	Very High
High	High	Very High	Very High

The previous value of SAP is from the fuzzy TSK model.

- Baseline SAP < 120 mmHg:
  - *Low*: SAP < 90 mmHg;
  - *Medium*: SAP between 90 and 120 mmHg;
  - *High*: SAP > 120 mmHg.

The effect of the perceived stimulus on the HR is described as a change in the HR value. A small change represents a change of 5% from the previous value, and a large change could be as big as 10%. Table 6.5 shows the rule base for the change in HR as a result of the perceived stimulus.

The previous value of HR is from the fuzzy TSK model.

The HR has lower baseline values when using remifentanil than with other opioids. The HR class values were classified by the anesthesiologist as follows:

- *Low*: HR < 70% of baseline;
- *Medium*: HR between 70% and 90% of baseline;
- *High*: HR > 90% of baseline.

Figure 6.12 shows the actual SAP versus the value of SAP arising from the stimulus model. The results of the model including the stimulus effect give a reasonable approximation to the observed value of SAP considering the multiple disturbances that are inherent in the cardiovascular system. Figure 6.13 shows the result of the patient model after the inclusion of the stimulus effect for the HR. The HR model, similarly, appears to be averaging out disturbances in the signal.

Table 6.5 Rule Base for the New Value of HR After the Stimulus

Perceived stimulus	*Previous Value of HR*		
	Low	*Medium*	*High*
Zero	Zero	Zero	Zero
Low	Zero	Zero	Little
Medium	Zero	Little	Large
High	Little	Little	Large

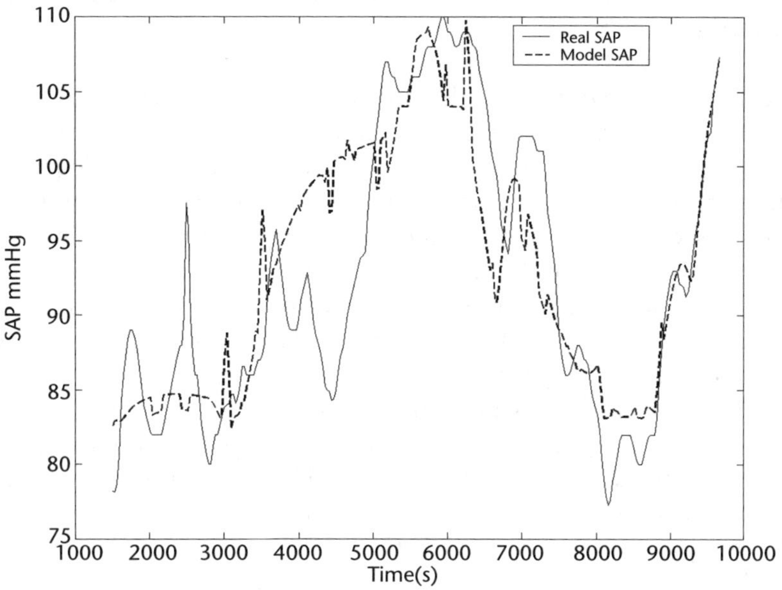

Figure 6.12 Actual SAP (solid line) versus SAP from patient model (dashed line) including the model of stimuli.

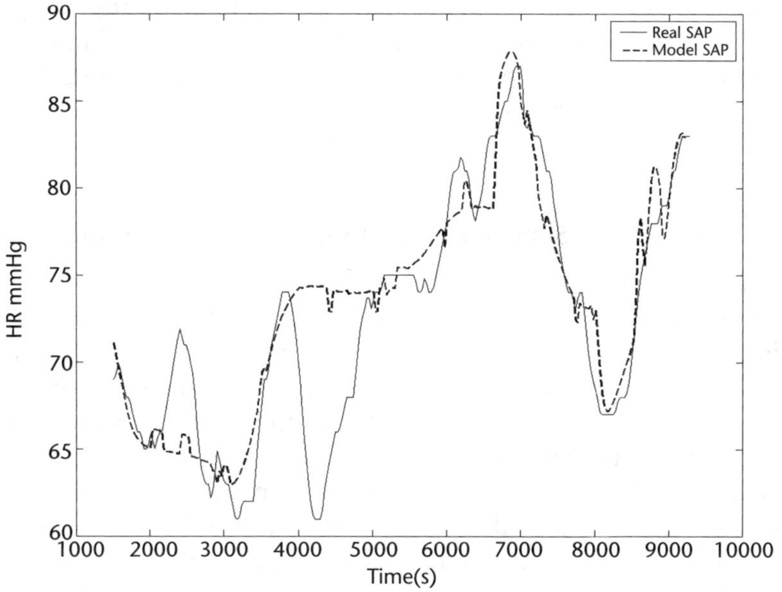

Figure 6.13 Actual HR (solid line) versus HR from patient model (dashed line) including the model of stimuli.

6.6 Exploitation of the Patient Model for Closed-Loop Drug Administration

The infusion rate of the anesthetic drug is titrated according to the patient's requirements, so as to maintain a certain level of DOA. The patient's clinical signs and/or brain signals are used by the anesthetist to determine the adequate infusion rate. In addition, the anesthetist also establishes the required infusion rate of the analgesic drug, based on the patient's response to surgical stimuli. A closed-loop control system of DOA will help the anesthetist adjust simultaneously the infusion rates of the anesthetic and analgesic drugs. The majority of the researches in the area are mainly concerned with the automatic control of the anesthetic drug, whereas the analgesic is controlled manually by the anesthetist.

However, in this research the objective is a multivariable control structure for both drugs. The patient model presented previously described adequately the effects and interactions of the two drugs in the presence of surgical stimuli. This model will be used to develop a control algorithm relating to the administration of both drugs. The study of the interactions between the anesthetic propofol and the analgesic remifentanil helps to determine the ideal combination of infusion rates. This study will also represent a practical guide for the anesthetist, which would help him or her learn how to adjust the amount of drug infused and hence improve the patient's comfort.

First, the patient model will be tested using a series of open-loop simulations with different infusion profiles. Second, the closed-loop structure will be presented, showing the links between the patient model, the DOA classifier (i.e., the FRC presented previously), and the controller. Third, a multivariable fuzzy controller, developed with the anesthetist's cooperation, is also described. This controller establishes the infusion rates of propofol and remifentanil simultaneously based on the level of DOA, the concentrations of the drugs, and the surgical stimuli. The performance of the controller is tested under different conditions.

6.6.1 Open-Loop Simulation Results Using the Patient Model

The patient model previously described was tested in open-loop simulations with different infusion profiles for propofol and remifentanil. Although three different infusion profiles were studied, only one profile will be presented and discussed here, expressing lessons learned from the experiment. The simulations were performed for 7,200 seconds (120 minutes) with a sampling time of 30 seconds. The first 1,500 seconds relate to the induction phase, followed by the maintenance phase. It is worth noting that the recovery phase is not simulated. The FRC for DOA developed previously is applied to the wavelet extracted AEP features, as determined by the maintenance-phase model. In the induction phase, the FRC uses only the cardiovascular parameters, a change in systolic arterial pressure (ΔSAP) and a change in heart rate (ΔHR), for the classification of DOA, since the AEP features are not modeled during this phase.

Figure 6.14 shows the infusion rate profiles for propofol and remifentanil used in the first simulation. The propofol infusion rate is very similar to the profile of

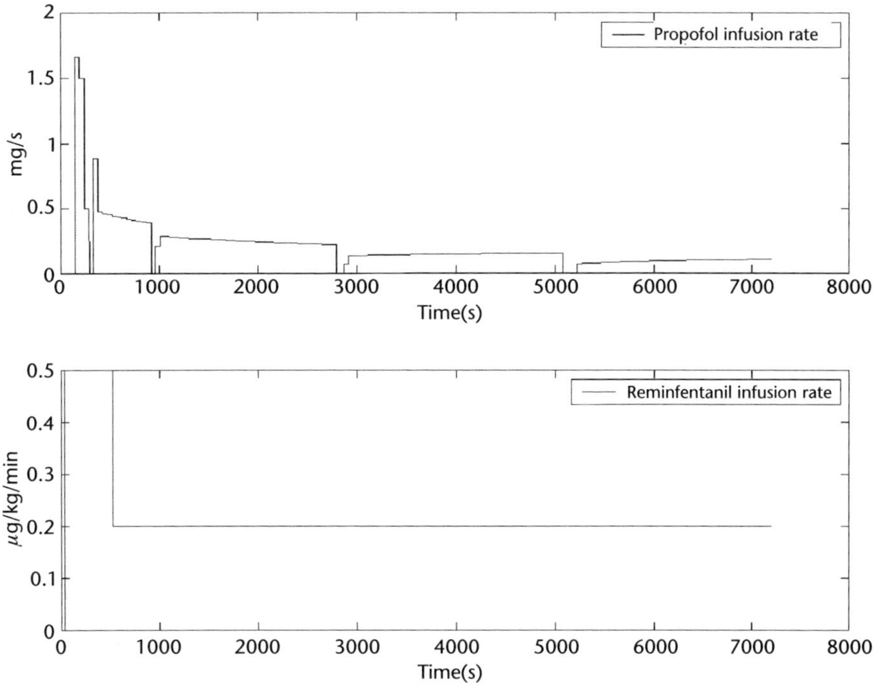

Figure 6.14 Propofol and remifentanil infusion rate profiles.

patient Pat 1, while the remifentanil infusion rate follows a typical profile during a surgical procedure (i.e., high at induction and constant during maintenance).

Figure 6.15 shows the HR and SAP as simulated by the patient model, using the Infusion Profile 1. The wavelet extracted AEP features as simulated by the patient model are used to classify the DOA level by the FRC during the maintenance phase. The DOA level is shown in Figure 6.16. The model performs adequately, describing the effects of the stimulus level on SAP and HR. The effect concentrations are within the ranges of the maintenance phase and this is reflected in the model's response. The DOA level as determined by the FRC cannot be compared with the anesthetist's

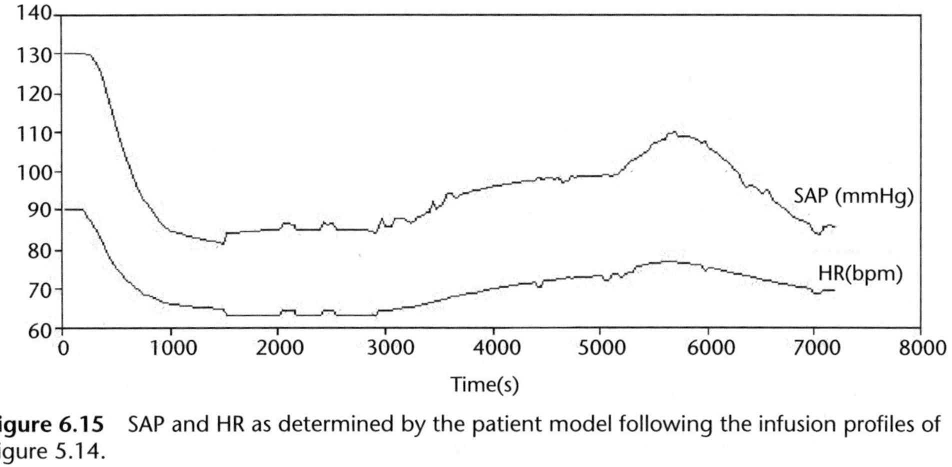

Figure 6.15 SAP and HR as determined by the patient model following the infusion profiles of Figure 5.14.

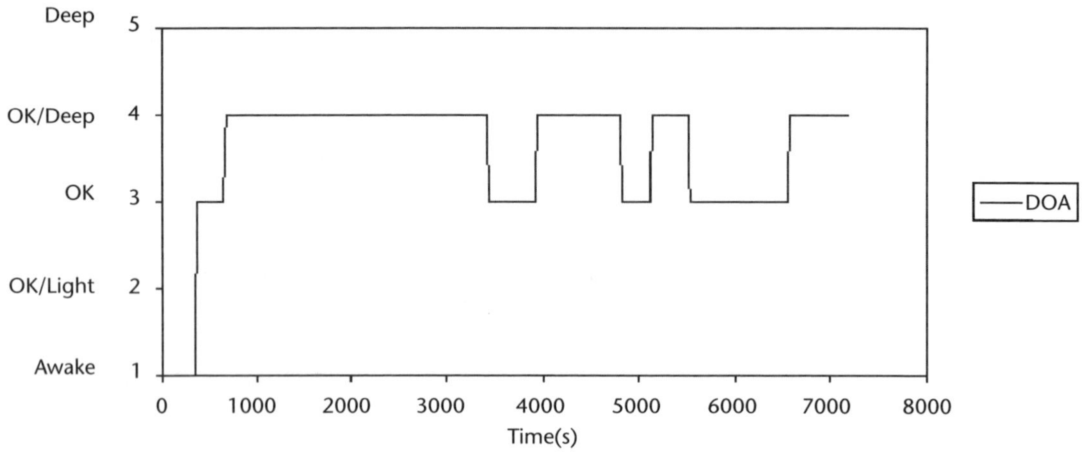

Figure 6.16 DOA as classified by the FRC using the wavelet extracted AEP features.

classification, since this is a simulated infusion profile. However, the classification is not unreasonable as it reflects the AEP features from the model.

6.6.2 Closed-Loop Control Structure

The closed-loop simulation system links the patient model, the FRC of DOA, and the control algorithm. Figure 6.17 shows the block diagram comprising the different components of the closed-loop system during the maintenance phase of anesthesia. The FRC uses the parameters from the patient model to classify DOA. Finally, a control structure maintains an adequate DOA level by adjusting the infusion rates of propofol and remifentanil, which are the inputs of the patient model. The fuzzy controller algorithm is discussed in the next section.

Multivariable Fuzzy Control

In general, propofol is used for maintenance of anesthesia in combination with an opioid; hence, the anesthetist is confronted with the dilemma of whether to vary

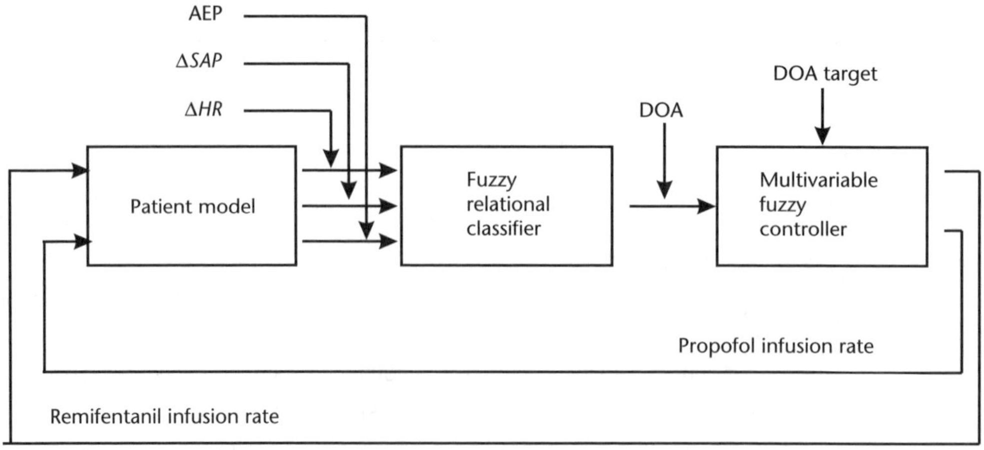

Figure 6.17 A schematic diagram relating to the closed-loop control system.

propofol or the opioid. In [40], Zhang and coworkers reported on a closed-loop system for total intravenous anesthesia by simultaneously administering propofol and fentanyl. They studied the interaction between propofol and fentanyl for loss of response to surgical stimuli using an unweighted least squares nonlinear regression analysis with human data. A look-up table of optimal and awakening combinations of concentrations was built, and it was used to determine the fentanyl set point according to the propofol set point. To our knowledge, this is one of the only studies in simultaneous control of anesthetic and analgesic drugs.

Multivariable fuzzy controllers are not easy to design. There are several studies on the design of multivariable controllers; however, there are very few target biomedical systems. In [41], King and coworkers described an integrated development system that permits the controller designer to test hypotheses, examine the effect of changes in the controller parameters, and perform a complete off-line simulation of a proposed multivariable fuzzy controller. This system was applied to industrial processes and tested off-line with mathematical models. In [42] Linkens and Nyongesa developed a hierarchical multivariable fuzzy controller for learning with GAs [43]. This controller was applied to a simulation case study in anesthesia, using mathematical models describing the action of atracurium and isoflurane (i.e., a muscle relaxant and an inhalational anesthetic). This structure decomposes a complex multivariable fuzzy controller into several simple fuzzy controllers.

The anesthesiologist is confronted with the dilemma of whether to vary propofol or the opioid. Anesthesiologists' knowledge and experience is incorporated into the fuzzy control system as a set of linguistic rules. In addition, the interaction between remifentanil and propofol provides information that can be used to determine an appropriate combination of the two drugs. The controller is activated only at the start of the maintenance phase.

The infusion rates of both drugs are determined according to the DOA level and the surgical stimulus. The controller comprises three different blocks, corresponding to three possible values of DOA: (1) the DOA level is at the target level (*ok*), (2) the DOA is lighter than that desired, or (3) the DOA is deeper than that desired. The controller acts differently according to these three stages. The linguistic scheme of the controller is described as follows:

- If DOA is *ok* then *no change*.
- If DOA is *light* and if:
  - Stimulus present, then increase remifentanil (via remifentanil rule base 1);
  - No stimulus, then increase propofol (via SISO fuzzy PI controller).
- If DOA is *deep* and if:
  - No stimulus and remifentanil *high*, then *decrease* remifentanil;
  - No stimulus and remifentanil *normal* and propofol *high*, then *decrease* propofol;
  - No stimulus and remifentanil *normal* and propofol *normal*, then *decrease* remifentanil;
  - Stimulus and propofol *high*, then *decrease* propofol (via SISO fuzzy PI controller) and *increase* remifentanil (via remifentanil rule base 2);

- Stimulus and propofol *normal*, then *decrease* remifentanil.

The above multivariable controller is based on linguistic rules that interact with three decision tables, two of which are rule bases for a change in remifentanil infusion rate (rule base 1 and rule base 2), and the other represents a SISO fuzzy PI controller for a change in propofol infusion rate. These rule bases are described in the next sections.

It is important to note that a minimum effect concentration of propofol and remifentanil is required at all times in order to ensure unconsciousness and prevent arousal. The minimum values were established as 2,250 ng/ml and 3.5 ng/ml for propofol and remifentanil, respectively. If the concentrations decrease to these minimum values, then the infusion rate of propofol will be increased by 0.2 mg/s and remifentanil will be increased by 0.025 μg/kg/min.

Remifentanil Rule Base 1

This determines the increment in the remifentanil infusion rate when the DOA level is *light* in the presence of a stimulus. The rule base is shown in Table 6.6, and it uses the perceived stimulus and the change in DOA error as inputs. The change in remifentanil infusion rate is normalized between [0,1], but the maximum value of the variable was established as 0.03 μg/kg/min.

Remifentanil Rule Base 2

This is used when the DOA is *deep*, but there is a stimulus present and an increment in the remifentanil infusion rate is necessary. Rule base 2 is shown in Table 6.7. The perceived stimulus level and the change-in-DOA error are used to determine the change in remifentanil infusion rate, similarly to rule base 1.

Table 6.6 Rule Base 1 Describing the Change in Remifentanil Infusion Rate Using the Stimulus Level and the Change in Error of DOA

Stimulus	*Change in DOA Error*			
	Negative	*Zero*	*Positive Small*	*Positive Big*
Low	Small	Small	Medium	Big
Medium	Small	Medium	Medium	Big
High	Medium	Big	Big	Big

Table 6.7 Rule Base 2 Describing the Change in Remifentanil Infusion Rate Using the Stimulus Level and the Change in Error of DOA

Stimulus	*Change in DOA Error*			
	Negative Big	*Negative Small*	*Zero*	*Positive*
Low	Small	Small	Small	Small
Medium	Small	Small	Medium	Medium
High	Small	Medium	Medium	Big

In rule base 2, the smaller increment in the infusion rate is predominant, since the DOA is deep and the remifentanil is increased for its analgesic properties. In rule base 1, the DOA is *light*, and so, remifentanil is increased as a response to the stimulus and for its synergistic properties in order to increase the DOA level.

6.6.3 SISO Fuzzy Proportional Integral Controller for Propofol

The fuzzy PI controller was designed to control the change in the infusion rate of propofol, in order to achieve and maintain a steady level of DOA and to minimize the amount of drug infused. The fuzzy PI controller uses the error (target DOA minus measured DOA) and the change of error as inputs. The controller's rule base is presented in Table 6.8.

The variables are normalized between [–1,1]. The maximum level of change in the infusion rate was set to 4,000 mg/h considering the maximum conditions preset by the anesthesiologist. The input scaling factors of the controller were optimized using a GA [43] (see Appendix 6B for more details on a GA-based optimization sequence) with a performance index given by the following equation:

$$PI_{Index} = \lambda_1 \sum_{k=0}^{N-1} k|e(k)| + \lambda_2 \sum_{k=0}^{N-1} u(k) \tag{6.11}$$

where $e(k)$ is the error, $u(k)$ is the propofol infusion rate, and N is the number of simulation samples. The weighting parameters λ_1 and λ_2 were chosen to place more emphasis on the error or on the infusion rate, so that an ideal balance between them can be reached. The GA was implemented using MATLAB for a population of 40 strings each of length 20, with a probability of crossover of 0.95 and a probability of mutation of 0.06. The optimization was run with parameters $\lambda_1 = 0.4$ and $\lambda_2 = 0.6$, which were found to be representative of the specifications for the control system, giving a bigger weight to the infusion profile. The GA optimization led to the value of 0.375 for the error scaling factor and 7.77 for the change-of-error scaling factor.

Table 6.8 Rule Base of the Fuzzy PI Controller for DOA

DOA error	*Change of DOA Error*				
	NB	*NS*	*ZE*	*PS*	*PB*
NB	NB	NB	NB	NS	ZE
NS	NB	NS	NS	ZE	PS
ZE	NB	NS	ZE	PS	PB
PS	NS	ZE	PS	PS	PB
PB	ZE	PS	PB	PB	PB

The inputs are the error and the change in error, and the output is the change in propofol infusion rate. The membership functions are labeled: negative big (NB), negative small (NS), zero (ZE), positive small (PS), and positive big (PB).

6.6.4 Simulation Results

The multivariable fuzzy controller was used in the closed-loop simulations with the patient model. Note that the controller only starts acting at 1,500 seconds (i.e., after the induction phase).

Simulation 1

In the first simulation shown here, a typical infusion profile of both drugs was used for the induction phase. The DOA level for this simulation is shown in Figure 6.18. The *ok* DOA level was achieved at 1,740 seconds. Figure 6.19 shows the infusion rates of propofol and remifentanil for this simulation.

The *ok* DOA level was achieved rapidly, and the controller maintained efficiently a stable DOA level by keeping both infusion rates constant. In fact, the controller only changed the propofol infusion rate in this simulation. Figure 6.20 shows the variations in the cardiovascular variables during this simulation.

Next, the same conditions as in Simulation 1 were considered, but with a set-point change to *ok/deep* level at 3,000 seconds. Figure 6.21 shows the infusion

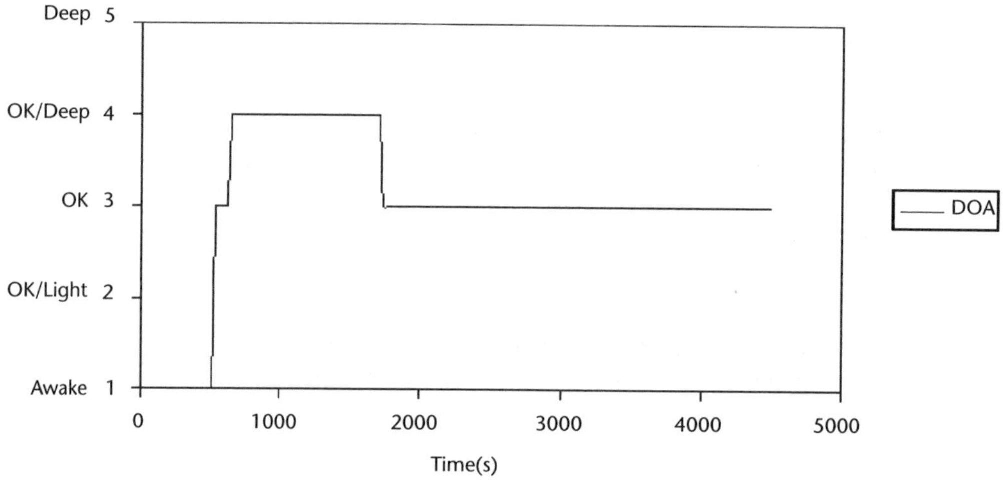

Figure 6.18 DOA level using the multivariable controller in Simulation 1.

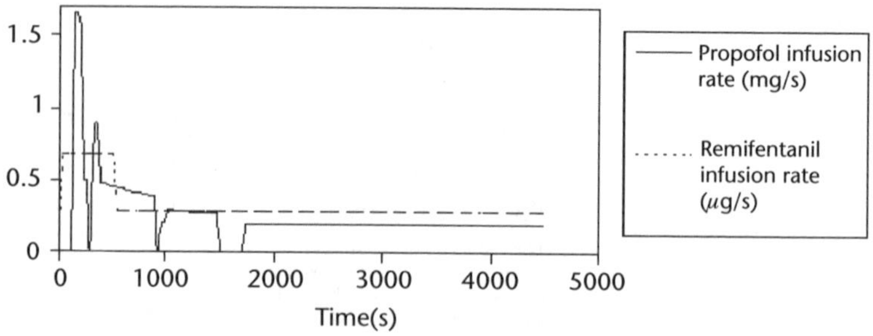

Figure 6.19 Propofol and remifentanil infusion rates as determined by the multivariable controller in Simulation 1.

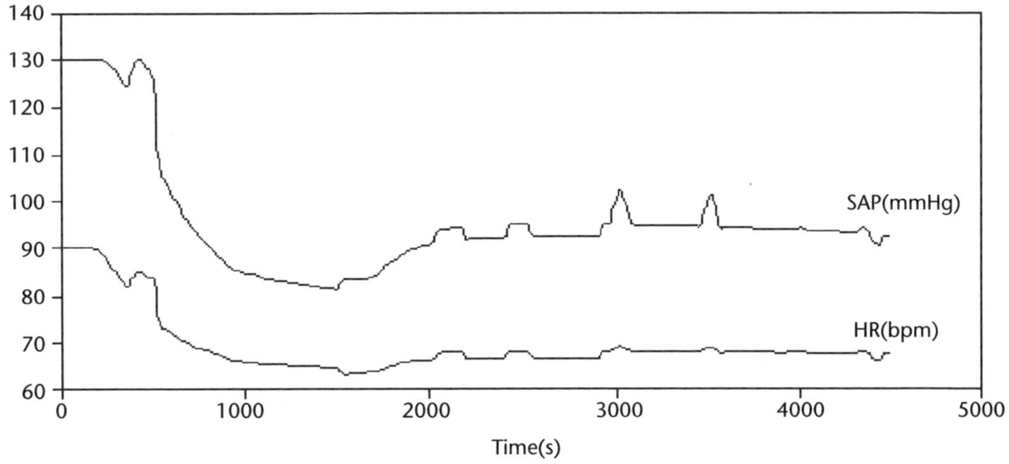

Figure 6.20 SAP and HR signals during Simulation 1.

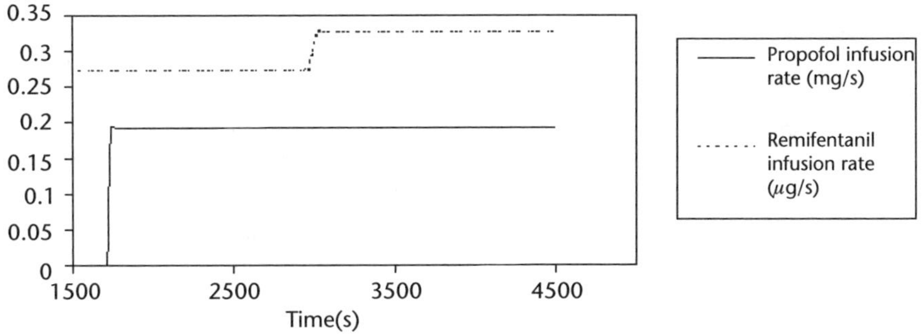

Figure 6.21 Propofol and remifentanil infusion rates as determined by the multivariable controller during the maintenance phase. DOA target change to *ok/deep* at 3,000 seconds.

rates of propofol and remifentanil as determined by the multivariable controller during the maintenance phase. The DOA level is shown in Figure 6.22.

The *ok/deep* DOA level was reached at 3,060 seconds (i.e., only 60 seconds after the set-point change). The multivariable controller reacted to the set-point change by increasing the remifentanil infusion rate, due to the stimulus level present in the system. This increase in the remifentanil infusion rate increased the level of analgesia and also potentiated the effect of propofol. Figure 6.23 shows the SAP during the maintenance phase. The decrease in SAP following an increase in the remifentanil infusion rate shows how a deeper level of depression was achieved, leading to the *ok/deep* DOA level. Therefore, the multivariable controller is taking advantage of the synergism between propofol and remifentanil for efficient control of DOA.

Simulation 2

Simulation 2 considered a different remifentanil infusion profile during the induction phase, while the propofol infusion profile was the same as the one used in Simulation 1, thus producing different initial conditions for the controller. Figure 6.24

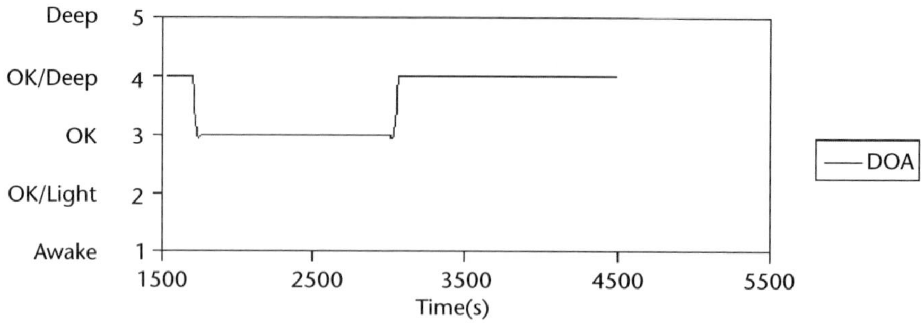

Figure 6.22 DOA level using the multivariable controller during the maintenance phase. DOA target change to *ok/deep* at 3,000 seconds.

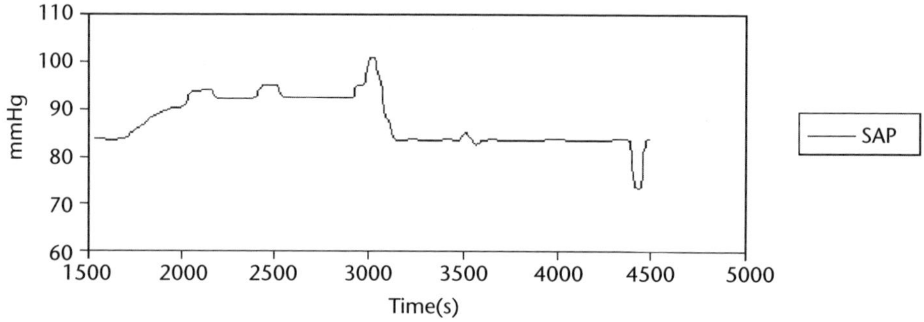

Figure 6.23 SAP during the maintenance phase. DOA target change to *ok/deep* at 3,000 seconds.

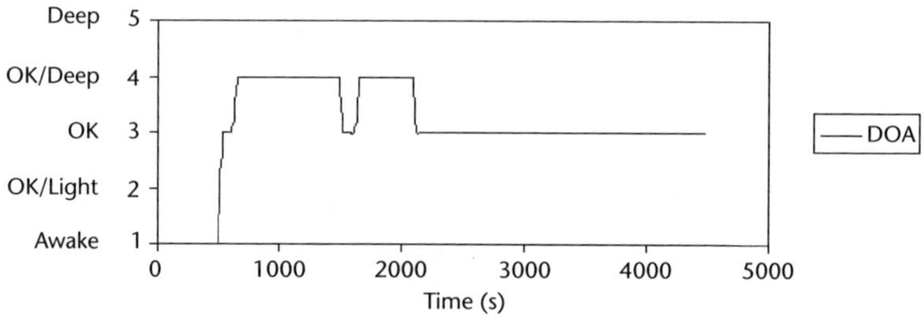

Figure 6.24 DOA level using the multivariable controller in Simulation 2.

shows the DOA level in Simulation 2, while the propofol and remifentanil infusion rates are shown in Figure 6.25. In contrast to the previous simulation, the multivariable controller changed both infusion rates in response to the *ok/deep* DOA level at approximately 1,700 seconds.

Figure 6.26 shows the infusion rate of both drugs during the maintenance phase. The multivariable controller decreased the propofol infusion rate first and then gradually decreased the remifentanil infusion rate. The controller was able to determine an adequate combination of the two drugs, thus achieving and maintaining the *ok* DOA level.

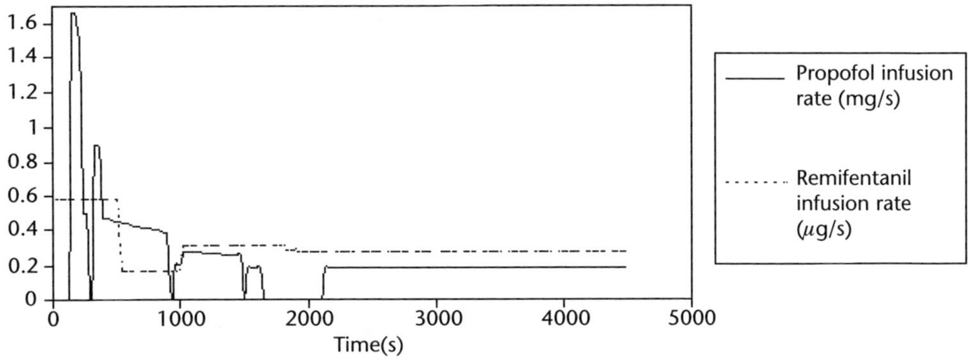

Figure 6.25 Propofol and remifentanil infusion rates as determined by the multivariable controller in Simulation 2.

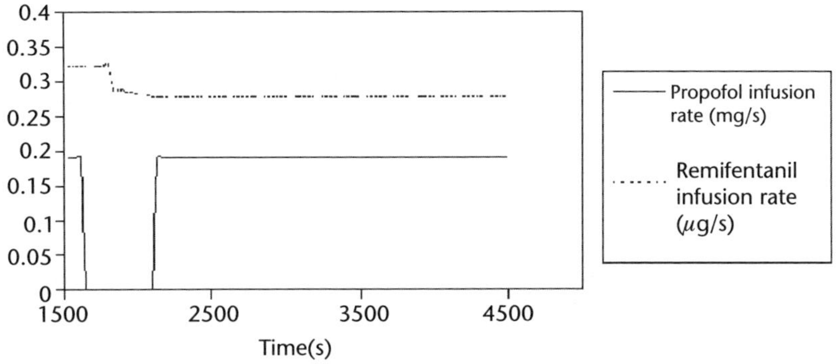

Figure 6.26 Propofol and remifentanil infusion rates as determined by the multivariable controller in Simulation 2 during the maintenance phase (zoom on Figure 6.25).

6.7 Discussions and Conclusions

A simulation platform for DOA has been developed, linking a patient model and a FRC system. This structure is suitable for use in the operating room as a monitoring system for the anesthesiologist, providing information about the patient's vital signs. The first step was the classification of DOA. A set of 10 wavelet extracted AEP features was found to provide the necessary information to distinguish between the DOA levels. The FRC was developed in order to process the AEP features and classify them into DOA levels. An automatic system is required to process the AEPs and translate the relevant information into a reference value, to which the anesthesiologist can relate. Online AEP processing provides the anesthesiologist with information about the CNS depression, and hence, the degree of unconsciousness of the patient. The nature of the FRC makes it adequate for high-dimensional vector features and complex classification processes when a small amount of data is available; this is reflected in the classification results.

A patient model was developed to determine the effects of the drugs propofol and remifentanil on the cardiovascular variables and on the AEP features. The objective was to establish the effects of the interaction of the two drugs and how this affects the required infusion rates to maintain a stable DOA level. It is worth noting that the two drugs interact only in the presence of stimuli, hence the focus on the maintenance phase of anesthesia.

The effect concentration governs the overall behavior, since it is at the effect site that the drug exerts its action. Therefore, a fuzzy structure of models was built to relate the effect concentrations of both drugs with patients' vital signs. Only the pharmacodynamic interactions were considered, since they have a higher clinical importance than the pharmacokinetic interactions, which are obscured by the interpatient variability.

ANFIS was used to adapt TSK fuzzy models for ΔSAP, ΔHR, and the AEP features based on clinical data. The fuzzy models led to acceptable performance and correctly reflected the drugs effect and synergistic interaction. The model has a good performance on the testing data sets and shows how SAP and HR respond to the drug related synergism. Anesthesiologists' experience was used to construct a Mamdani-type fuzzy model describing the effect of surgical stimuli. The results of the stimuli model give a reasonable approximation to the observed values. This was expected, since the stimulus level affects the cardiovascular variables. However, there is a small delay between the model results and the observed SAP and HR. This happens because in reality it takes a few minutes for the stimulus effect to be reflected in the cardiovascular parameters, meaning that the influence of the stimulus is not instantaneous. This time delay has not been included in the model. The observed time delay is different for SAP and HR, and to our knowledge there are no studies defining its value. It was found that (1) the DOA level has some influence on the stimulus effect, (2) the cardiovascular variables include a time delay in response to surgical stimulus, and (3) the stimulus effect is proportional to its intensity.

The AEP features were then analyzed according to stimuli effects. The brain signals responded immediately to stimulus, indicating the degree of arousal. The low and high frequency AEP features responded differently to different stimuli. It is important to establish that there is a relationship between these two signals, and one needs to be aware of its influence on the state of the patient. However, the effects of stimuli on the AEPs were not modeled in detail due to insufficient information (further details can be found in [44–46]).

Overall, the developed patient model performed well, reflecting the effect of the drugs and of the surgical stimuli. This model also represents a typical patient's behavior during a surgical procedure, and should be recognized as such when exploiting it for feedback control of drug administration. Indeed, the objective of a control system for DOA would be to determine the best infusion rates of the anesthesiologist and analgesic drugs, helping the anesthetist to decide which drug should be changed in response to different events. As a result, the developed patient model was used to construct and test a multivariable controller for simultaneous administration of remifentanil and propofol during the maintenance phase. Anesthesiologists' experience was incorporated into the control structure using linguistic rules. According to the different profiles for the DOA level and for the surgical stimuli, the multivariable controller defines the required change in the infusion rates of the two drugs.

The controller was able to adjust the remifentanil infusion rate according to the stimulus intensity, and was able to take advantage of the synergistic interaction to change appropriately the propofol infusion rate. Propofol is titrated to lower infusion rates, decreasing the amount of drug infused, and speeding up recovery. In addition, the controller ensures adequate analgesia by titrating the remifentanil according to the stimulus. The multivariable fuzzy controller was tested under different simulations, and responded efficiently to different induction profiles and set-point changes (further details can be found in [47, 48]).

Fuzzy logic techniques have been shown to be efficient in incorporating human knowledge for better solutions in biomedicine. The complexity of DOA and the unavailability of large data sets make this an ideal application for fuzzy logic based concepts. The FRC, the Takagi-Sugeno-Kang pharmacodynamic models, and the multivariable fuzzy controller were combined successfully in an integrated structure to provide a closed-loop simulation platform for future anesthesia monitoring and control in the operating theater during surgery.

References

[1] Thornton C., and J. Jones, "Evaluating Depth of Anesthesia: Review of Methods," *Int. Anesthesiol. Clin.*, Vol. 31, 1993, pp. 67–88.

[2] Stanski. D. R., "Monitoring Depth of Anesthesia," in R. D. Miller, (ed.), *Anesthesia*, New York: Churchill-Levingstone, 1990, pp. 1127–1159.

[3] Nayak A, and R. J. Roy, "Anesthesia Control Using Midlatency Auditory Evoked Potentials," *IEEE Trans. on Biomedical Engineering*, Vol. 45, 1998, pp. 409–421.

[4] Kenny, G. N., and H. Mantzaridis, "Closed-Loop Control of Propofol Anaesthesia," *Br. J. Anaesth.*, Vol. 83, 1999, pp. 223–228.

[5] Kumar A., A. Bhattacharya, and N. Makhija, "Evoked Potential Monitoring in Anesthesia and Analgesia," *Anesthesia*, Vol. 55, 2000, pp. 225–241.

[6] Thornton C., et al., "The Auditory Evoked Response as an Indicator of Awareness," *Br. J. Anaesth.*, Vol. 63, 1989, pp. 113–115.

[7] Thorntonm, C., and R. M. Sharpe, "Evoked Responses in Anaesthesia," *Br. J. Anaesth.*, Vol. 81, 1998, pp. 771–781.

[8] Jones, P. G., "Perception and Memory During General Anaesthesia," *Br. J. Anaesth.*, Vol. 73, 1994, pp. 31–37.

[9] Berenbaum, M., "What Is Synergy?" *Pharmacological Reviews*, Vol. 41, 1989, pp. 93–141.

[10] Minto, C. F., et al., "Response Surface Model for Anesthetic Drug Interactions," *Anesthesiology*, Vol. 92, 2000, pp. 1603–1616.

[11] Shieh, J. S., D. A. Linkens, and J. E. Peacock, "Hierarchical Rule-Based and Self-Organizing Fuzzy Logic Control for Depth of Anesthesia," *IEEE Trans. on Systems, Man and Cybernetics*, Part-C, 1999, Vol. 29, pp. 98–109.

[12] Vuyk, J., "Drug Interactions in Anesthesia," *Minerva Anestesiol*, Vol. 65, 1999, pp. 215–218.

[13] Vuyk, J., "The Effect of Opioids on the Pharmacokinetics and Pharmacodynamics of Propofol," in J. Vuyk, F. Engbers, and S. Groen-Mulder, (eds.), *On the Study and Practice of Intravenous Anesthesia*, Boston, MA: Kluwer Academic Publishers, 2000, pp. 99–111.

[14] Olivier, P., et al., "Continuous Infusion of Remifentanil and Target-Controlled Infusion of Propofol for Patients Undergoing Cardiac Surgery: A New Approach for Scheduler Early Extubation," *J. Cardiothorac. Vasc. Anesth.*, Vol. 14, 2000, pp. 29–35.

[15] Takagi, T., and M. Sugeno, "Fuzzy Identification of Systems and Its Applications to Modeling and Control," *IEEE Trans. Syst. Man. Cybern.*, Vol. SMC-15, 1985, pp. 116–132.

[16] Uhl, R. R., et al., "Effect of Halothane Anesthesia on the Human Cortical Visual Evoked Response," *Anesthesiology*, Vol. 53, No. 4, 1980, pp. 273–276.

[17] Thornton, C., "Mesure du Niveau de Conscience et de la 'Profondeur de l'Anesthésie'," *Cahiers d'Anesthésiologie*, Vol. 38, No. 7, pp. 457–475.

[18] Thornton, C., "Evoked Potentials in Anesthesia," *European Journal of Anesthesiology*, Vol. 8, No. 2, 1991, pp. 89–107.

[19] Sebel, P. S., P. J. Flynn, and D. A. Ingram, "Effect of Nitrous Oxide on Visual, Auditory and Somatosensory Evoked Potentials," *British Journal of Anaesthesia*, Vol. 56, 1984, pp. 1403–1407.

[20] Samra, S. K., et al., "Remifentanil- and Fentanyl-Based Anesthesia for Intraoperative Monitoring of Somatosensory Evoked Potentials," *Anesthesia Analgesia*, 2001, Vol. 92, No. 6, 2001, pp. 1510–1515.

[21] Thornton, C., and D. E. F. Newton, "The Auditory Evoked Response: A Measure of Depth of Anesthesia," *Baillière's Clinical Anesthesiology*, Vol. 3, No. 3, 1989, pp. 559–85.

[22] Thornton, C., et al., "Enflurane Anaesthesia Causes Graded Changes in the Brainstem and Early Cortical Auditory Evoked Response in Man," *British Journal of Anaesthesia*, Vol. 55, 1983, pp. 479–485.

[23] James, M. F. M., C. Thornton, and J. G. Jones, "Halothane Anaesthesia Changes the Early Components of the Auditory Evoked Response in Man," *British Journal of Anaesthesia*, Vol. 54, 1982, p. 787.

[24] Navaratnarajah, M., et al., "Effect of Etomidate on the Auditory Evoked Response in Man," *British Journal of Anaesthesia*, Vol. 55, 1983, pp. 1157P–1158P.

[25] Navaratnarajah, M., et al., "Effects of Isoflurane on the Auditory Evoked Responses in Man," *British Journal of Anaesthesia*, Vol. 57, 1985, p. 352P.

[26] Thornton, C., et al., "Effects of Halothane or Enflurane with Controlled Ventilation on Auditory Evoked Potentials," *British Journal of Anaesthesia*, Vol. 56, 1984, pp. 315–323.

[27] Thornton, C., et al., "Effect of Etomidate on the Auditory Evoked Response in Man," *British Journal of Anaesthesia*, Vol. 57, 1985, pp. 554–561.

[28] Thornton, C., et al., "Selective Effect of Althesin on the Auditory Evoked Response in Man," *British Journal of Anaesthesia*, Vol. 58, 1986, pp. 422–427.

[29] Aunon, J. I., C. D. McGillem, and D. G. Childers, "Signal Processing in Evoked Potential Research: Averaging and Modeling," *CRC Critical Reviews in Bioengineering*, Vol. 5, 1981, pp. 323–367.

[30] Bertrand, O., J. Bohorquez, and J. Pernier, "Time-Frequency Digital Filtering Based on an Invertible Wavelet Transform: An Application to Evoked Potentials," *IEEE Trans. on Biomedical Engineering*, Vol. 41, No. 1, 1994, pp. 77–88.

[31] Daubechies, I., "Orthonormal Bases of Compactly Supported Wavelets," *Communications on Pure and Applied Mathematics*, Vol. 41, 1988, pp. 909–996.

[32] Setnes, M., and R. Babuska, "Fuzzy Relational Classifier Trained by Fuzzy Clustering," *IEEE Trans. on Systems, Man and Cybernetics*, Part-B, Vol. 29, 1999, pp. 619–625.

[33] Xie, X. L., and G. Beni, "A Validity Measure for Fuzzy Clustering," *IEEE Trans. on Pattern Analysis and Machine Intelligence*, Vol. 13, 1991, pp. 841–847.

[34] Gepts, E., et al., "Disposition of Propofol Administered as Constant Rate Intravenous Infusions in Humans," *Anesth. Analg.*, Vol. 66, 1987, pp. 1256–1263.

[35] Marsh, B., M. N. M. White, and G. N. C. Kenny, "Pharmacokinetic Model Driven Infusion of Propofol in Children," *Br. J. Anaesth.*, Vol. 67, 1991, pp. 41–48.

[36] Egan T. D., et al., "The Pharmacokinetics of the New Short-Acting Opioid Remifentanil (GI87084B) in Healthy Adult Male Volunteers," *Anesthesiology*, Vol. 79, 1993, pp. 881–892.

[37] Mahfouf, M., D. A. Linkens, and J. E. Peacock, *Propofol Induced Anaesthesia: A Comparative Control Study Using a Derived Pharmacokinetic/Pharmacodynamic Model—Part II: Results of Experiments,* Research Report, Department of Automatic Control and Systems Engineering, The University of Sheffield, 1995.

[38] Vuyk, J., et al., "Propofol Anesthesia and Rational Opioid Selection: Determination of Optimal EC50-EC95 Propofol-Opioid Concentrations That Assure Adequate Anesthesia and a Rapid Return of Consciousness," *Anesthesiology*, Vol. 87, 1997, pp. 1549–1562.

[39] Jang, J. R., "ANFIS: Adaptive-Network-Based Fuzzy Inference System," *IEEE Trans. on Systems, Man, and Cybernetics*, Vol. 23, 1993, pp. 665–685.

[40] Zhang, X. S., R. J. Roy, and J. W. Huang, "Closed-Loop System for Total Intravenous Anesthesia by Simultaneously Administering Two Anesthetic Drugs," in H. K. Chang, and Y. T. Zhang, (eds.), *Proceedings of the 20th Annual International Conference of the IEEE Engineering in Medicine and Biology Society*, Vol. 20, Biomedical Engineering Towards the Year 2000 and Beyond, Hong Kong, October 29–November 1, 1998, Piscataway, NJ: IEEE Service Center, pp. 3052–3055.

[41] King, R. E., G. D. Magoulas, and A. A. Stathaki, "Multivariable Fuzzy Controller Design," *Control Eng. Practice*, 1994, Vol. 2, pp. 431–437.

[42] Linkens, D. A., and H. O. Nyongesa, "A Hierarchical Multivariable Fuzzy Controller for Learning with Genetic Algorithms," *Int. J. Control.*, 1996, Vol. 63, pp. 865–883.

[43] Goldberg, D. E., *Genetic Algorithms in Search, Optimization, and Machine Learning*, Reading, MA: Addison-Wesley, 1989.

[44] Nunes, C. S., M. Mahfouf, and D. A. Linkens, "A Fuzzy Relational Classifier Applied to Depth of Anaesthesia," *Proc. of the 7th UK Workshop on Fuzzy Systems*, Whirlow Grange Conference Centre, Sheffield, U.K., October 26–27, 2000, Vol. 2, pp. 100–105.

[45] Nunes, C. S., M. Mahfouf, and D. A. Linkens, "Fuzzy Logic Monitoring in Anesthesia." *Proc. of the Sixth Portuguese Conference on Biomedical Engineering, BioEng 2001,* Faro, Portugal, June 11–12, 2001, pp. 71–72.

[46] Nunes, C. S., M. Mahfouf, and D. A. Linkens, "Neuro-Fuzzy Modeling in Anesthesia," *Proc. of the Fifth Portuguese Conference on Automatic Control, CONTROLO2002*, Aveiro, Portugal, September 5–7, 2002.

[47] Mahfouf, M., C. S. Nunes, and D. A., Linkens, "Closed Loop Control for Depth of Anesthesia Using a Fuzzy PI Structure," *Proc. of the International Conference on Neural Networks and Expert Systems in Medicine and Health Care, NNESMED 2001*, Milos Island, Greece, June 20–22, 2001, pp. 342–346.

[48] Nunes, C. S., M. Mahfouf, and D. A. Linkens, "Fuzzy Control of Depth of Anesthesia," *Proc. of 2001 UK Workshop on Computational Intelligence, UKCI-01*, Edinburgh-First Conference Centre, Edinburgh, U.K., September 10–12, 2000, pp. 99–104.

[49] Babuska, R., *Fuzzy Modeling and Control*, Boston, MA: Kluwer Academic Publishers, 1998.

[50] Wang, L. X., *A Course in Fuzzy Systems and Control,* Englewood Cliffs, NJ: Prentice Hall International, 1997.

[51] Holland, J. H., *Adaptation in Natural and Artificial Systems*, Reading, MA: Addison-Wesley, 1975.

[52] Nyongesa, H., "Genetic Based Machine Learning Allied to Multi-Variable Fuzzy Control of Anaesthesia," Ph.D. thesis, University of Sheffield, Dept. of Automatic Control and Systems Engineering, 1994.

Appendix 6A: Fuzzy Clustering—The Fuzzy C-Means Algorithm [49]

A large family of fuzzy clustering algorithms is based on minimization of the *fuzzy c-means functional* formulated as follows:

$$J(Z,U,V) = \sum_{i=1}^{c} \sum_{k=1}^{N} (\mu_{ik})^m \|z_k - v_i\|^2 \tag{6A.1}$$

where:

$U = [\mu_{ik}] \in M_{fc}$ is the fuzzy partition matrix of the original data set Z, μ_{ik} is the degree of membership of a particular data point from Z in the fuzzy partitioning.

$V = [v_1, v_2, \ldots, v_c], v_i \in R^n$ is the vector of cluster prototypes (cluster centers or cluster means).

$D_{ik}^2 = \|z_k - v_i\|^2 = (z_k - v_i)^T (z_k - v_i)$ is the squared distance norm.

$m \in [1, \infty]$ is the weighting exponent ($m = 1$ means that the partition is "hard," whereas $m \to \infty$ implies that the partition is fuzzy).

The minimization of the *c-means functional* (6A.1) represents a nonlinear optimization problem that can be solved using a variety of methods, such as Genetic Algorithms. However, using Lagrange multipliers (not given here), objective (6A.1) is minimized only if

$$\mu_{ik} = \frac{1}{\sum_{j=1}^{c} \left(\frac{D_{ik}}{D_{jk}} \right)^{\frac{2}{(m-1)}}} \quad 1 \le i \le c \quad 1 \le k \le N \tag{6A.2}$$

$$v_i = \frac{\sum_{k=1}^{N} (\mu_{ik})^m z_k}{\sum_{k=1}^{N} (\mu_{ik})^m} \quad 1 \le i \le c \tag{6A.3}$$

Equation (6A.2) gives v_i as the weighted mean of the data points that belong to the cluster, where the weights are the membership degrees (that is why the algorithm is called *c-means*). The FCM algorithm iterates through (6A.2) and (6A.3) as follows.

Fuzzy C-Means (FCM) Algorithm

Given the data set Z, choose the number of clusters $1 < c < N$, the weighting exponent $m > 1$, the termination tolerance $\varepsilon = 0$. Initialize the partition matrix randomly, such that $U^{(0)} \in M_{fc}$.

Repeat for $l = 1, 2, \ldots$.

Step 1. Compute the cluster prototypes (centers):

$$v_i^{(l)} = \frac{\sum_{k=1}^{N} \left(\mu_{ik}^{(l-1)} \right)^m z_k}{\sum_{k=1}^{N} \left(\mu_{ik}^{(l-1)} \right)^m} \quad 1 \le i \le c$$

Step 2. Compute the distances:

$$D_{ik}^2 = \left(z_k - v_i^{(l)}\right)^T \left(z_k - v_i^{(l)}\right)$$
$$1 \le i \le c$$
$$1 \le k \le N$$

Step 3. Update the partition matrix U:

$$\text{If } D_{ik} > 0 \; \mu_{ik}^{(l)} = \frac{1}{\sum_{j=1}^{c} \left(\frac{D_{ik}}{D_{jk}}\right)^{\frac{2}{(m-1)}}}$$

$$\text{If } D_{ik} = 0 \; \mu_{ik} = 0, \; \mu_{ik} \in [0,1], \sum_{i=1}^{c} \mu_{ik}^{(l)} = 1$$

Until $\left\| U^{(l)} - U^{(l-1)} \right\| < \varepsilon$.

Example of Two-Dimensional Data [50]

Suppose that X consists of 15 points in R^2, as shown in Figure 6A.1. These data points look well distributed in this two-dimensional space, except for x_8. Indeed, in the case of two clusters, for instance, x_1 to x_7 would form the first grouping, x_9 to x_{15} would form the second grouping, and x_8 is the bridge between the two groupings. The hard c-means algorithm would group x_8 either with the first grouping or the second grouping, leading hence to an asymmetric matrix U.

Choosing

$$c = 2,\; m = 1.30,\; \varepsilon = 0.001, \quad \text{and } U^{(0)} = \begin{bmatrix} 0.8 & 0.2 & 0.8 & \ldots & 0.8 \\ 0.2 & 0.8 & 0.2 & \ldots & 0.2 \end{bmatrix}_{2\,X15}$$

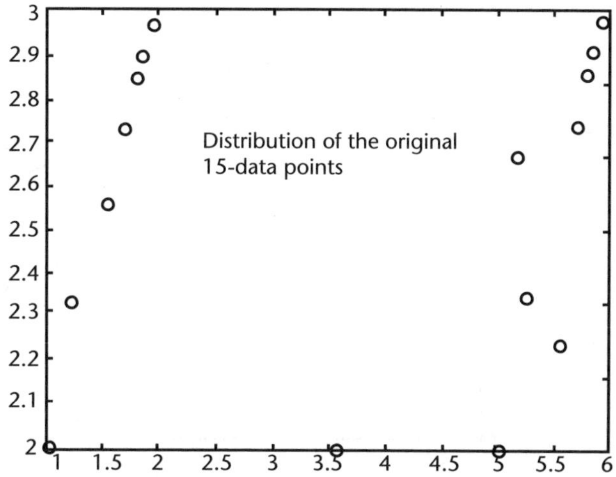

Figure 6A.1 Distribution of the original data in a two-dimensional space.

The FCM algorithm, written in C-Code, stopped at iteration $l = 8$ with the following membership values:

$$U^{(7)} = \begin{bmatrix} 0 & 0 & 0 & 0 & 0 & 0 & 0 & 0.56 & 1 & 1 & 1 & 1 & 1 & 1 & 1 \\ 1 & 1 & 1 & 1 & 1 & 1 & 1 & 0.44 & 0 & 0 & 0 & 0 & 0 & 0 & 0 \end{bmatrix}$$

Centers were at v_0 (1.68, 2.58) to the left of the plane and v_1 (5.46, 2.53) to the right of the plane.

The data appears to be well classified, with the bridge x_8 belonging to both clusters with the similarly close degrees of membership (see Figure 6A.2).

Appendix 6B: Genetic Algorithms

Genetic Algorithms are exploratory search and optimization methods that were devised on the principles of natural evolution and population genetics. Holland first developed the technique of GA [51], and several other research studies provided a comprehensive review and introduction of the concept [43]. Unlike other optimization techniques, GA does not require gradients, but instead relies on a function, better known as a *fitness function*, in order to assess the fitness of a particular solution to the problem in question. Possible solution candidates are represented by a population of individuals (generation), and each individual (chromosome) is encoded as a binary string containing a well-defined number of genes (1s and 0s) as Figure 6B.1 shows.

Initially, a population of individuals is generated and the fittest individuals are chosen by ranking them according to a priori-defined fitness-function, which is evaluated for each member of this population. In order to create another better population from the initial one, a mating process is carried out among the fittest individuals in the previous generation, since the relative fitness of each individual is used as a criterion for choice. Hence, the selected individuals are randomly combined in pairs to

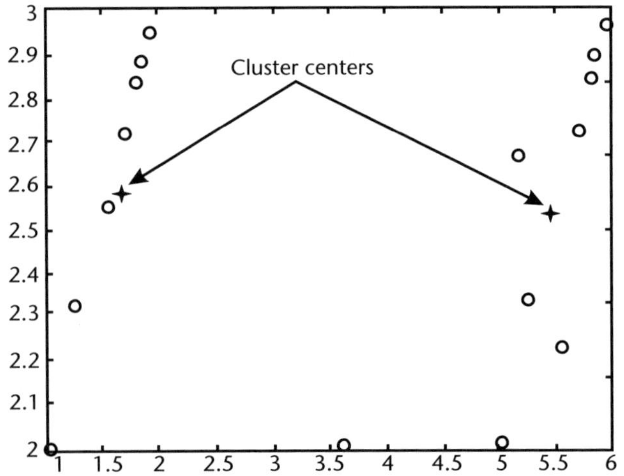

Figure 6A.2 Coordinates of the two cluster centers following optimization via the fuzzy c-means algorithm.

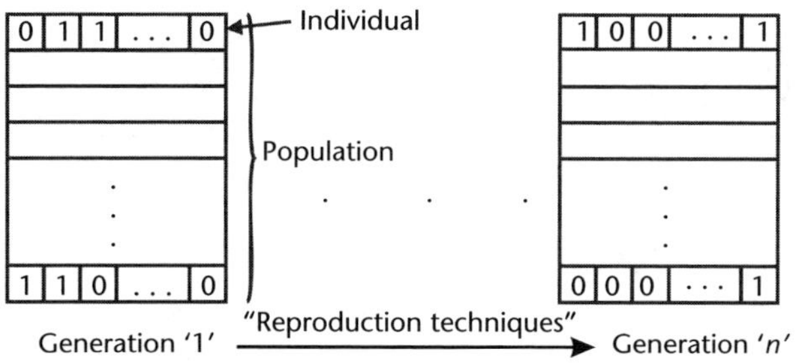

Figure 6B.1 A schematic diagram illustrating the philosophy behind a generation, a population, an individual, and the binary coding of each individual in a GA.

produce an offspring by *crossover* parts of their chromosomes at a randomly chosen position of the string. The new offspring is supposed to represent a better solution to the problem. In order to provide extra excitation to the process of generation, randomly chosen bits in the strings are inverted (0s to 1s and 1s to 0s). This mechanism is known as *mutation* and helps to speed up convergence and prevents the population from being predominated by the same individuals. All in all, it ensures that the solution set is never empty. A compromise should be reached, however, between too much excitation and none by choosing a small probability of mutation (Figure 6B.2).

It was pointed out that the successful running of a GA involves having to set a number of control parameters, which include population size; the nature and rates of the recombination operators; and crossover, mutation, and reproduction. Reproduction is defined as the process through which *parent structures* are selected to form new offspring, by applying the above genetic operators, which can then replace members of the old generation. The method of selecting an individual to produce offspring (or to be deleted from the population) determines its life span and

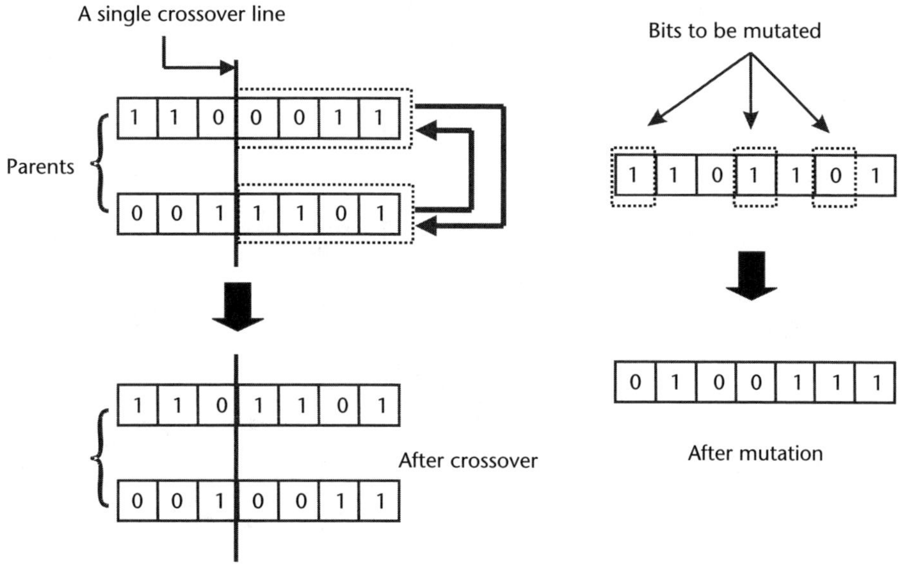

Figure 6B.2 A schematic diagram illustrating the various GA operators.

the number of its offspring. The most common reproduction techniques are Generational Replacement (GR), Steady-State (SS), Generational Gap (GG), and Selective Breeding (SB) [52]. Figure 6B.3 illustrates a GA-based optimization sequence.

Appendix 6C: The ANFIS Architecture

ANFISs are used for tuning the rules of fuzzy systems automatically, given sets of data that relate to the mapping of inputs and outputs under investigation. To explain this particular architecture, let us consider the simple example of two inputs

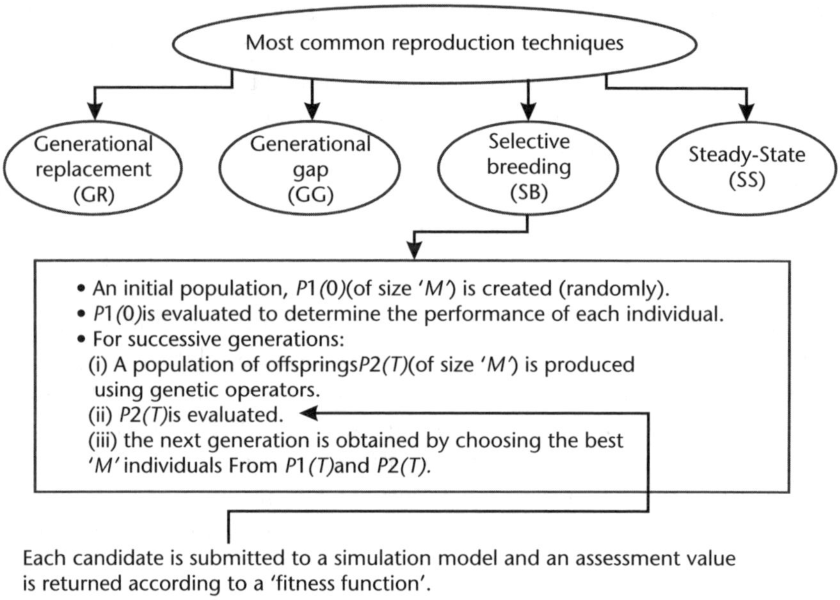

Figure 6B.3 Successive operations involved in a GA calculation sequence.

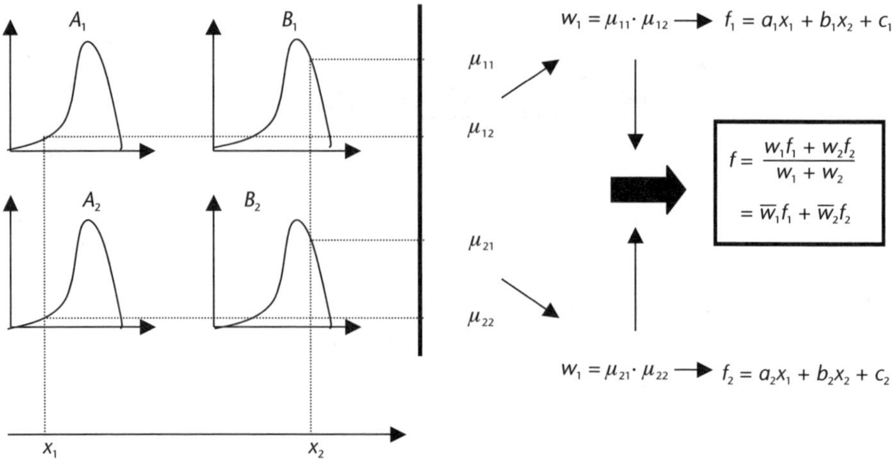

Figure 6C.1 The fuzzy inference mechanism using TSK models.

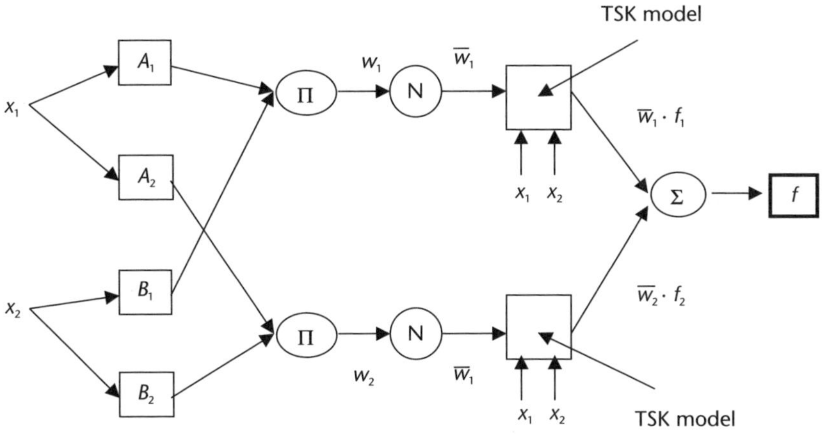

Figure 6C.2 The ANFIS architecture, which uses TSK models.

x_1 and x_2, and one output f, with the rules expressed using the TSK model rather than the Mamdani model (see Chapter 5):

$$\text{Rule 1: If } x_1 \text{ is } A_1 \text{ and } x_2 \text{ is } B_1\text{, then } f_1 = a_1x_1 + b_1x_2 + c_1 \tag{6C.1}$$

$$\text{Rule 2: If } x_1 \text{ is } A_2 \text{ and } x_2 \text{ is } B_2\text{, then } f_2 = a_2x_1 + b_2x_2 + c_2 \tag{6C.2}$$

Graphically, the above rules can be illustrated by Figure 6C.1. All parameters relating to the fuzzy membership functions A_i, B_i and those relating to the TSK models [i.e., a_i, b_i, and c_i—see (6C.1) and (6C.2)] can be optimized using the neural network architecture of Figure 6C.2.

CHAPTER 7

Intelligent Modeling and Decision Support in the General Intensive Care Unit

With the aim of developing a hybrid knowledge and model-based intelligent advisory system for intensive care ventilators that will be implemented clinically, this chapter proposes to shed more light into the complex dynamics involved in the interactions between patient and ventilator. In addition to introducing some of the components relating to the model of patient/ventilator system, this chapter reviews the remaining components of the decision support architecture, which are formulated using a hybrid approach combining the themes of fuzzy and neural-fuzzy systems as well as mechanistic formulations. The end product is called SIVA: An Intelligent Advisory System for Intensive Care Ventilators.

7.1 Introduction

Critically ill patients in the ICU often develop respiratory failure, and they are assisted with their breathing by mechanical ventilation (also referred to as artificial ventilation), which is a procedure that replaces or supplements the function of the respiratory muscles during respiration via mechanical means. Most commonly, this is the result of a lung disease such as pneumonia, or as a result of lung damage secondary to a process that began elsewhere in the body but has led to lung dysfunction as part of multiorgan failure. The patients' requirements will vary depending on the severity of their lung disease, their degree of sedation, and whether it is appropriate to reduce[1] and eventually withdraw their ventilatory support to allow them to breathe normally once again and without assistance. Failure to adequately adjust the mode or parameters of ventilatory support will endemically result in further problems. These will include:

- Failure to oxygenate;
- Failure to produce adequate lung expansion;
- Overexpansion of the lung and exposure of the lung to high-pressure damage.

Once the patients reach a stage in their critical illness where, although the load on their respiratory muscles is high or they are relatively weak [1], it is appropriate for them to trigger assistance from the ventilator with their own breaths, at which

1. This is also known as the "weaning" period in which assistance with mechanical ventilation is gradually suppressed.

time it is important to tailor the amount of respiratory support they receive to their requirements. On the one hand, too much support will switch off their own muscles, leading to atrophy,[2] and on the other, too little of such support will result in fatigue, slowing the weaning from the ventilator and promoting areas of collapse and consolidation of the lungs.

The normal process of setting the ventilator is labor-intensive and represents a feedback decision support environment, requiring manual adjustment(s) of ventilator settings. In a previous study conducted by Morgan [2], only in 22% to 33% of cases did the clinical team respond to an abnormal blood oxygen partial pressure result within 30 minutes. This is hardly surprising since the ICU clinical team has to simultaneously look after other equally important patients' needs. The ventilator settings include the mode of ventilation, depending on the nature of the respiratory failure, and inspired oxygen, depending on arterial oxygen or pulse-oximeter readings. Respiration rate, positive end-expiratory pressures (PEEP), and peak pressures are adjusted in relation to oxygenation, carbon dioxide levels, and tidal volume. Inspiratory flow rate influences the patient's spontaneous breathing rate [3, 4]. The ratio of inspiratory to expiratory time depends on expiratory flow, oxygenation, and the nature of the lung disease. The inspiratory time of the ventilator and that of the patient need to be matched to allow efficient support for the patient [5, 6]. The degree of pressure assistance during breathing will be determined by the following variables:

- Respiratory rate;
- Tidal volume;
- Indicators of likely impending fatigue, such as the pressure change by the patient in the first 0.1 second of a breath [7].

There have been numerous studies examining predictors of weaning success [8–10] based on the above indicators. Some of these studies have allowed one to extrapolate which aspects of ventilatory support should be altered in response to changes in the respiratory system. However, the ventilator settings are frequently delayed, and hence, patient ventilation is suboptimized, hence the need for an optimal control strategy, which was described as the most pressing practical need in ventilation at several meetings and symposia in Europe and in the United States.

Gradual advances in treating the underlying critical illness are slowly being made. However, advances in attempting to preserve lung function and avoid the adverse effects of mechanical ventilation are progressing more rapidly. Unfortunately, it may be difficult for busy clinical staff to either keep abreast of, or implement these developments, and to compound these problems, much still remains unvalidated by proper clinical trials. Despite this, Ely [11] showed that consistent procedures, which are based on medical knowledge and directed towards good weaning protocols, can reduce the period of ventilation by more than 24 hours! However, given the constraints under which ICUs operate, ventilation cannot be customized for every patient at 1-minute intervals. Therefore, the need for an adap-

2. This refers to the decrease in size and wasting of muscle tissue. Muscles that lose their nerve supply can atrophy and waste away.

tive advisory support system to ensure optimal ventilation in both the acute phase of respiratory failure and during weaning from mechanical ventilation is crucial. Such a system has the potential to impact on morbidity as well as mortality by reducing the duration of mechanical ventilation [12] or by ensuring that lung protective strategies for mechanical ventilation are always employed. Indeed, it is estimated that one single day saved in terms of mechanical ventilation will reduce morbidity, particularly the number of pneumonias that occur due to the susceptibility of the critically ill to additional opportunistic infections, which increase by 10% per day while patients are artificially ventilated. Moreover, such a system could have an impact on mortality or death rate in critically ill patients, which is currently between 20% and 30%.

The little research work hitherto conducted on the development of these systems by several groups worldwide is already starting to bear fruit [12–15]. Such understanding and hence behavior prediction will also lead to the design of more effective strategies for preventing lung collapse and subsequent barotraumas, which currently include the open-lung approach [16] and prone ventilation [17], while restricting overdistension [18] during the acute phase of mechanical ventilation. The development and application of feedback control to mechanical ventilation is still in its early stages. Particular work in this area has focused mainly on knowledge-based systems [12], which can potentially provide a framework for application and synthesis of the latest respiratory and therapeutic knowledge to actual ventilatory care [13]. Early work in this area focused on the control of individual decoupled variables, such as oxygen levels [19] with systems being in the potentially less complicated environment of the operating theater in patients with a "more normal" physiology than those normally encountered in the ICU. Later, such work expanded into other areas such as decision support for weaning from mechanical ventilation. For instance, Adlassnig has integrated this type of system into the clinical care of one hospital via fuzzy open-loop systems such as FuzzyARDS and FuzzyKB [20]. Nemoto et al. [21] have shown that automatic control can at least match the human operator when controlling pressure support ventilation. However, trials to determine the clinical advantages of knowledge-based weaning systems are rare, as only a few workers have demonstrated the possibility of optimizing patient ventilation in a clinical environment [15]. Moreover, little is known about specific aspects of the weaning problem, for example, the patient-ventilator interactions that limit the development of knowledge-based systems.

The study presented in this chapter will comprise two facets, which consist of actual milestones in a major research project initiated in 2000. The first facet relates to the hybrid[3] modeling of the mechanisms involved in the ventilator-human *interactions*; the second facet concerns the exploitation of this model to design a decision support system to *automate* the ventilator settings given certain physiological observations.

In addressing the above issues, the project was deliberately steered towards investigating neural fuzzy modeling and control techniques for several reasons:

3. The author is aware that this word has somewhat sadly been abused over the years, but by "hybrid" here it is meant a combination of mechanistic model(s) together with linguistic descriptions, for instance using fuzzy logic–based statements.

- General ICU environments are nonlinear.
- They can be uncertain.
- Intra- and interpatient parameter variability mean that generalization over a wide range of patients' dynamics is crucial.
- Data relating inputs to outputs can be sparse.
- Model maintenance and incremental learning are essential ingredients for the future as new data regions are explored continuously.

It is hoped that a system built around this strong flavor of computational intelligence and able to address the two facets of modeling and decision support will affect morbidity and mortality. Indeed, it is estimated that one single day saved in terms of mechanical ventilation will reduce the risk of pneumonias that occur due to infection, which in turn will reduce mortality. Also, a single day reduction in mechanical ventilation will save as much as £500,000 in costs per year for an ICU with more than 400 admissions per year. Similarly, any progress in the design of optimal "recipe(s)" for patient ventilation, in the form of an advisory system, similar to the one proposed in this study, will represent an ideal training tool for junior as well as skilled staff who care for the intensive-care patients. This will help them to (1) care for the patients more effectively, and (2) allow them to concentrate on other related duties without compromising the patients' immediate health.

The starting point of the study was a simulator based upon a process model of patients on artificial ventilation (SOPAVENT[4]), which was developed using a series of differential equations to describe the exchange of gases in the lungs and tissues together with their transport through the circulatory system [17]. The model was later to be used to develop an expert system for the manipulation of ventilator settings.

7.2 Description of the Original SOPAVENT Model

The SOPAVENT model equations are based on respiratory physiology, and they have been used by several researchers in this domain, as will be outlined next. The prototype model equations pertain to five categories:

1. Oxygen transport equations;
2. Oxygen dissociation function and its inverse;
3. Carbon dioxide transport equations;
4. Carbon dioxide dissociation function and its inverse;
5. Airway mechanics and ventilator equations.

It is apparent from the above that the model should include two sets of equations: oxygen transport equations and carbon dioxide transport equations. These should describe the passage of oxygen and carbon dioxide in the pulmonary and circulatory systems. They are dynamic equations and their analytical solutions are not trivial; however, because ICU clinicians are more interested in obtaining blood-gas

4. SOVAPENT stands for simulation of patients under artificial ventilation.

level information at steady state, such equations will be analyzed during this period and the results obtained used to see how the model responds to changes in the following settings:

- Respiratory rate (RR);
- Tidal volume (VT);
- Inspired O_2 gas fraction (FIO_2);
- Positive end-expiratory pressure (PEEP);
- Inspiratory-expiratory ratio (I:E).

The transport of O_2 and that of CO_2 are both described by sets of five linked differential equations targeting the lung and circulatory systems. As Figure 7.1 shows, the lung system comprises three compartments [22], which are the alveolus (where all gas exchanges take place), the dead space (representing the ventilated lungs), and the shunt (representing the perfused area). The circulatory system, however, comprises four compartments, which include the arterial pool (where shunted venous blood mixes with the oxygenated blood from the lungs), the venous pool, and the pulmonary capillary bed and the systemic tissue capillary bed (where O_2 is consumed and CO_2 is produced). Because diffusion across alveolar membranes is triggered by pressure gradients, O_2 and CO_2 contents are converted into partial pressures via inverse gas dissociation functions (GDFs). It is worth noting, however, that gas transport is described in the model by concentrations.

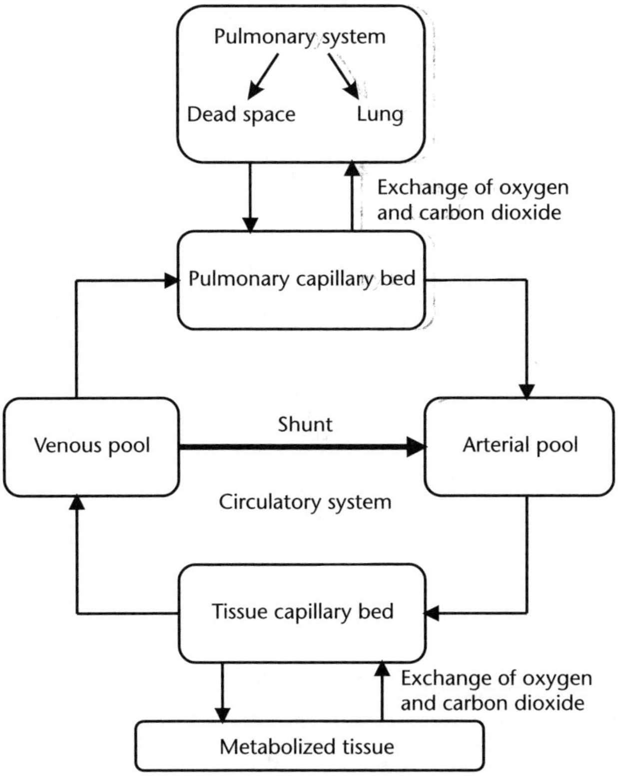

Figure 7.1 A schematic diagram of the patient model. (*After:* [17].)

7.2.1 Oxygen Transport Equations

The transport dynamics relating to O_2 can be described by the following set of differential equations, which target the alveolar, arterial pool, tissue, venous pool, and pulmonary compartments:

$$\frac{\partial C_a O_2}{\partial t} \cdot V_a = \dot{Q}_t \cdot \left[X \cdot C_v O_2 + (1 + X) \cdot C_p O_2 - C_a O_2 \right] \tag{7.1}$$

$$\frac{\partial C_t O_2}{\partial t} \cdot V_t = \dot{Q}_t \cdot \left[C_a O_2 - C_t O_2 \right] - \dot{V}o_2 \tag{7.2}$$

$$\frac{\partial C_v O_2}{\partial t} \cdot V_v = \dot{Q}_t \cdot \left[C_t O_2 - C_v O_2 \right] \tag{7.3}$$

$$\frac{\partial C_p O_2}{\partial t} \cdot V_p = \dot{Q}_t \cdot (1 - X) \cdot \left[(C_v O_2 - C_p O_2) + O_2 Diff \right] \tag{7.4}$$

$$\frac{\partial C_A O_2}{\partial t} \cdot V_A = RR \cdot (V_T - V_D) \cdot (F_I O_2 - C_A O_2 / 1{,}000) - \dot{Q}_t (1 - X) \cdot O_2 Diff \tag{7.5}$$

$$O_2 Diff = B_{O_2} \cdot \left(P_B \cdot (C_A O_2 / 1{,}000) - P_p O_2 \right) \tag{7.6}$$

$$P_P O_2 = f_{inv} \left(C_p O_2 \right) \tag{7.7}$$

where:

$V_x (x = A, a, t, v, p)$ is volumes of alveolar, arterial, tissue, venous, and pulmonary compartments.

$\dot{Q}_t$ is the cardiac output.

X is the fraction of blood shunted past the lungs.

$\dot{V}_{O_2}$ is the oxygen consumption by tissues.

V_D is the alveolar dead-space volume.

V_T is the tidal volume.

RR is respiratory rate.

$C_A O_2$ is alveolar O_2 content.

$C_x O_2 (x = A, a, t, v, p)$ is the alveolar, arterial, tissue, venous, and pulmonary O_2 content.

$P_p O_2$ is the pulmonary partial pressure of O_2.

t is time.

$F_I O_2$ is the inspired O_2 gas fraction.

P_B is the barometric pressure.

B_{O_2} is the diffusion constant.

The following notes apply as well:

- $\dot{V}_A = RR \cdot (V_T - V_D)$ is the alveolar ventilation which depends on the size of the dead space VD; see (7.5).
- Fick's first law of diffusion [23, 24] (see also Chapter 4) is reflected in (7.6) after derivation. This law can be formulated as follows: $V = \dfrac{D \cdot A \cdot (P_1 - P_2)}{\Delta X}$ where $\dot{V}$ is the volume gradient of gas across the membrane, A is the membrane area, $P_1 - P_2$ is the partial pressure gradient on either side of the membrane, D is the diffusion coefficient relating to the gas, and ΔX is the membrane thickness. The previous expression can be rearranged to reflect the diffusion capacity of the gas across the alveolar capillary membrane, D_L, expressed as follows:

$$D_L = \frac{D \cdot A}{\Delta X} = \frac{\dot{V}}{\left(P_A - P_p\right)} \tag{7.8}$$

where P_A and P_p are the alveolar and pulmonary capillary partial pressures respectively.
- The rate of O_2 diffusion is given by the ratio of the diffusion coefficient of O_2 to the pulmonary blood flow; that is,

$$B_{O_2} = \frac{D_{O_2}}{\dot{Q}_l(1 - X)} \tag{7.9}$$

- The alveolar O_2 partial pressure P_AO_2 is given by the following expression:

$$P_AO_2 = P_B \cdot \frac{C_AO_2}{1{,}000} \tag{7.10}$$

- The calculation of the pulmonary partial pressure of O_2, P_pO_2, is carried out via the inverse of the gas dissociation function relating to O_2, as will be explained next:

7.2.2 The Oxygen Gas Dissociation Function (GDF) and Its Inverse

The following equation expresses the fact that the O_2 content in blood is made up of oxygen and hemoglobin in addition to a fraction dissolved in plasma:

$$C(O_2) = \beta_b \cdot H_b \cdot SO_2 + \alpha_b \cdot P_{O_2} \tag{7.11}$$

where:

H_b is the hemoglobin concentration.

SO_2 is the O_2 saturation capacity.

PO_2 is the O_2 partial pressure.

β_b is the hemoglobin/O_2 combining capacity.

α_b is the O_2 carrying capacity of blood plasma.

Sharanm and coworkers [25] suggested the following formulation of the O_2 saturation capacity SO_2, which in fact is a modified version of Kelman's original formula [26]:

$$SO_2 = 0.003683x + 0.000584x^2 \quad x < 10\,\text{mmHg} \tag{7.12}$$

where x is the virtual PO_2 given by the following expression:

$$\begin{aligned} x &= \left[\frac{3.5774}{P_{50}} \cdot PO_2\right] \cdot 10^{(A+B+C)} \cdot \left[\frac{760}{101{,}325}\right] \\ A &= 0.024(37 - T) \\ B &= 0.40(pH - 7.4) \\ C &= 0.06 \cdot \log\left(\frac{5.3329}{PCO_2}\right) \end{aligned} \tag{7.13}$$

The constants A, B, and C are modifiers which account for shifts in the curve[5] due to the blood pH, PCO_2, and temperature T. P_{50} is the 50% saturation normal operating point of the curve (this is normally taken to be 3.5774 kPa).

As already stated in Section 7.2.1, we are interested in obtaining the O_2 partial pressure in the pulmonary compartment rather the O_2 content. Hence, (7.11), which is now nonlinear, needs to be inverted using an iterative algorithm—the secant gradient approach for instance [27]—to obtain an estimate of such pressure. An estimate of the O_2 partial pressure, P_{est}, is obtained given two initial values of PO_2, P_1, and P_2, and their corresponding O_2 contents, C_1 and C_2. Using the secant method, one obtains the following pressure estimate:

$$P_{est} = \left[(C_f - C_1) \cdot \frac{(P_2 - P_1)}{(C_2 - C_1)}\right] + P_1 \tag{7.14}$$

Iterations around (7.14) are repeated until the difference between the current estimate P_{est} and the previous one reaches a predefined tolerance *Err*.

7.2.3 Carbon Dioxide Transport Equations

Similarly to Section 7.2.1, the CO_2 gas transport equations are given as follows:

$$\frac{\partial C_a CO_2}{\partial t} \cdot V_a = \dot{Q}_l \cdot \left[X \cdot C_v CO_2 + (1 - X) \cdot C_p CO_2 - C_a CO_2\right] \tag{7.15}$$

$$\frac{\partial C_t CO_2}{\partial t} \cdot V_t = \dot{Q}_t \cdot \left[C_a CO_2 - C_t CO_2\right] + \dot{V}_{CO_2} \tag{7.16}$$

$$\frac{\partial C_v CO_2}{\partial t} \cdot V_v = \dot{Q}_t \cdot \left[C_t CO_2 - C_v CO_2\right] \tag{7.17}$$

5. A normal curve is obtained for T = 37°C, pH = 7.4, PCO2 = 5.3329 kPa (40 mmHg).

$$\frac{\partial C_pCO_2}{\partial t}\cdot V_p = \dot{Q}_t\cdot(1-X)\cdot\left[\left(C_vCO_2 - C_pCO_2\right) - CO_2Diff\right] \tag{7.18}$$

$$\frac{\partial C_ACO_2}{\partial t}\cdot V_A = RR\cdot(V_T - V_D)\cdot\left(F_ICO_2 - \frac{C_ACO_2}{1{,}000}\right) + \dot{Q}_t\cdot(1-X)\cdot CO_2Diff \tag{7.19}$$

$$CO_2Diff = B_{CO_2}\cdot\left(P_B\cdot\left(\frac{C_ACO_2}{1{,}000}\right) - P_pCO_2\right) \tag{7.20}$$

$$P_pCO_2 = f_{inv}\left(C_pCO_2\right) \tag{7.21}$$

where:

$\dot{V}_{CO_2}$ is the carbon dioxide production by tissues.
C_ACO_2 is the alveolar CO_2 content.
C_xCO_2 ($x = A, a, t, v, p$) is the alveolar, arterial, tissue, venous, and pulmonary CO_2 content.
P_pCO_2 is the pulmonary partial pressure of CO_2.
F_ICO_2 is the inspired CO_2 gas fraction.
B_{CO_2} is the diffusion constant.

All the other parameters are as defined in Section 7.2.1.

7.2.4 The Carbon Dioxide Gas Dissociation Function and Its Inverse

Similarly to the O_2 GDF, the CO_2 GDF is derived using a nonlinear equation. This latter relates the total CO_2 content (or concentration) of the plasma from its pH and P_{Co_2} as follows:

$$[CO_2]_{PLASMA} = \alpha\cdot P_{CO_2}\cdot\left(1+10^{(pH-pK)}\right) \tag{7.22}$$

where:

$\alpha =$ solubility of CO_2 in plasma

$$=0.0307 + 00057\cdot(37-T) + 0.00002\cdot(37-T)^2\cdot\left(\frac{760}{101.325}\right) \tag{7.23}$$

$T =$ blood temperature

$pK = 6.086 + 0.042\cdot(7.4 - pH) + (38 - T)\cdot(0.0047 + 0.0014\cdot(7.4 - pH))$

The CO_2 concentration in the blood is a combination of that in the plasma and the cells such that:

$$[CO_2]_{BLOOD} = 22.2([CO_2]_{CELLS}\cdot pcv + [CO_2]_{PLASMA}\cdot(1-pcv)) \tag{7.24}$$

where pcv = packed cell volume fraction.

The ratio of $[CO_2]_{CELLS}$ to $[CO_2]_{PLASMA}$ is given by the following index:

$$\begin{aligned} d &= ((reduced\ ratio - oxygenated\ ratio) \cdot SO_2) + reduced\ ratio \\ reduced\ ratio &= 0.664 + 0.2275 \cdot (7.4 - pH) - 0.0938 \cdot (7.4 - pH)^2 \\ oxygenated\ ratio &= 0.590 + 0.2913 \cdot (7.4 - pH) - 0.0844 \cdot (7.4 - pH)^2 \end{aligned} \tag{7.25}$$

Hence, (7.24) becomes

$$[CO_2]_{BLOOD} = 22.2 \cdot [CO_2]_{PLASMA} \cdot (d \cdot pcv + (1 - pcv)) \tag{7.26}$$

Similarly to Section 7.2.2, in order to obtain an estimate of the CO_2 partial pressure, (7.26) needs to be inverted numerically via a linear iterative process—the secant method for instance [27]. Figure 7.2 summarizes the way the O_2 and CO_2 GDFs work. Table 7.1 lists the various parameters that SOPAVENT should accept in order to predict the above pressures [17].

To summarize the sequence of calculations performed by SOPAVENT, the following steps describe the movement of oxygen and carbon dioxide:

1. Venous blood (in the venous compartment) lacks oxygen and therefore passes to the pulmonary compartment where it is oxygenated.

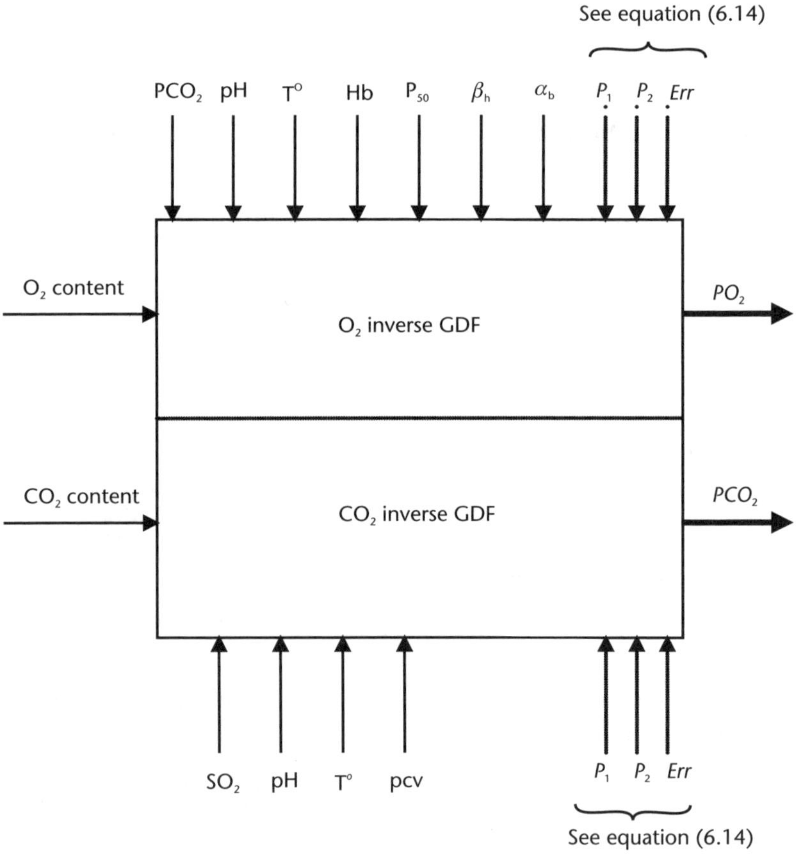

Figure 7.2 Diagram showing the input/output mappings in SOPAVENT. (*After:* [17].)

Table 7.1 Parameters Needed by SOPAVENT

Blood Parameters	*Patient Parameters*	*Ventilator Parameters*
Temperature (T_O) and pH	O_2 consumption ($\dot{V}O_2$)	Inspired O_2 fraction (FiO_2)
Hemoglobin (H_b)	Cardiac output ($\dot{Q}_t$)	Respiratory rate (RR)
P_{50}: 50% saturation point	Dead space (V_D)	Tidal volume (V_T)
	Shunt ($\frac{\dot{Q}_S}{\dot{Q}_t}$)	Positive end-expiratory pressure (PEEP), peak inspiratory pressure (PIP), inspiratory/expiratory ratio (I:E)
	O_2 diffusion rate (DO_2)	
	Age, height, weight, sex	
	Compartmenal volumes: Venous, Arterial, Tissue, Pulmonary, Alveolar	

2. Not all the venous blood passes to the pulmonary compartment. Some of it may bypass the lung and pass directly into the arterial compartment, which is called the shunt.
3. Oxygen diffuses from the alveolar compartment to the pulmonary compartment, and excess carbon dioxide diffuses from the pulmonary compartment to the alveoli, the rate of which is governed by Fick's Law [24].
4. Oxygenated blood is then mixed with the deoxygenated blood in the shunt (the anatomical shunt) and together they pass into the arterial compartment.
5. The blood is carried by the arterial compartment to the tissue compartment where the oxygen is taken up. The amount taken up is determined by the oxygen consumption in the tissues.
6. Carbon dioxide diffuses to the blood stream from the tissues and passes into the venous compartment. The oxygen and carbon dioxide content of the venous compartment is determined by the arterial gas content, oxygen consumption, and carbon dioxide production. The gases are carried in the venous blood back to the lungs.

It should be noted that the overall shunt, which is called the physiological shunt, is normally greater than the anatomical shunt. The physiological shunt includes the shunt effect caused by a ventilation-perfusion mismatch in the lungs, and this effect usually dominates in pathological cases. The inputs to the model include FiO_2, PEEP, respiratory rate (RR), tidal volume, peak inspiratory pressure (PIP), inspiratory time, and inspiratory to expiratory ratio (I:E ratio). The outputs of the model include the arterial pressure of oxygen (PaO_2) and carbon dioxide ($PaCO_2$). The dynamic relationship between the inputs and the outputs depends on the model parameters and constants. The model parameters are patient-specific and the model can therefore be matched to each patient provided the parameters are known. These parameters include age, gender, weight, height, body temperature, hemoglobin level

(Hb), arterial pH, bicarbonate concentration, cardiac output (CO), oxygen consumption rate (VO_2), carbon dioxide production rate, respiratory quotient, shunt, and dead space. The first eight parameters are routinely measured clinically. The CO, VO_2, and shunt can be measured clinically but they are not available routinely. A pulmonary arterial catheter (PAC), which is considered an invasive monitoring equipment, is usually required for the measurements of these three parameters. The gas pressures in the mixed venous blood have to be known for the calculation of VO_2 and shunt.

7.2.5 Model Implementation and Exploitation

The model was implemented using MATLAB-SIMULINK, and a GUI, as shown in Figure 7.3, was produced to enable the user to observe the relevant data that is vital for monitoring and eventually decision-making. These include the demographic data, the ventilator settings, the blood gases, the blood pressures, the heart rate, the cardiac output, the oxygen consumption, the carbon dioxide production, the hemoglobin level, and the body temperature of the current simulated event. The GUI also provides a graphical display of the PaO_2, $PaCO_2$, FiO_2, PEEP, and PIP trends. The user can give advice as to the desirable range of the PaO_2, $PaCO_2$, pH, and tidal volume. He or she can change the ventilator settings (including FiO_2, PEEP, PIP, ventilatory rate, and the I:E ratio) based upon the evidence presented. He/she can also choose from a range of other nonventilatory advice from the pull-down menu. This advice includes the administration of bicarbonate, initiation of continuous venous-venous hemodialysis, prone ventilation, and blood transfusion.

As it stands the model configuration allows one to simulate patients' scenarios, which can be extracted from the Royal Hallamshire Hospital (RHH) Patient Data Management System (PDMS). Since the SOPAVENT model required knowledge of the cardiac output and oxygen consumption, only data from patients who were ventilated and had a pulmonary arterial catheter inserted were suitable for the construction of simulated patients. The records in the PDMS between December 1999 and

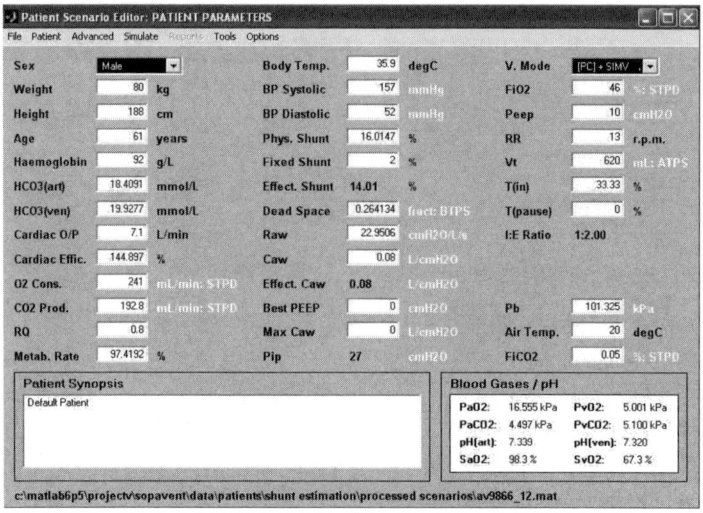

Figure 7.3 Graphical user interface relating to SOPAVENT.

June 2000 were searched and 11 patients were identified. For each patient, there were between 7 and 44 sets of data available for the construction of the simulated events. This data provided 11 simulated patients, one of which was used as a test case so that the user could familiarize himself or herself with the simulator before the actual simulations. The patient's demographic data (include the gender, age, weight, and height), ventilator settings, cardiac output, hemoglobin, blood pressures, oxygen consumption, carbon dioxide production, pH, bicarbonate, shunt, and dead space were used to characterize each simulated event. The carbon dioxide production was often not measured. Therefore, it was necessary to estimate it using the oxygen consumption and the respiratory quotient (assumed to be 0.8). The dead space and the shunt at each simulated event were calculated based on the patient's blood gases, ventilator settings, oxygen consumption, carbon dioxide production, and cardiac output using the secant method on the SOPAVENT model.

Any system's model can, in general, be exploited in two ways: in a forward fashion purely for prediction purposes, and/or in a "reverse engineering" way to control the corresponding output properties via the manipulation of input variables. The ultimate objective of this patient's model is to help design a decision support system for use in the ICU by incorporating expert's knowledge, so it was decided to invite hospital consultants to undergo a study in which they would be asked to use patients' scenarios to decide on the adequate treatment.

The simulation procedure is shown in Figure 7.4. Four intensive care consultants from the Royal Hallamshire Hospital were invited to undertake the study. Before the simulations, the procedure was explained to the clinician. Then he/she had an opportunity to familiarize himself/herself with the GUI using the test case. Afterwards, the clinician was presented with the first set of data relating to one of the simulated patients. The derived parameters were not shown on the GUI. Only the data directly retrieved from the PDMS were shown. The clinician was asked to

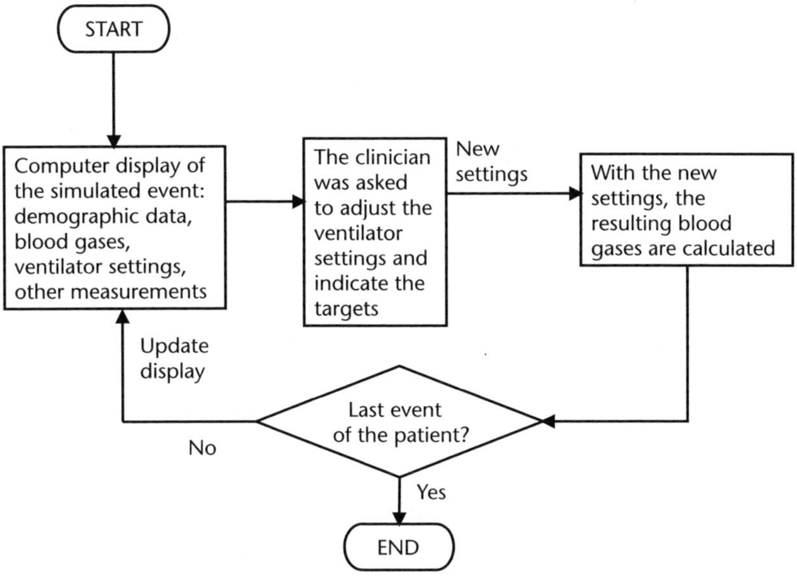

Figure 7.4 Flow chart illustrating the steps included in the patient simulator as used by the clinicians.

key in the target range of PaO_2, $PaCO_2$, pH, and tidal volume that he/she felt appropriate for the simulated patient. Then the clinician would need to key in the advised new ventilator settings. From the advice given by the clinician and the internal parameters of the patient, the simulator calculated the predicted blood gases and the resulting blood gases were shown to the clinician. After the clinician was shown the resulting blood gases, the simulator would calculate the blood gases of the next case event based upon the advised ventilator settings and the internal parameters of the new case event. Then the screen was updated using the data, the advised ventilated settings and the calculated blood gases of the new case event. The clinician was asked to repeat the exercise again until the clinician had completed the simulation on all the case events of the simulated patient. Then, he/she could move on to the next simulated patient. The results obtained, which will be discussed in later sections, will form the basis for the elicitation of the rules embedded in the decision support system.

7.3 Noninvasive Estimation of Shunt

As already stated in Section 7.2, when the shunt is not known, it can be estimated using the secant method on the SOPAVENT model. The secant method [27] is a numerical algorithm designed to solve nonlinear equations when the Jacobian (the first derivative) of the equation is not known. It starts with two arbitrarily chosen values of the independent variable—in this case, the shunt. At each iteration, the estimated root of the equation is determined by joining the two points by a straight line, then determining where the line intersects the target value of the dependent variable—in this case P_aO_2. It is computationally expensive, and occasionally no solution can be found. Therefore, a simple and noninvasive method of shunt estimation is required. Zetterstrom [28] showed that some indices of hypoxemia correlated well with the venous admixture. In a previous study conducted in my research group, the correlation between four indices of hypoxemia and the shunt estimated by the SOPAVENT model in nine ICU patients were compared and the respiratory index was found to correlate best with the shunt. In the present paper, two models of the relationship between the respiratory index (RI) and shunt are compared. They are a linear regression model and an adaptive neuro-fuzzy inference system (ANFIS) model [29].

7.3.1 Method

The data was collected retrospectively from the patient data management system (PDMS) in a General ICU of the Royal Hallamshire Hospital (RHH) teaching hospital in Sheffield. The local medical research ethics committee was consulted and the use of the patient data was approved. Records of patients admitted between December 1999 and November 2001 were reviewed. Patients who were mechanically ventilated and had a PAC inserted were identified for the study. The diagnosis of each patient was retrieved from the ICU audit database. The patients were divided into a septic group and a nonseptic group. The septic group included those patients whose diagnosis included septicemia or septic shock. The remaining patients were in the nonseptic group. This is by no means the only way to classify the patients. However, it was considered an important one because septic patients behave differently from nonseptic patients. They generally have a higher metabolic rate and a higher chance of getting acute respiratory distress syndrome (ARDS). Data from both groups of patients were used to derive the CI, CO, and VO_2 median values (Part II), whereas data from septic patients were used to estimate the shunt (Part I) and the effect of PaO_2 (Part III). All the statistical analyses were performed using SPSS v11.0.

7.3.2 Estimation of Shunt Using the Respiratory Index

Data from the septic group was used in this part of the study. These patients were randomly allocated into two groups. The randomization was done using the MS EXCEL function RANDBETWEEN. Data from one group (Group A) was used to derive the models of relationship between the shunt and the respiratory index. Data from the other group (Group B) was used in the simulator study described later.

The CO measurements of each patient were retrieved. The ventilator settings, the blood gas measurements, the body temperature, and the hemoglobin level when the CO was measured were recorded. The VO_2 was calculated as part of the CO measurements. The carbon dioxide production was not available. It has also been shown [17] in the sensitivity analysis that the PaO_2 was relatively insensitive to the carbon dioxide production. Therefore, it was estimated from the VO_2 assuming a respiratory quotient (RQ) of 0.8. For each data set, the ideal alveolar partial pressure of oxygen (PAO_2) was calculated using the alveolar gas equation [28]:

$$PAO_2 = FiO_2\left(P_B - P_{H_2O}\right) - PaCO_2\left(FiO_2 + (1 - FiO_2)RQ\right) \tag{7.27}$$

where P_B is the barometric pressure and P_{H_2O} is the water vapor pressure. The RI was calculated using the formula

$$RI = \frac{PAO_2 - PaO_2}{PaO_2} \tag{7.28}$$

The effective shunt and total shunt in the SOPAVENT model from each data set were estimated using the secant method based on the patient's measurements. During the process, the patient's age, gender, weight, height, body temperature, *Hb*, arterial pH, bicarbonate concentration, CO, and VO_2 were used as the respective parameter values of the SOPAVENT model. The shunt and dead space were adjusted according to the model output errors in the PaO_2 and $PaCO_2$. The shunt estimated was the effective shunt, which includes the effect of the PEEP. The total shunt was calculated by assuming a linear relationship between the PEEP and effective shunt. The correlations between the RI and the two shunts were calculated and the one with a higher correlation was used for the study. A linear regression model and an ANFIS model were fitted to the paired RI and shunt data. Since the number of samples available from one individual patient varied from 4 to 19, if all the data from each patient was used in the modeling process, some patients might have had more influence on the results than the others. Therefore, four data sets were chosen randomly from each patient to form the training data set (Group A1), and the rest of the data (Group A2) was used for cross-validation and testing. The linear regression model and the ANFIS model were used in Part III of the study.

The linear regression model was represented by the following linear function:

$$shunt = a \cdot RI + b \tag{7.29}$$

where a and b are constants. The input to the ANFIS model was the RI, and the output was the effective shunt. The input space of the ANFIS model was divided into three fuzzy partitions. The input membership functions were bell shaped. The output memberships were linear and each of the output linear function[6] was in the following form:

$$output = c \cdot input + d \tag{7.30}$$

where c and d are constants. The ANFIS model was trained using the MATLAB Fuzzy Logic Toolbox. Half of the data from Group A2 were used for the cross-validation and the other half were used for testing. During the ANFIS training, the mean squared error for the training set and the cross-validation set were monitored continuously and the training was stopped when the mean squared error for the cross-validation set started to increase. The optimization was done using the hybrid Levenberg-Marquardt back propagation [30] and gradient descent algorithm.

7.3.3 Results

Forty patients were identified as being mechanically ventilated and having a PAC inserted during their ICU admissions; 27 patients were diagnosed as having septicemia or septic shock. The age, gender, weight, height, and body surface area of the septic and nonseptic groups of patients are shown in Table 7.2. None of the differences were statistically significant.

Ten patients were randomly allocated to Group B. Therefore, 17 patients were chosen as Group A, from whom the data was used in this part of the study. As shown in Table 7.2, the demographic data of the two groups of patients were not significantly different. From Group A, 202 data sets were retrieved for analysis. However,

Table 7.2 The Demographic Data of the Three Groups of Patients

	Septic		
Group	*Group A (n=17)*	*Group B (n=10)*	*Nonseptic (n=13)*
Age (years)	66 (45.5–75)	64.5 (46.8–72)	66 (63.5–80)
F:M	7:10	1:1	4:9
Height (cm)	178 (168.0–183.5)	164 (158.8–178.5)	174 (163.5–179.5)
Weight (kg)	75 (62.5–80.0)	79.5 (61.9–97)	70 (63.5–80)
BSA (m^2)	1.90 (1.76–1.98)	1.91 (1.64–2.14)	1.81 (1.77–1.95)

The median values are shown in the table and the interquartile ranges are shown in the brackets.
None of the differences between the three groups of patients was statistically significant.

6. It is worth reiterating here that in fuzzy TSK model, although the output function is linear, the overall fuzzy mapping is still nonlinear.

in 43 cases no solution for the shunt could be found using the secant method. Two patients had to be excluded because no solution could be found for any of their data. The estimated shunts of each patient were plotted against the RIs. Three data points, which were outliers, were excluded. Of the remaining 156 data points, 60 were chosen to be the training set. The correlation coefficient between the RI and the total shunt was 0.839 and that of the effective shunt was 0.849. The effective shunt was chosen for use in the latter part of the study not only because of the slightly higher correlation coefficient but also because of the different PEEP that the patients are likely to be on.

Figure 7.5 shows the relationship between the RI and the shunt. The linear regression model derived was as follows:

$$effective\ shunt = 4.5236RI + 9.3767 \quad (7.31)$$

The relationship between the effective shunt predicted by the ANFIS model and the RI is shown in Figure 7.6 and the resulting rule base is shown in Table 7.3. The shunt estimation errors of the linear regression model and the ANFIS model on the Group A1 data, Group A2 data, and the data used for Part III were compared and are shown in Table 7.4. The mean absolute estimation error of the ANFIS model was significantly lower than that of the linear regression model for the Group A2 data ($p = 0.028$) and for the data used in Part III ($p = 0.003$).

7.4 The Sheffield Intelligent Ventilator Advisor (SIVA): Design Concepts

When a clinical decision is made for a change in ventilator settings, the decision can be divided into two parts. The first part is the *qualitative* aspect of the change—for example, the inspired fraction of oxygen (FiO_2) should be increased. The second part is the *quantitative* aspect, which defines by how much the particular setting should be changed. To achieve a target blood gas level, there are often several options available. For example, to increase the arterial partial pressure of oxygen

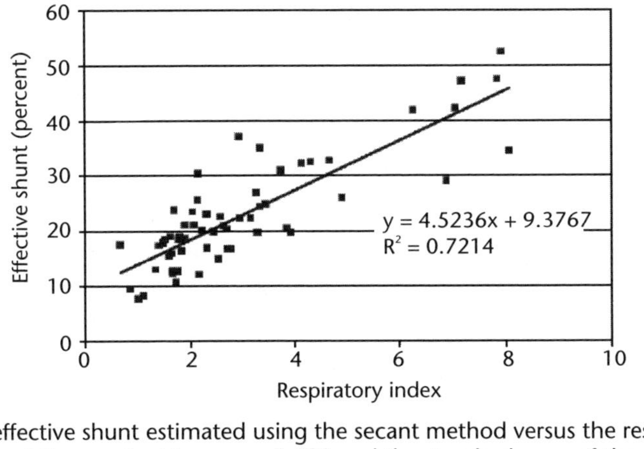

Figure 7.5 The effective shunt estimated using the secant method versus the respiratory index. The standard error of the constant term was 1.288 and the standard error of the regression coefficient was 0.369.

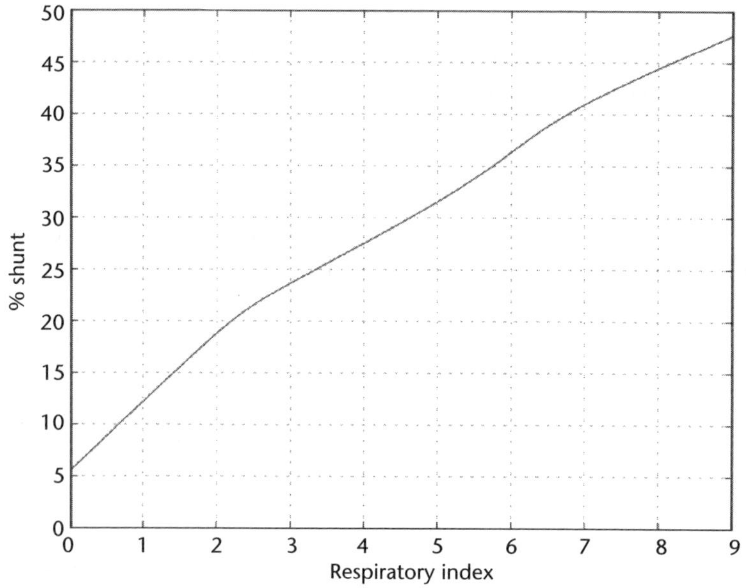

Figure 7.6 The shunt characterization using the ANFIS model.

Table 7.3 The Rule Base Elicited by ANFIS Model for Shunt Estimation

Antecedence	*Consequence*
When the RI is low	Shunt = 6.499 RI + 5.293
When the RI is moderate	Shunt = 3.787 RI + 11.89
When the RI is high	Shunt = 3.119 RI + 19.61

Table 7.4 The Mean Squared Shunt Estimation Error and the Mean Absolute Shunt Estimation Error of the Linear Regression Model and ANFIS Model for the Three Groups of Data

	Group A1		*Group A2**		*20 Samples from Group B***	
Shunt Estimation Model	*Linear Regression*	*ANFIS*	*Linear Regression*	*ANFIS*	*Linear Regression*	*ANFIS*
m.s.e	26.52	25.99	27.24	25.48	20.65	12.12
Mean absolute estimation error	4.06 ± 3.19	4.07 ± 3.01	4.23 ± 3.07	4.04 ± 3.04	4.08 ± 2.04	3.05 ± 2.04

Group A1 is the training data set. Group A2 is the cross-validation and testing set. The 20 samples from Group B were used in Part III of this study. The paired t-test was done on the mean absolute estimation error of the two models.
*$p = 0.028$. **$p = 0.003$.

(PaO_2), one can increase the FiO_2 or increase the PEEP or both. One can also start prone ventilation and prescribe other respiratory therapies. Humans are very good at pattern recognition and can often come quickly to a reasonable conclusion as to which ventilator setting should be changed. On the other hand, due to the complexity of the problem and the lack of a comprehensive model, it is difficult and time consuming to use a computer algorithm to find the optimal solution in the domain of possible solutions. However, compared to computers, humans are often less capable of making good quantitative decisions. In clinical practice, how much a ventilator setting should be changed is very often arbitrarily determined. This is the reason why we have adopted the combined knowledge-and-model-based computer-assisted approach.

There is another problem that was mentioned earlier: inconsistency and lack of information. In most ICUs, the blood gases are measured every few hours. The calculation of shunt, which is an important parameter, requires the measurement of mixed venous blood gases via a PAC, which is considered invasive and therefore is only used in a small number of patients. The indications for its use are related to management of the cardiovascular system. A PAC would not be inserted specifically for management of the ventilation. Therefore any advisory system must be able to function with a variety and limited number of inputs. This leads to another aspect of our design, which is to offer the user different modes of control. The system can be operated in invasive mode or noninvasive mode. Operations on the invasive mode require data from invasive cardiovascular measurements, which are usually acquired through a PAC. In the noninvasive mode, the cardiovascular measurements and the related parameters are estimated noninvasively. Sometimes an intensivist may delegate part of the ventilator management to nurses by defining blood gases and other measurement targets. Similarly, the SIVA can be operated under full advisory mode or clinician-directed mode. In the full advisory mode, the therapeutic goals (target blood gases) and the type of ventilator setting(s) to be changed are determined by the system, whereas in the clinician-directed mode, the clinician can define the target blood gases and/or choose which ventilator setting(s) should be changed.

SIVA is designed to advise the clinicians on four ventilator settings: FiO_2, PEEP, inspiratory pressure (Pinsp) and ventilatory rate (Vrate). The intensive care unit that was involved in this research is a general ICU. During most of the acute phase of the disease, the ventilation mode adopted is the biphasic positive airway pressure (BiPAP) ventilation. This is a form of pressure control ventilation where the airway pressure switches between two levels. Therefore, the system was confined to deal only with BiPAP ventilation. SIVA concentrates on dealing with patients who suffer from pneumonia, septicemia, or ARDS, which are considered to be commonly occurring diseases in the ICU. The inputs to the system include the patient's demographic data such as age and gender, routine measurements, blood gases, ventilator settings, and the respiratory measurements. In the initial prototype, the data is keyed in by the user. However, in the future, data will be automatically retrieved from the PDMS and the routine and respiratory measurements will be automatically logged into the system.

In order to make the system easy to test and maintain, a modular approach has been adopted. The architecture is shown in Figure 7.7. It is divided into two main parts: the *top-level knowledge-based module* and a *lower level model-based module*. Each module is divided into a FiO_2/PEEP subunit that controls the oxygenation related settings and a Pinsp/Ventilatory rate subunit that controls the settings relating to the minute ventilation. The top-level module advises the type of the ventilator settings to be changed and the target PaO_2 and $PaCO_2$, while the lower-level module derives the amount of change required in each setting.

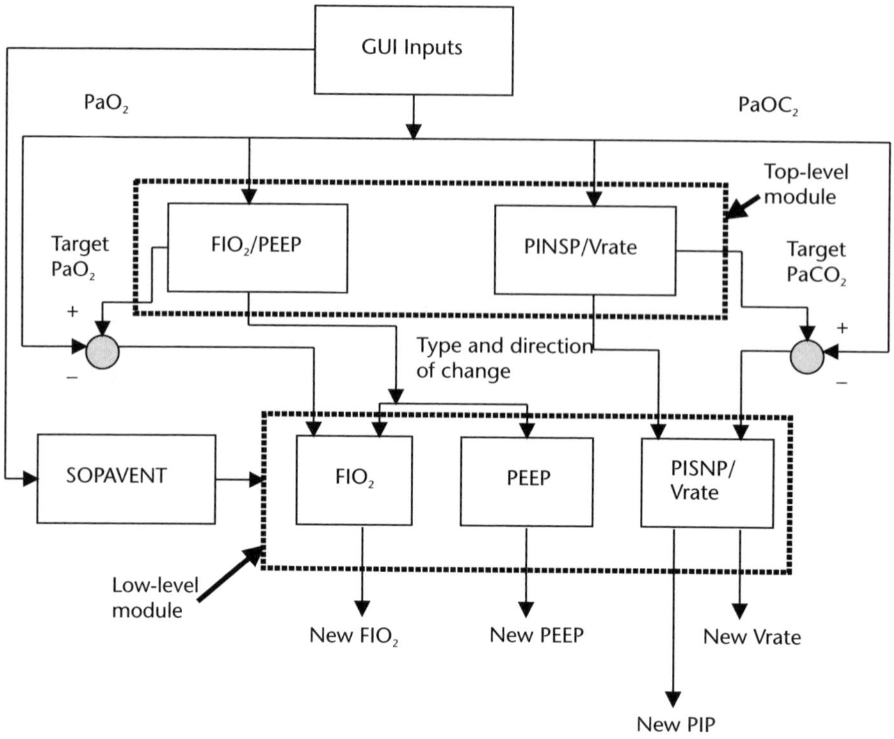

Figure 7.7 The architecture of the Sheffield Intelligent Ventilator Advisor (SIVA).

7.5 Design of the Knowledge-Based Levels

As shown in Figure 7.7, the top-level module consists of two subunits: the FiO_2/PEEP subunit and Pinsp/Vrate subunit. Each subunit generates advice for two ventilator settings and is implemented in two fuzzy rule bases, one for each ventilator setting. The primary input variables to the FiO_2/PEEP subunit include the PaO_2, the previous PaO_2, FiO_2, previous FiO_2, and PEEP. The primary input variables to the Pinsp/Vrate subunit include the $PaCO_2$, previous $PaCO_2$, pH, Pinsp, and Vrate. There are three input fuzzy partitions for each primary variable, and grid-partitions are used. There are five input variables in each rule base, and therefore, 243 rules would be required. In order to reduce the number of rules, three secondary variables are derived from the primary variables, as will be described next.

7.5.1 Top-Level FiO_2/PEEP Subunit

The structure of the fuzzy rule bases in the FiO_2/PEEP subunit is shown in Figure 7.8. The five input variables are summarized to give three input variables, which are the PaO_2, the patient's condition, and the level of ventilatory support. There are three input partitions for PaO_2: low, normal, and high. The patient's condition is determined by the percentage change in the hypoxemia index (PaO_2/FiO_2) and there are three fuzzy partitions: deteriorating, static, and improving. The ventilator support level is derived from the normalized FiO_2 and PEEP. The therapeutic range of FiO_2 was found to vary between 0.25 to 1, and the therapeutic range of PEEP ranged

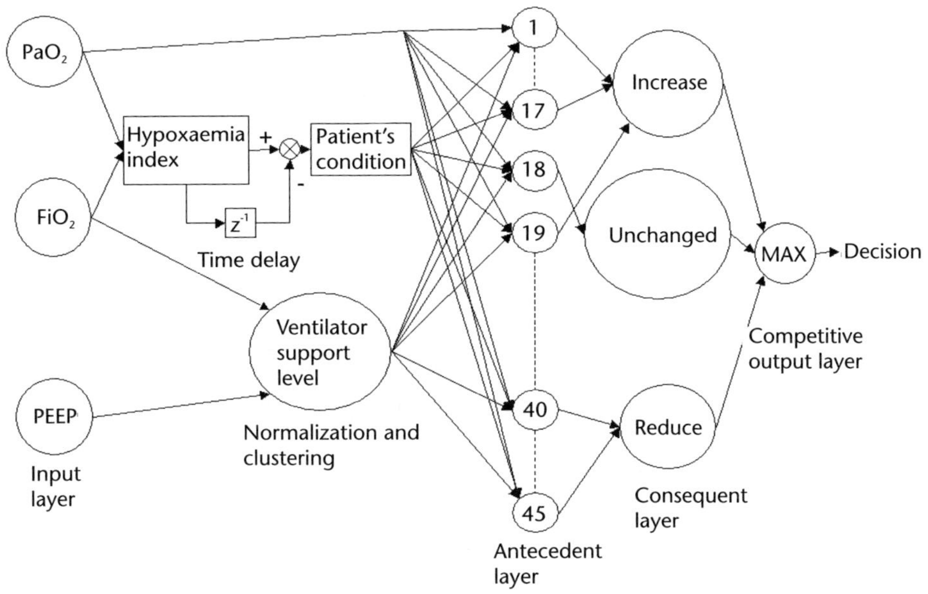

Figure 7.8 The structure of the fuzzy rule base in the top level FiO_2/PEEP subunit.

from 0 to 20 cmH_2O. Both the FiO_2 and PEEP were normalized to the range [0,1]. There are five fuzzy clusters (see Chapter 6, Appendix 6A for details about fuzzy clustering) defined for the ventilatory support level, and their corresponding cluster centers are shown in Table 7.5. This resulted in 45 rule antecedents. It is worth noting that grid partitions were used for the fuzzy rule antecedents and a modified "multiplication" was chosen as the inference method. The consequent part of the rules can be *reduce*, *unchanged*, or *increase*. The output member with the maximum membership value is chosen as the output.

7.5.2 Top-Level Pinsp/Vrate Subunit

The structure of the rule bases in the Pinsp/Vrate subunit is shown in Figure 7.9. The five primary input variables are amalgamated into the previous $PaCO_2$, metabolic status, and the ventilator support level. There are three fuzzy partitions for the previous $PaCO_2$: low, normal, and high. There are three fuzzy partitions for the current

Table 7.5 The Cluster Centers of the Support Level Used as the Input to the FiO_2/PEEP Control of the Top-Level Module

Support Level	*Cluster Center*
Minimal	0.00, 0.00
Moderate with PEEP dominance	0.25, 0.75
Moderate	0.50, 0.50
Moderate with FiO_2 dominance	0.75, 0.25
Maximal	1.00, 1.00

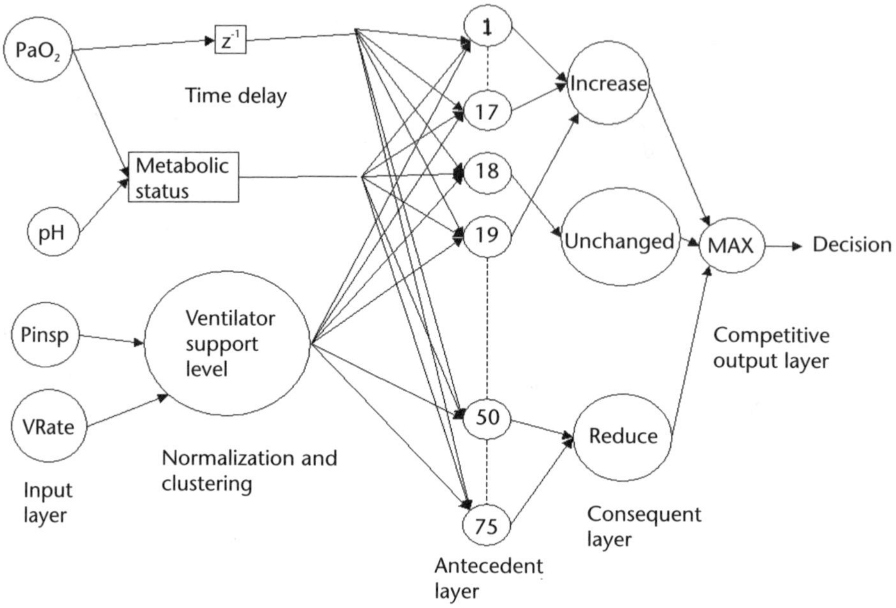

Figure 7.9 The structure of the rule bases in the top-level Pinsp/Vrate subunit.

pH and $PaCO_2$: low, normal, and high. The metabolic status was derived from the pH and $PaCO_2$ according to Table 7.6. The ventilator support level was derived from Pinsp and Vrate using a clustering approach similar to that used in the FiO_2/PEEP subunit. The therapeutic range of the Pinsp is from 10 to 40 cmH_2O and that of the Vrate is from 4 min^{-1} to 20 min^{-1}. Both the Pinsp and Vrate were normalized to [0,1] and the cluster centers are shown in Table 7.7.

7.5.3 Parameters Assigned to the Input Membership Functions

The fuzzy partitions for four of the primary inputs (FiO_2, PEEP, Pinsp, and Vrate) have already been defined using the clustering technique outlined previously using the parameters of Tables 7.5 and 7.7.

The fuzzy partitions derived for the primary input variables PaO_2, $PaCO_2$, and pH are shown in Figures 7.10 to 7.12. The secondary inputs include the patient's condition and the ventilatory support level for the FiO_2/PEEP subunit, and the acid-base status and the ventilatory support level for the PINSP/Ventilatory rate

Table 7.6 The Definition of the Metabolic Status

Metabolic Status	*pH*	*$PaCO_2$*
Metabolic acidosis	Low	Low
Respiratory acidosis	Low	High
Normal	Normal	Normal
Metabolic alkalosis	High	High
Respiratory alkalosis	High	Low

Table 7.7 The Cluster Centers of the Support Levels Used as the Input to the Pinsp/Vrate Subunit of the Top-Level Module

Support Level	*Cluster Center*
Minimal	0.00, 0.00
Moderate with Vrate dominance	0.25, 0.75
Moderate	0.50, 0.50
Moderate with Pinsp dominance	0.75, 0.25
Maximal	1.00, 1.00

subunit. For the FiO_2/PEEP subunit, the patient's condition is determined by the percentage change in the hypoxemia index (PaO_2/FiO_2). Its fuzzy partitions were derived from the probability distribution of the percentage change in the hypoxemia index among the simulated patients. The fuzzy partitions are shown in Figure 7.13. The ventilatory support level is determined by the FiO_2 and PEEP levels. The therapeutic range of FiO_2 was found to vary between 0.25 to 1, and the therapeutic range of PEEP varied from 0 to 20 cmH_2O. Both the FiO_2 and PEEP were normalized to the range [0,1]. Five cluster centers were defined as shown in Table 7.8. The degree of membership in each ventilatory support category for a particular set of inputs was determined by the Euclidean distance between the normalized input FiO_2 and PEEP levels, and the cluster centers. Arbitrarily, a degree of membership of 0.5 was assigned to any data point at a distance of 0.25 from the cluster center. That gives the following membership function:

$$\mu_c(x) = \exp\left(-\frac{\|x - c\|^2}{0.1803}\right) \tag{7.32}$$

where $\mu_c(x)$ is the membership of x in cluster c, x = [normalized FiO_2, normalized PEEP], and $\|x - c\|$ is the Euclidean distance of x from the cluster center c.

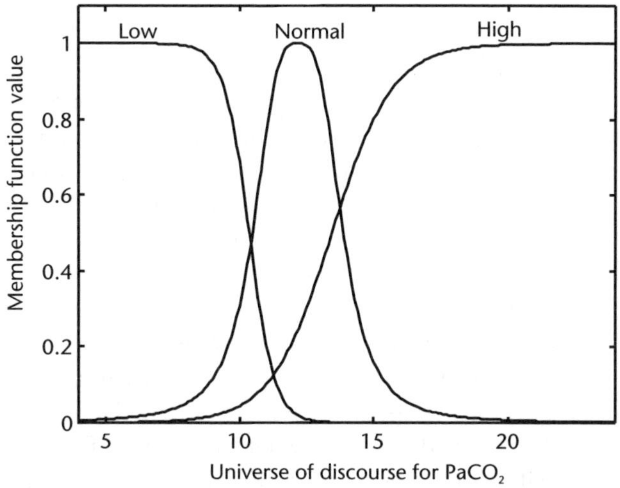

Figure 7.10 The fuzzy partitions for PaO_2.

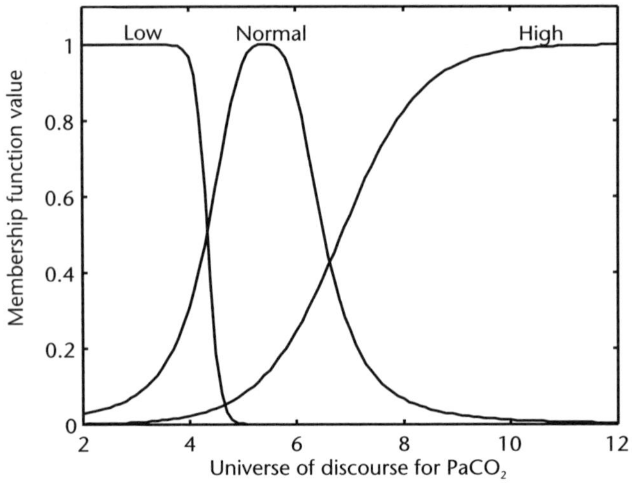

Figure 7.11 The fuzzy partitions for $PaCO_2$.

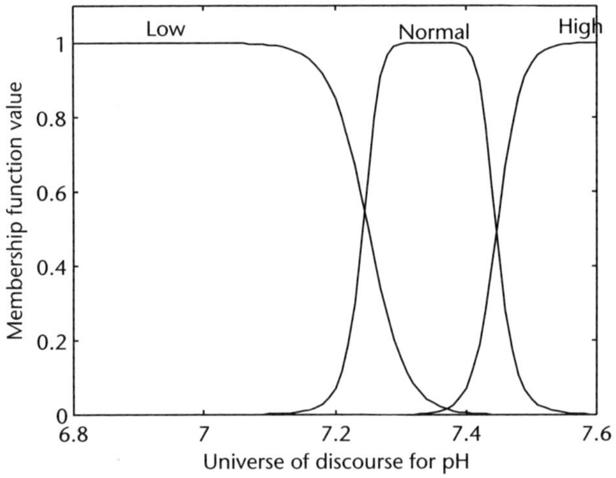

Figure 7.12 The fuzzy partitions for pH.

The targets set by the clinicians during the study were used in the derivation of parameters for the fuzzy membership functions of the primary inputs. The cumulative distributions of the target range set by the clinicians were used to derive the parameters for the fuzzy membership functions. The derivation was based on the assumption that the fuzzy membership value represents the probability that an input will be classified as belonging to the respective fuzzy set by an intensive care consultant. For example, Figure 7.14 shows the histogram and the cumulative frequency for the upper limit of the acceptable tidal volume level set by the consultants. From the cumulative frequency curve, one could deduce that if the tidal volume was more than 11.5 ml/kg, all of the consultants would regard it to be too high, and therefore, it should be regarded as high and it will be assigned a fuzzy membership value of 1 in the fuzzy set "high" and 0 in fuzzy set "normal." Therefore, from the cumulative frequency distribution of the upper and lower limits, one could derive the fuzzy membership values. Also, if the upper limit of the target PaO_2 range in 50% of the

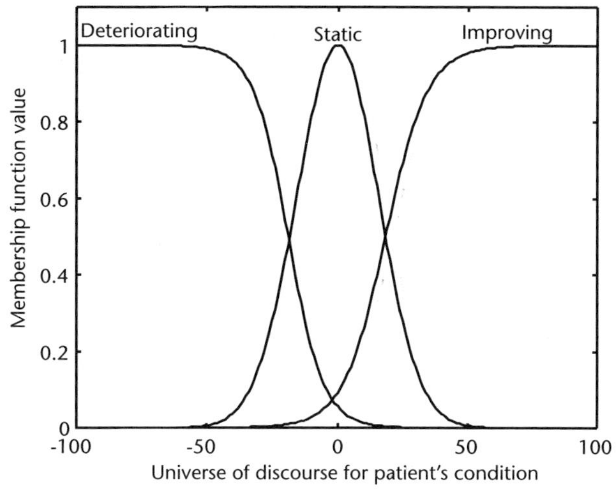

Figure 7.13 The fuzzy partitions for the patient's condition in the FiO_2/PEEP subunit.

Table 7.8 The Fuzzy Cluster Centers of the Support Levels in the FiO_2/PEEP Subunit

Support Level	*Cluster Center*
Minimal	0, 0
Moderate with PEEP dominance	0.25, 0.75
Moderate	0.5, 0.5
Moderate with FiO_2 dominance	0.75, 0.25
Maximal	1, 1

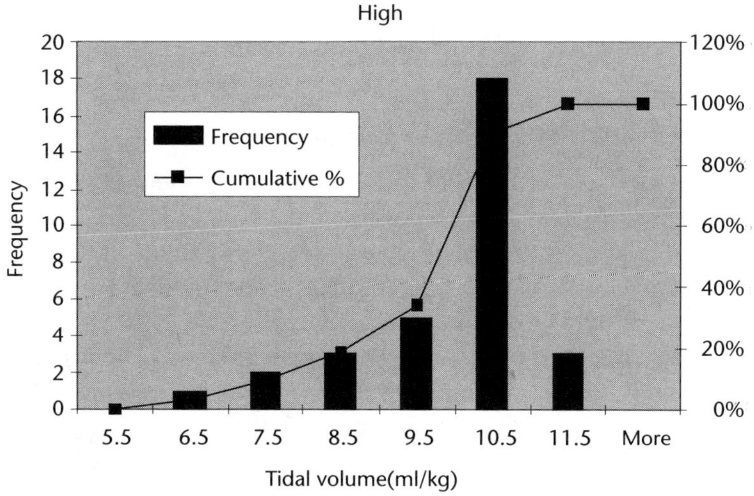

Figure 7.14 The histogram and cumulative frequency of the upper limit of tidal volume set by the clinicians.

cases was less than 15 kPa, then this PaO_2 level should have a degree of membership of 0.5 in both the *normal* and *high* fuzzy partitions.

Sigmoidal membership functions were used for the fuzzy sets *low* and *high*, whereas bell-shaped membership functions were used for the fuzzy sets *normal*. The sigmoidal membership function is given by:

$$\mu(x) = \frac{1}{1 + e^{\left(\frac{-a}{x-c}\right)}} \tag{7.33}$$

where x is the value of the input, $\mu(x)$ is the membership function value, and a and c are constant parameters. The bell-shaped membership function is defined as follows:

$$\mu(x) = \frac{1}{1 + \left|\frac{x-c}{a}\right|^{2b}} \tag{7.34}$$

where b is a constant parameter.

It is worth noting that the parameters of the fuzzy sets were derived using the experimental data and calculated using the nonlinear least squares program provided by the MATLAB Statistics Toolbox.

For the secondary variable *patient condition*, the percentage changes in the hypoxemia index for the case events during the simulation were analyzed. The parameters of the membership functions are based on the statistical distribution of the changes during the simulations. A Gaussian membership function was used for the *static* patient condition, the parameters of which were derived from the mean and standard deviation of the change in hypoxemia index. Sigmoidal membership functions were used for the *deteriorating* and *improving* patient condition, the parameters of which were derived based on the cumulative frequency curve.

7.5.4 Derivation of the Initial Rule Base

From each case event in the simulation study, the degrees of membership in the rule antecedents were calculated. The clinicians' advice in each ventilator settings for the associated case event was considered the target output of the fuzzy rule bases. During the analysis, each set of data consisted of the degree of membership in a rule antecedent and the advice on the ventilator setting (reduce, unchanged or increase). Therefore, for each rule antecedent, there existed a collection of the degree of membership and advised outputs. The range of the degree of membership for each rule antecedent was divided into 10 intervals. The frequency of each category of the consultants' advice in each degree of membership interval was determined. If a rule antecedent should result in a particular advice being given, one should find a positive correlation between the probability (the relative frequency) of the advice being given and the degree of membership. Hence, the output (rule consequent) associated with a rule antecedent was determined using correlation analysis. Not all the fuzzy rules could be derived from the correlation analysis because some of the results were equivocal. Therefore, an intensive care consultant with a special interest in ventilator management, who was also an expert in the behavior of the lung in the critically ill patient and respiratory physiology, was asked to review the initial rule-base and to supplement it where necessary.

Grid partitioning was used for the fuzzy rule–based system. There are 45 rules in each of the FiO_2 and PEEP rule bases and 75 rules in each of the Pinsp and Ventilatory rate rule bases. The rule consequent of 6 FiO_2 rules, 7 PEEP rules, 27 Pinsp rules, and 28 Ventilatory rate rules could not be decided due to a lack of statistical significance. For the undetermined rules, possible consequents were determined by ruling out those with significantly negative correlation coefficients. The candidate consequents of these rules were then presented to the intensive care consultant for the final verdict. The fuzzy rule bases for the four ventilator settings are shown in Tables 7.9 through 7.12.

7.5.5 Validation of the Initial Rule Bases

The initial rule bases were validated from the simulation data and retrospective clinical data. The retrospective clinical data were collected from the PDMS of the ICU. Ten ventilated patients were randomly chosen from the database. For each patient, the blood gas records from the time when artificial ventilation began to the time the weaning process started were retrieved. Due to the retrospective nature of the data collection, it was not always clear-cut when the weaning process started. It was decided that it would be defined as the time when the ventilatory mode was changed

Table 7.9 The FiO_2 Rule Base

When the PaO_2 is low	*Condition*		
Support	*Deteriorating*	*Static*	*Improving*
Minimal	Increase	Increase	Maintain
Moderate (PEEP dominant)	Increase	Increase	Maintain
Moderate	Increase	Increase	Maintain
Moderate (FiO2 dominant)	Increase	Increase	Maintain
Maximal	Increase	Increase	Increase

When the PaO_2 is normal	*Condition*		
Support	*Deteriorating*	*Static*	*Improving*
Minimal	Maintain	Maintain	Maintain
Moderate (PEEP dominant)	Increase	Maintain	Maintain
Moderate	Maintain	Maintain	Maintain
Moderate (FiO2 dominant)	Increase	Maintain	Maintain
Maximal	Increase	Maintain	Reduce

When the PaO_2 is high	*Condition*		
Support	*Deteriorating*	*Static*	*Improving*
Minimal	Maintain	Maintain	Reduce
Moderate (PEEP dominant)	Maintain	Reduce	Reduce
Moderate	Maintain	Reduce	Reduce
Moderate (FiO2 dominant)	Maintain	Reduce	Reduce
Maximal	Reduce	Reduce	Reduce

Table 7.10 The PEEP Rule Base

When the PaO_2 *is low*	*Condition*		
Support	*Deteriorating*	*Static*	*Improving*
Minimal	Increase	Increase	Increase
Moderate (PEEP dominant)	Increase	Increase	Increase
Moderate	Increase	Increase	Maintain
Moderate (FiO_2 dominant)	Increase	Increase	Increase
Maximal	Increase	Increase	Increase

When the PaO_2 *is normal*	*Condition*		
Support	*Deteriorating*	*Static*	*Improving*
Minimal	Maintain	Maintain	Maintain
Moderate (PEEP dominant)	Increase	Maintain	Reduce
Moderate	Increase	Increase	Reduce
Moderate (FiO_2 dominant)	Increase	Increase	Reduce
Maximal	Maintain	Increase	Reduce

When the PaO_2 *is high*	*Condition*		
Support	*Deteriorating*	*Static*	*Improving*
Minimal	Maintain	Maintain	Maintain
Moderate (PEEP dominant)	Maintain	Reduce	Reduce
Moderate	Maintain	Maintain	Reduce
Moderate (FiO_2 dominant)	Maintain	Maintain	Maintain
Maximal	Maintain	Maintain	Maintain

from BiPAP to CPAP/ASB/IPPV (Continuous Positive Airway Pressure/Assisted Spontaneous Breath/Intermittent Positive Pressure Ventilation). For the validation of the FiO_2/PEEP subunit, the FiO_2 and PEEP when the blood gas sample was taken were recorded. The FiO_2 and the PaO_2 in the last set of blood gas measurements were also recorded. (It should be noted that the blood gas sampling in the ICU was irregular and therefore, the time interval between the PaO_2 and the last PaO_2 could range from 30 minutes to 4 hours.) These provided data for the input to the fuzzy system. Each set of inputs consisted of PaO_2, the last PaO_2, FiO_2, the last FiO_2, and PEEP. For the target outputs, the patient record in the 1 hour following each blood gas result was examined to look for any changes in the FiO_2 or PEEP settings.

For the Pinsp/Vrate subunit, the PIP, the ventilatory rate, and the PEEP when the blood gas sample was taken were recorded, and the $PaCO_2$ in the last set of blood gas was noted. The Pinsp was calculated from the difference between the PIP and the PEEP. The $PaCO_2$, pH, the last $PaCO_2$, Pinsp and Vrate formed the input data to the fuzzy system. The patient record in the 1 hour following each blood gas result was examined to look for any changes in Pinsp or Vrate.

Three types of matching were defined. An *exact* match was defined to be when the rule base's output was the same as the clinician's advice or action. A *partial* match was defined to be when the rule base's output was different from the clinician's advice or action but they were not in opposite directions (for example, one is *unchanged* and the other one is *increase*). A *conflicting mismatch* was defined to be

Table 7.11 The Pinsp Rule Base

When the past $PaCO_2$ is low	*Metabolic Acidosis*	*Respiratory Acidosis*	*Normal*	*Metabolic Alkalosis*	*Respiratory Alkalosis*
Minimal support	Maintain	Increase	Maintain	Reduce	Reduce
Moderate support (PINSP dominant)	Maintain	Increase	Reduce	Reduce	Reduce
Moderate support	Increase	Increase	Maintain	Reduce	Reduce
Moderate support (Ventilatory rate dominant)	Increase	Increase	Maintain	Reduce	Reduce
Maximal support	Maintain	Maintain	Reduce	Reduce	Reduce

When the past $PaCO_2$ is normal	*Metabolic Acidosis*	*Respiratory Acidosis*	*Normal*	*Metabolic Alkalosis*	*Respiratory Alkalosis*
Minimal support	Maintain	Increase	Maintain	Reduce	Reduce
Moderate support (PINSP dominant)	Maintain	Increase	reduce	Reduce	Reduce
Moderate support	Increase	Increase	Maintain	Reduce	Reduce
Moderate support (Ventilatory rate dominant)	Maintain	Increase	Maintain	Reduce	Reduce
Maximal support	Maintain	Increase	Reduce	Reduce	Reduce

When the past $PaCO_2$ is high	*Metabolic Acidosis*	*Respiratory Acidosis*	*Normal*	*Metabolic Alkalosis*	*Respiratory Alkalosis*
Minimal support	Increase	Increase	Maintain	Maintain	Reduce
Moderate support (PINSP dominant)	Maintain	Increase	Reduce	Reduce	Reduce
Moderate support	Increase	Increase	Maintain	Reduce	Reduce
Moderate support (Ventilatory rate dominant)	Increase	Increase	Maintain	Reduce	Maintain
Maximal support	Maintain	Maintain	reduce	Reduce	Reduce

when the rule base's output was in the opposite direction to the clinician's advice or action. The number and percentage of the exact matches, partial matches, and conflicting mismatches for the simulation data and the retrospective clinical data are shown in Table 7.13.

7.5.6 Further Tuning of the Initial Rule Bases

Table 7.13 shows that although the FiO_2, PEEP, and the Vrate rule bases outputs matched the clinician's advice or action in the majority of cases, the Pinsp rule base performance was less than satisfactory. Moreover, there were seven cases of conflicting mismatches for the PEEP rule base with the simulation data. Therefore, it was necessary to adjust the initial rule bases. To adjust the parameters of the rule bases, the clinicians' simulation results were used as the training data. For each data set, the patient's data became the training inputs and the clinician's advice was used as the targets. In the initial rule bases, the membership in the consequent associated with a rule antecedent was either 1 or 0. This was adjusted during the training. The output from the fuzzy inference system was compared to the target during the train-

Table 7.12 The Ventilator Rule Base

When the past $PaCO_2$ is low	*Metabolic Acidosis*	*Respiratory Acidosis*	*Normal*	*Metabolic Alkalosis*	*Respiratory Alkalosis*
Minimal support	Increase	Increase	Maintain	Reduce	Reduce
Moderate support (PINSP dominant)	Maintain	Increase	Maintain	Reduce	Maintain
Moderate support	Maintain	Increase	Maintain	Reduce	Reduce
Moderate support (Ventilatory rate dominant)	Maintain	Increase	Maintain	Reduce	Reduce
Maximal support	Maintain	Maintain	Maintain	Maintain	Reduce

When the past $PaCO_2$ is normal	*Metabolic Acidosis*	*Respiratory Acidosis*	*Normal*	*Metabolic Alkalosis*	*Respiratory Alkalosis*
Minimal support	Maintain	Increase	Maintain	Reduce	Reduce
Moderate support (PINSP dominant)	Increase	Increase	Maintain	Maintain	Reduce
Moderate support	Increase	Increase	Maintain	Maintain	Reduce
Moderate support (Ventilatory rate dominant)	Maintain	Increase	Maintain	Maintain	Reduce
Maximal support	Maintain	Increase	Reduce	Reduce	Reduce

When the past $PaCO_2$ is high	*Metabolic Acidosis*	*Respiratory Acidosis*	*Normal*	*Metabolic Alkalosis*	*Respiratory Alkalosis*
Minimal support	Maintain	Increase	Maintain	Maintain	Maintain
Moderate support (PINSP dominant)	Increase	Increase	Maintain	Maintain	Reduce
Moderate support	Maintain	Increase	Maintain	Reduce	Reduce
Moderate support (Ventilatory rate dominant)	Maintain	Maintain	Maintain	Reduce	Reduce
Maximal support	Maintain	Maintain	Maintain	Reduce	Reduce

ing. If the output was output member i and the target was output member j, the weight of output member i was adjusted using the formula

$$w_i(n+1) = w_i(n) - \mu_A \cdot u \cdot l_r \tag{7.35}$$

The weight of output member j was adjusted using the formula

$$w_j(n+1) = w_j(n) - \mu_A \cdot u \cdot l_r \tag{7.36}$$

where $w_i(n+1)$ and $w_j(n+1)$ are the new weight vectors for output members i and j, respectively, $w_i(n)$ and $w_j(n)$ are the old weight vectors for output members i and j, respectively, μ_a is the membership values in the antecedents for the particular data set, u is the input vector, and l_r is the learning rate, which is always positive and is set to be between 0 and 1.

During the tuning, a *performance index* was introduced to monitor the progress. The performance index was calculated based on the number of exact matches and

Table 7.13 The Validation Results of the Initial Rule Bases on the Simulation Data (S) and the Retrospective Data (C)

	Exact Match		*Partial Match*		*Conflicting Mismatch*	
	S	C	*S*	C	*S*	C
FiO_2	528 (70.2%)	119 (78.8%)	224 (29.8%)	32 (21.2%)	0 (0.0%)	0 (0.0%)
PEEP	486 (64.6%)	107 (70.9%)	259 (34.4%)	44 (29.1%)	7 (0.9%)	0 (0.0%)
Pinsp	413 (54.8%)	49 (32.5%)	316 (41.9%)	99 (65.6%)	25 (3.3%)	3 (2.0%)
Vrate	623 (82.6%)	123 (81.5%)	130 (17.2%)	24 (15.9%)	1 (0.1%)	1 (0.7%)

conflicting mismatches between the targets and outputs for all the data sets. The performance index (*Perf*) is given by

$$Perf = n(exact\_match) - n(conflicts) \tag{7.37}$$

where *n*(*exact_match*) is the number of exact matches and *n*(*conflicts*) is the number of conflicting mismatches. The tuned rule bases were validated using the retrospective clinical data. Table 7.14 shows that the performance of the FiO_2 and PEEP rule bases improved after the tuning, but that of the Pinsp and Vrate rule bases deteriorated slightly.

To retune the initial Pinsp and Vrate rule bases, the data from the PDMS was used for tuning so that the final rule bases would match the clinical decision-making better. The Pinsp/Vrate subunit was revalidated using new clinical data from the PDMS. Data from six patients were used with a total of 125 data sets. For the Pinsp rule base, there were 75.2% perfect matches. There were 86.4% perfect matches from the Vrate rule base. Before the retuning, the proportion for perfect match for the Pinsp rule base and the Vrate rule base was 30.5% and 73.5%, respectively. Therefore, there were substantial improvements in the performance of both rule bases as a result of the retuning.

7.5.7 Assessment of the Final Fuzzy Rule Bases by an Independent Clinician

In clinical decision support, one can never achieve complete and perfect matching between an advisory system advice and the clinician's advice. The final fuzzy rule bases were assessed by comparing the degree of mismatches between the fuzzy rule bases' outputs and an independent clinician's advice and the degree of mismatches

Table 7.14 The Number of Exact Matches, Partial Matches, and Conflicting Mismatches Between the Clinician's Action as Recorded in the PDMS (Validation Data) and the Rule Base Outputs Before and After Tuning

	Exact Match		*Partial Match*		*Conflicting Mismatch*	
	Before	*After*	*Before*	*After*	*Before*	*After*
FiO2	119	123	32	28	0	0
PEEP	107	134	44	17	0	0
Pinsp	49	46	99	103	3	2
Vrate	123	111	24	39	1	1

$p < 0.01$ using the Chi-Square Test.

between this independent clinician's advice and the clinicians' actions as recorded in the PDMS.

An intensive care consultant was invited to assess the acceptability of each ventilator control action for the data. For each patient, he was presented with the patient's diagnosis and background data. Then, the data sets were presented to him in random order. This was to prevent him from guessing what the actual clinician's action was. Each data set included the blood gas measurements, the previous blood gas measurements, the ventilator settings and measurements, and the previous FiO_2. He was asked to comment on each of the 12 possible ventilator control actions (a reduction, maintenance, or increase in FiO_2, PEEP, Pinsp, and Vrate) as to whether it was *preferable*, *acceptable*, or *unacceptable*. In this way, he was blinded to both the PDMS record and the fuzzy rule bases' outputs. Using his comments, the number of preferred actions, acceptable actions, and unacceptable actions by the clinicians recorded in the PDMS and the fuzzy rule bases' outputs were compared. The results are shown in Table 7.15 and it shows that the advice generated by the rule base was comparable to that of the clinician's actions recorded in the PDMS. This provides further support for the validity of the top-level module knowledge base.

7.6 Integration of SOPAVENT with the Knowledge-Based Levels

The lower-level model-based module generates the *quantitative* component of the advice. It receives inputs from the top-level module, which include the target blood gas levels and the type and direction of the change required for each ventilator setting. Then it calculates the amount of change required in each ventilator setting based on a mathematical model of the respiratory system. However, due to the limitation of the model, the change in PEEP was not derived using a model-based method. The mathematical model chosen was SOPAVENT [31]. It is a model of blood gas exchange during artificial ventilation based on respiratory physiology and mass balance equations. The gas transport system is divided into five compartments: an alveolar compartment, a pulmonary compartment, an arterial compartment, a tissue compartment, and a venous compartment. The lungs are assumed to be composed of three types of alveolar, each with a different ventilation/perfusion ratio (V/Q): normal alveolar (V/Q = 1), dead space (V/Q = ∞), and shunt (V/Q = 0). This transport of gases between the compartments is described by 10 linked first-order

Table 7.15 The Number of Clinician's Actions Recorded in the PDMS (C) and the Number of the Rule Base Advice (F) That Were Judged "Preferable," "Acceptable," and "Unacceptable" by an Independent Consultant

	FiO_2		*PEEP*		*Pinsp*		*Vrate*	
	C	F	C	F	C	F	C	F
Preferable	76	84	77	80	84	85	112	103
Acceptable	50	47	46	47	40	32	32	40
Unacceptable	25	20	28	20	27	34	7	8

None of the differences were statistically significant using the Chi-Square Test.

differential equations: five for oxygen transport and five for carbon dioxide transport. These equations are implemented in MATLAB-SIMULINK. The inputs to the model are the ventilator settings and the outputs of the model include pH, PaO_2, and $PaCO_2$. The model outputs depend on the patient's demographic data (age, gender, weight, and height), the patient's routine measurements (body temperature, systolic and diastolic blood pressure), hemoglobin (Hb) and cardiovascular measurements [cardiac output (CO), oxygen consumption (VO_2) and carbon dioxide production (VCO_2)]. In addition to these variables, the model outputs are determined by the derived internal parameters including shunt and dead space (or dead space to tidal volume ratio, K_D). The shunt can be defined as "the degree of admixture of mixed venous blood with pulmonary end-capillary blood which would be required to produce the observed difference between the arterial and the pulmonary end-capillary partial pressure of oxygen (usually taken to equal ideal alveolar PO_2)" [32]. The dead space is the volume of lungs that needs to be depleted of perfusion in order to give rise to the observed blood gases (mainly $PaCO_2$) given the minute volume ventilation. Only when the patient has a PAC inserted and the model variables are known, can the shunt and dead space be estimated numerically.

7.6.1 Control of FiO_2

By evaluating the oxygen transport equations at steady state, one can derive the Jacobian, which is the first derivative of PaO_2 to FiO_2. The Newton's method can then be used to calculate the change in FiO_2 required to achieve the target PaO_2. The following equation needs to be solved:

$$f(\mathrm{FiO}_2) = SOPAVENT(\mathrm{FiO}_2, \theta) - t\mathrm{PaO}_2 \tag{7.38}$$

where θ is the parameters vector which includes the shunt, CO, VO_2, hemoglobin, body temperature, arterial pH, $PaCO_2$ and bicarbonate; and the $tPaO_2$ is the target PaO_2. The iterative formula can be written as follows:

$$\mathrm{FiO}_2(n+1) = \mathrm{FiO}_2(n) - \frac{SOPAVENT(\mathrm{FiO}_2(n), \theta) - t\mathrm{PaO}_2}{\frac{\partial \mathrm{PaO}_2}{\partial \mathrm{FiO}_2}\Big|\mathrm{FiO}_2(n)} \tag{7.39}$$

where FiO_2 denotes the nth approximation of FiO_2. The iteration stops when the error is within the error tolerance level. In clinical practice, a PaO_2 error of 0.5 kPa is considered acceptable because the error one expects from the blood gas analysis is more than 0.5 kPa due to the measurement errors of the blood gas machines and errors from the contamination and reduction of the PaO_2 in the sample during transit. The algorithm is implemented in MATLAB. The default starting point of the FiO_2 for the iterative procedure is the current FiO_2 the patient is receiving. Theoretically, sometimes a solution can only be found if the FiO_2 is allowed to be of negative value or a number greater than unity. Thus, to prevent the delivery of dangerously low levels of FiO_2 or unachievably high FiO_2, the output FiO_2 is limited to the range from 0.3 to 1.

Figure 7.15 shows how the lower-level FiO_2/PEEP subunit works in closed-loop. It calculates the FiO_2 required to achieve the target PaO_2 level using Newton's method. The calculation is based on the SOPAVENT model, which has been matched to the individual patient. The parameters vector θ in (7.38) is known if the patient has a PAC (i.e., the advisor is operating under the invasive mode). However, in the noninvasive mode, the shunt is estimated using the RI while the CO and VO_2 are estimated using the population median values.

7.6.2 Control of Pinsp and Vrate

The $PaCO_2$ is mainly affected by the minute volume ventilation, which in turn is determined by the Pinsp and Vrate. From steady-state analysis of the carbon dioxide transport equations, when the minute volume is changed from MV_0 to MV, the change in $PaCO_2$, $\Delta PaCO_2$, is given by

$$\Delta PaCO_2 = \frac{P_B \cdot \dot{V}_{CO_2}}{1000(1-K_D)}\left(\frac{1}{MV} - \frac{1}{MV_0}\right) \tag{7.40}$$

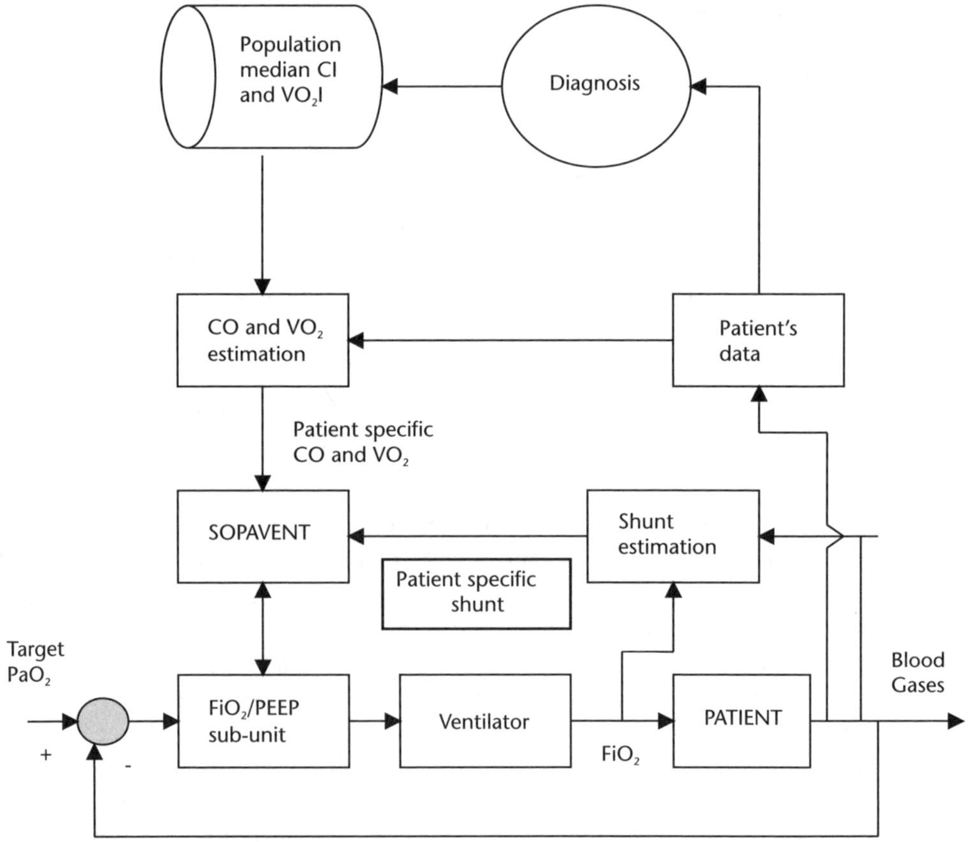

Figure 7.15 The FiO_2 control algorithm in the noninvasive mode.

In pressure-controlled ventilation, one can assume that the Pinsp is constant during inspiration, and therefore, a first-order RC model can be used:

$$MV = \frac{Vrate \cdot pinsp \cdot C}{1000}\left(1 - e^{\left(-\frac{60t_i}{Vrate \cdot RC}\right)}\right) \tag{7.41}$$

where t_i is the inspiratory time as a proportion of the whole respiratory cycle. This new minute volume ventilation can be achieved either by adjusting Vrate, Pinsp, or both. In clinical practice, there is a trade-off between achieving the target $PaCO_2$ level and minimizing the adverse effects of artificial ventilation brought about by excessive ventilatory support. Therefore, it is not always desirable to deliver this target minute volume ventilation. In the lower-level modules, this problem is solved by minimizing a cost function, which includes the error in minute volume ventilation, the change in Pinsp and/or Vrate. The cost function to be minimized is as follows:

$$J_{Cost} = \left(\Delta PaCO_2 - \Delta PaCO_{2\,Target}\right)^2 + \lambda\left(pinsp - pinsp_0\right)^2 + \lambda_2\left(Vrate - Vrate_0\right)^2 \tag{7.42}$$

where $\lambda 1 = 0$ and $\lambda = 0$. If only the Vrate is to be adjusted, the Pinsp is fixed at the initial value ($Pinsp = Pinsp_0$) in the calculation of the minute volume ventilation. If only the Pinsp is to be adjusted, the Vrate is fixed ($Vrate = Vrate_0$) in the calculation of the MV. For the design choice of λ_1 and λ_2, the maximum and minimum Pinsp and Vrate and the mean squared control errors to increase or reduce the $PaCO_2$ by 2 kPa in 20 simulated patients at different values of λ_1 and λ_2 were assessed. It was found that in order to keep the Pinsp and Vrate within the therapeutic range and at the same time achieve a small control error, when both Pinsp and Vrate are to be adjusted, λ_1 should be set to 0.025 and λ_2 should be set to 0.075. When only one of the settings is to be adjusted, the corresponding λ should be set to 0.025.

7.6.3 Setting the PaO_2 and $PaCO_2$ Targets

The lower-level model-based module can only execute if the target PaO_2 and $PaCO_2$ are defined. Therefore, the top-level module is responsible for defining the targets as well as generating advice on which ventilator settings should be changed. Although the targets can be derived using a knowledge-based approach via interviewing clinical consultants, one has to ensure that the targets generated do not conflict with the advice generated by the rule bases. To ensure consistency of the advice, a different approach was taken. We define the target range as a range of values where none of the ventilator settings needs to be changed. For the target range of PaO_2, it means a range of values when neither the FiO_2 nor the PEEP needs to be changed. Similarly, the target range of $PaCO_2$ is defined by the range of values where neither the Pinsp nor the Vrate needs to be changed.

The algorithm for the search of the target values uses the previous PaO_2 level, FiO_2, the previous FiO_2, and PEEP as inputs. The algorithm then searches for a range of in which the FiO_2/PEEP subunit will advise maintenance of both settings. The limit of this range is output target range.

For the $PaCO_2$ target, the bicarbonate level is first estimated from the $PaCO_2$ and pH using the Henderson-Hasselbach equation:

$$pH = pK' + \log\left(\frac{[HCO_3]}{\alpha PaCO_2}\right) \tag{7.43}$$

where pK' is the logarithm of the equilibrium constant of the reaction, $[HCO_3^-]$ is the bicarbonate concentration and α is a constant. Then, the algorithm searches for a range of $PaCO_2$ range within which the output advice of the Pinsp/Vrate subunit is *maintain*. During the search, for each presumed $PaCO_2$ level, the pH is estimated using (7.43) assuming the bicarbonate level is unchanged. The algorithms were tested using the same set of patient data as in Section 7.5.5. The results are shown in Table 7.16. The target ranges generated were considered to be clinically acceptable.

7.7 Implementation and Validation of SIVA

SIVA is implemented in MATLAB, the top-level module being integrated with the lower-level module. A GUI was implemented in LabVIEW to enable the clinicians to key in the data and observe the advice given by the SIVA (see Figure 7.16 for a snapshot of this GUI). In order for the advice to make clinical sense, we limited the output resolution of the system to be the same as the smallest change that a clinician would make to a ventilator setting.

The aim of the closed-loop validation was to assess the system's ability to deal with different scenarios that may occur in the actual clinical environment when the advisory system is used in closed-loop. In the actual clinical environment, the patient's condition may deteriorate or improve over time. There may be inaccuracies in the measurements and the patient's internal parameters may fluctuate naturally. Therefore, a good advisory system should be able to cope with these situations. In the closed-loop validation, the performance of the system was assessed within different clinical scenarios under simulation conditions. The scenarios included:

1. An acute increase in shunt and then the shunt returned to the baseline level after 2 hours;

Table 7.16 The Mean and Range of the Lower and Upper Limits of the Target PaO_2 and $PaCO_2$ Ranges Generated by the Top-Level Module on 151 Sets of Data

		Mean ± s.d. (kPa)	*Range (kPa)*
PaO_2	Lower	10.05 ± 0.40	9.50–12.00
	Upper	15.44 ± 1.27	13.00–20.00
$PaCO_2$	Lower	5.41 ± 0.96	3.20–9.20
	Upper	7.96 ± 0.91	5.20–20.00

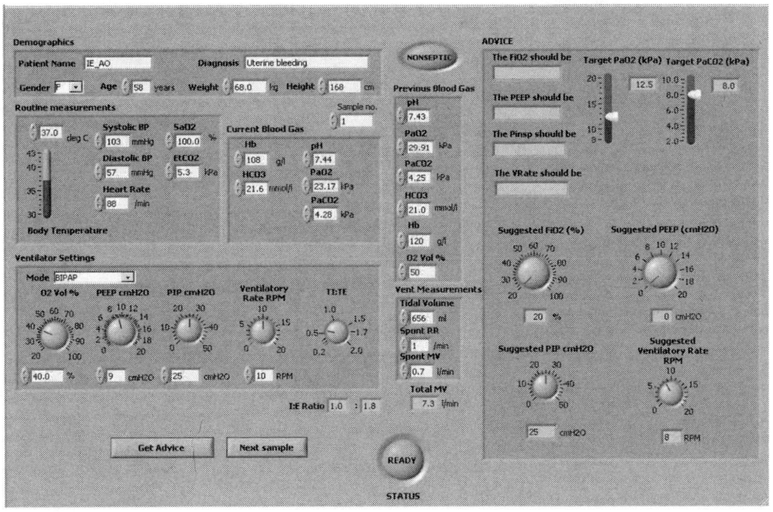

Figure 7.16 The GUI for SIVA.

2. An acute increase in K_D and the K_D returned to the baseline level after 2 hours;
3. A slowly increasing shunt;
4. A slowly increasing K_D.

The simulations emulated the intended use of the advisory system in the ICU. It was assumed that the blood gases were sampled every 30 minutes. The simulated patient was represented by a SOPAVENT model with model parameters initially matched to an ICU patient, and changed during the simulation process to represent the four different scenarios listed above. It was assumed that the advisory system did not know the internal parameters of the simulated patients and the parameters had to be estimated noninvasively if they were not part of the routine measurements. To simulate the natural fluctuations in the parameters, random noise (disturbances) was introduced to the internal parameters of the simulated patient. To assess the effect of K_D estimation errors, scenario 2 was repeated assuming a K_D estimation error of 0.10. The simulated patient was constructed based on a real ICU patient who was a 48-year-old male with septic shock and pneumonia. He was 1.78m tall and weighed 95 kg. His initial shunt and K_D were 0.20 and 0.35, respectively.

The results for the first two scenarios are shown in Figures 7.17 and 7.18. When the shunt was changed, only the PaO_2, FiO_2, and PEEP are shown, and when the K_D was changed, only the $PaCO_2$, Pinsp, and Vrate are shown because the shunt will mainly affect the oxygen transport and the K_D the carbon dioxide transport. For the septic simulated patient, when there was a ramped increase in shunt (Figure 7.17), as no blood gas sampling occurred during the 30 minutes when the shunt was increased, the PaO_2 reduced to 7.16 kPa. The advisory system then increased the FiO_2 from 0.51 to 0.76 and the PEEP was increased to 17 cmH_2O. As a result, the PaO_2 was restored successfully close to the baseline level. During the period when the shunt was reduced, the PaO_2 increased to a value 37.12 kPa because the ventilator settings remained the same (between two sampling points). At the sampling

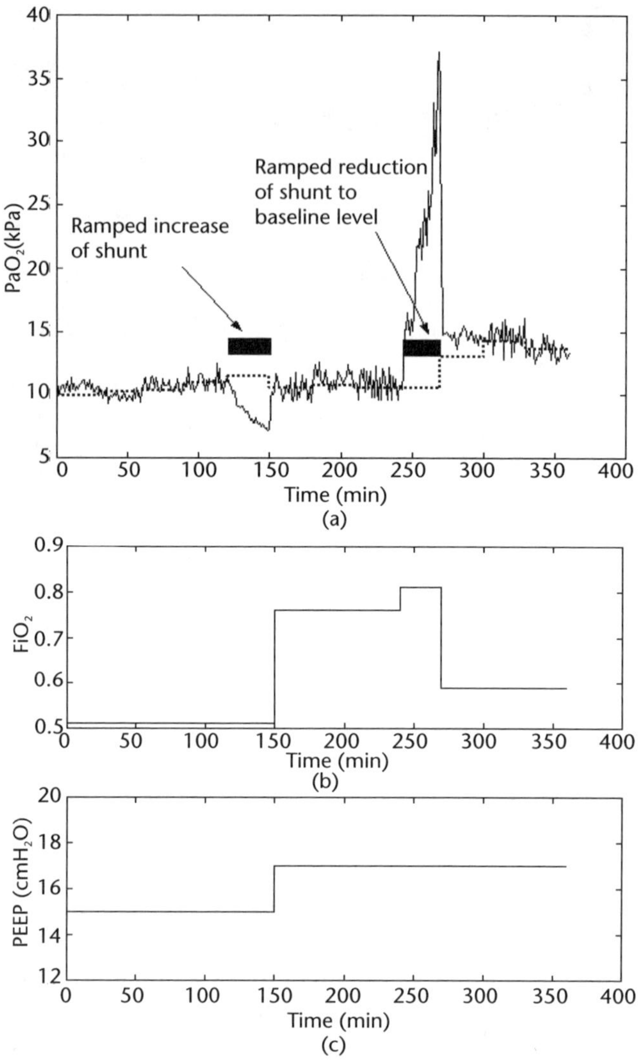

Figure 7.17 (a) PaO_2, (b) FiO_2, and (c) PEEP during a closed-loop simulation of the simulated patient. The dotted line on the PaO_2 plot is the target PaO_2 advised by the top-level module, and the solid line is the simulated patient output PaO_2. The shunt was increased from 0.2 to 0.33 over a period of 30 minutes at the 120th minute and was reduced to 0.2 over a period of 30 minutes at the 240th minute.

point immediately following this period, the advisory system responded by reducing the FiO_2 appropriately to bring the PaO_2 back to the target level. It should be noted that the target PaO_2 changed according to the situation although it was generally close to the normal range (12 to 15 kPa).

Figure 7.19 shows how the PaO_2, FiO_2, and PEEP changed in the simulated patient when the shunt was increased from 0.20 to 0.37 over 330 minutes. When there is an increase in shunt, the PaO_2 will decrease if the ventilator settings are unchanged. This is reflected by the slow decline in PaO_2 in between two sampling points in the graph. Every time a blood gas sample was available, the advisory sys-

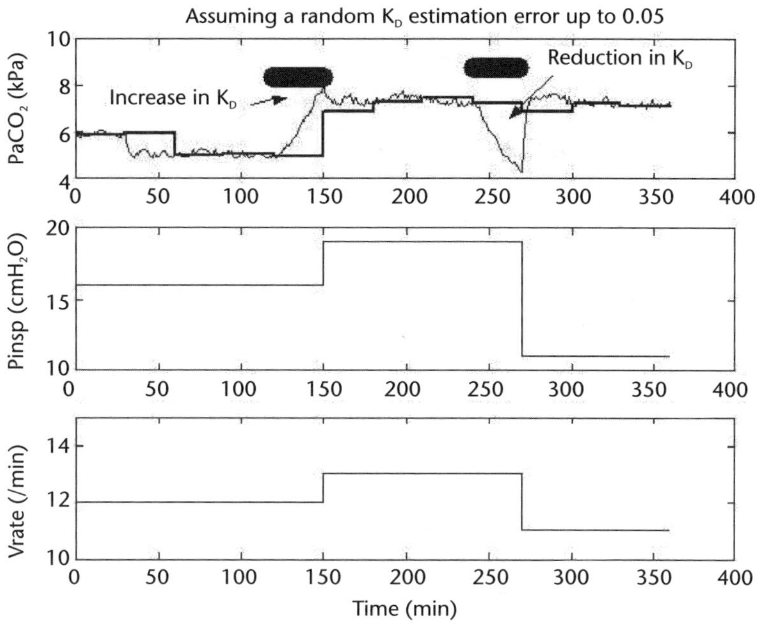

Figure 7.18 The $PaCO_2$, Pinsp, and Vrate of the simulated patient when the K_D was increased from 0.35 to 0.65 over 30 minutes from the 120th minute and was reduced to 0.35 over 30 minutes from the 240th minute. A K_D estimation error of up to 0.1 was assumed in the derivation of Pinsp and Vrate. On the $PaCO_2$ graph, the dotted line is the target $PaCO_2$ advised by the top-level module and the solid line is the simulated output $PaCO_2$.

tem responded appropriately by increasing FiO_2 and/or PEEP. This improved the PaO_2 and kept it close to the target PaO_2 level. At the end of the simulation, both FiO_2 and PEEP were at their maximal setting and if there were a further increase in shunt, the PaO_2 would continue to decline.

Figure 7.20 shows the $PaCO_2$, Pinsp, and Vrate of the simulated septic patient when the K_D was increased gradually from 0.35 to 0.68 over 330 minutes. Unlike the case when the shunt was increased gradually, the advisory system seemed to be slow to respond. It was only after 4.5 hours that the Pinsp was increased. The $PaCO_2$ at the time was 7.68 kPa (compared to the initial $PaCO_2$ of 5.90 kPa). The Vrate was not increased at all. The target $PaCO_2$ level was increased gradually. This is similar to what is being practiced clinically. To reduce the risk of barotrauma, clinicians in generally will accept a high $PaCO_2$ provided the pH is close to normal. It is not unusual for a $PaCO_2$ of 8 kPa to be accepted in ICU.

7.8 Conclusions

The study presented in this chapter utilized multiple sources of knowledge and modeling to develop a hybrid advisory system to mimic the actions of medical staff to treat seriously ill patients in general ICU. Among the serious challenges posed by such an ambitious project one can cite the difficulty in collecting meaningful data to be able to relate system's inputs to outputs, the complexity of the processes involved in patient-ventilator interactions including uncertainties, and the intrapatient and

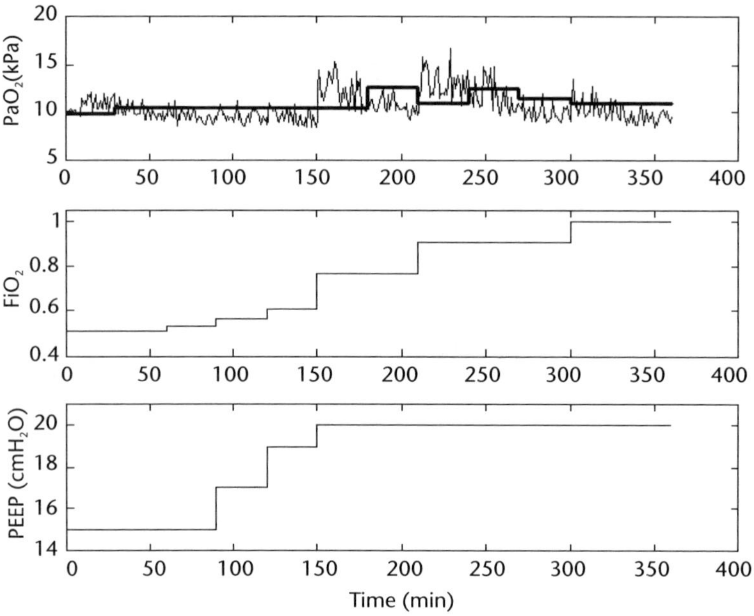

Figure 7.19 The PaO_2, FiO_2, and PEEP during the closed-loop simulation of the patient having a gradually increasing shunt (from 0.2 to 0.37 over 330 minutes). The dotted line on the PaO_2 plot is the target PaO_2 advised by the top-level module and the solid line is the simulated patient output PaO_2.

interpatient parameter variability. In relation to the last of these, one can mention the fact that different patients under different circumstances should have different target blood gas levels. A PaO_2 of 10 kPa may be acceptable for a patient with ARDS and maximum ventilatory support but may be unacceptable for a patient with relatively normal lungs who has always had a PaO_2 of 15 kPa in the previous few days.

Having said this, the research project was fortunate enough to have had a very good starting point, the SOPAVENT model, which was previously developed under a doctoral project by Dr. K. Goode [17]. The model was mechanistic but had the advantage of being easily open to amendments and extensions. The first amendment to the model was to alter the way in which the shunt was estimated. The proposed method used the respiratory index and a neuro-fuzzy model (ANFIS) to provide a better fit to the data than the linear regression model. The use of the respiratory index offers, in fact, a noninvasive and simple alternative to the current unreliable secant method in shunt estimation for the SOPAVENT model. Having strengthened the model to provide better and reliable predictions, the next step consisted of designing the decision support system.

One can easily and rightly argue that there are currently many advisory systems based on fuzzy logic theory (i.e., fuzzy systems), and one can ask what makes the present system unique or stand out, so to speak, from the crowd. The answer perhaps lies in the fact that the system, named SIVA throughout, is hybrid and also uses a combination of handcrafted rules and automatically generated fuzzy rules, via neural networks, which were also automatically fine-tuned to reflect the "true" environment of general ICU.

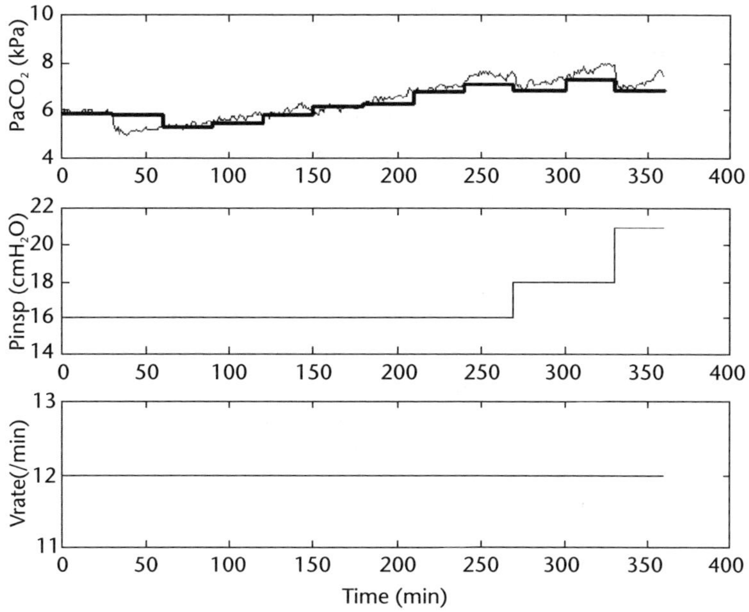

Figure 7.20 The $PaCO_2$, Pinsp, and Vrate during a closed-loop simulation of a septic patient with gradually increasing K_D (from 0.35 to 0.68 over 330 minutes). The dotted line on the $PaCO_2$ plot is the target $PaCO_2$ advised by the top-level module and the solid line is the simulated patient output $PaCO_2$.

SIVA's architecture includes two hierarchical levels, the top-level module, which by using retrospective clinical data was shown to match the clinicians' actions recorded in the PDMS to a high degree, with the degree of disagreement between the two being no worse than that between two clinicians. However, it must be stressed that the validation was performed using the data from the same institution as those used for the development of the system, therefore, the good result may be due to the rule base being biased towards one institution's protocol and policy. It will be interesting for the system to be validated by clinicians from other institutions.

For the model-based lower-level components, we have modified the SOPAVENT model for online adaptive decision support. This is a significant step forward from the work done by Rutledge [33], for instance. Rutledge and colleagues could only use the model for patients with normal lungs and the model had to be improved for patients with abnormal lungs and the computation time became unacceptable for them. The closed-loop validation has indicated that the system is tolerant to the effects of parameter estimation errors, disturbances, and noise. The PaO_2 and $PaCO_2$ targets generated by the top-level module changed as the simulated patient's status changed, and this phenomenon is commonly seen in clinical practice. The lower-level module generated appropriate changes in the ventilator settings to keep the simulated patient's PaO_2 and $PaCO_2$ close to the target and at the same time limit the Pinsp and Vrate to reduce the risk of barotrauma.

A "Conclusion" section to a chapter might suggest that the research study under question is finished and that all objectives have been ascertained. I am convinced that that is not the case and that a lot of work needs to be done in this multidisciplinary area not only in terms of variants of decision support architectures but also from the types of measurements that need to be routinely available. As a result, current research involving electrical impedance tomography (EIT) technique [34, 35], which will allow one to obtain detailed images of the lungs, is currently being investigated. It is envisaged that this technique will be integrated within SIVA's internal structure.

References

[1] Mills, G. H., et al., "Diaphragm Contractility and Lung Volume After Coronary Bypass Surgery," *Am. J. Respir. Crit. Care Med.*, Vol. A788, 1996, p. 153.

[2] Morgan, C. L., "Decision Support for Weaning Patients from Mechanical Ventilation," M.Phil. Thesis, The University of Sheffield, U.K., 2000.

[3] Ward, M. E., C. Corbeil, and W. Gibbons, "Optimization of Respiratory Muscle Relaxation During Mechanical Ventilation," *Anesthesiology*, Vol. 69, 1998, pp. 29–25.

[4] Georgapulos, G., I. Mitrousa, and Z. Bshouty, "Effect of N-REM Sleep on the Response of Respiratory Output to Varying Inspiratory Flow," *Am. J. Respir. Crit. Care Med.*, Vol. 153, 1996, pp. 1624–1630.

[5] Ranieri, V. M., et al., "Patient Ventilator Interaction During Acute Hypercapnia: Pressure Support v Proportional Assist Ventilation," *J. Appl. Physiol.*, Vol. 81, 1996, pp. 426–437.

[6] Nava, S., et al., "Respiratory Response and Inspiratory Effort During Pressure Support Ventilation in COPD Patients," *Int. Care Med.*, Vol. 21, 1995, pp. 871–879.

[7] Alberti, A., et al., "PO Is a Useful Parameter in Setting the Level of Pressure Support Ventilation," *Int. Care Med.*, Vol. 21, 1995, pp. 547–553.

[8] Tobin, M. J., "The Pattern of Breathing During Successful and Unsuccessful Trials of Weaning from Mechanical Ventilation," *Am. Rev. Respir. Dis.*, Vol. 135, 1986, pp. 1111–1118.

[9] Tobin, M. J., S. M. Guenther, and W. Perez, "Konno-Mead Analysis of Ribcage-Abdominal Motion During Successful and Unsuccessful Trials of Weaning from Mechanical Ventilation," *Am. Rev. Respir. Dis.*, Vol. 135, 1987, pp. 1320–1328.

[10] Yang, K. L., and M. J. Tobin, "Decision Analysis of Parameters Used to Predict Outcomes of a Trial of Weaning from Mechanical Ventilation," *Am. Rev. Respr. Dis.*, Vol. A98, 1998, p. 139.

[11] Ely, E. W., et al., "Effect on the Duration of Mechanical Ventilation of Identifying Patients Capable of Breathing Spontaneously," *New England Journal of Medicine*, Vol. 335, No. 25, 1996, pp. 1864–1869.

[12] Dojat, M., et al., "NeoGanesh: A Working System for the Automated Control of Assisted Ventilation in ICUs," *Artificial Intelligence in Medicine*, Vol. 11, 1996, pp. 97–117.

[13] Dojat, M., and L. Brochard, "Knowledge-Based Systems for Automatic Ventilatory Management," *Respiratory Care Clinic of North America*, Vol. 7, 2001, pp. 379–396.

[14] Dojat, M., et al., "Evaluation of a Knowledge-Based System Providing Ventilatory Management and Decision for Extubation," *Am. J. Respir. Crit. Care Med.*, Vol. 153, 1996, pp. 997–1004.

[15] Dojat, M., et al., "Clinical Evaluation of a Computer-Controlled Pressure Support Mode," *American Journal of Respiratory and Critical Care Medicine*, Vol. 161, 2000, pp. 1161–1166.

[16] Hartog, A., et al., "At Surfactant Deficiency, Application of the Open Lung Concept Prevents Protein Leakage and Attenuates Changes in Lung Mechanics," *Critical Care Medicine*, Vol. 28, No. 5, 2000, pp. 1450–1454.

[17] Goode, K. M., "Model-Based Development of a Fuzzy Logic Advisor for Artificially Ventilated Patients," Ph.D. Dissertation, The University of Sheffield, Sheffield, U.K., 2000.

[18] The Acute Respiratory Distress Syndrome (ARDS) Network, "Ventilation with Lower Tidal Volumes as Compared with Traditional Tidal Volumes for Acute Lung Injury and the Acute Respiratory Distress Syndrome," *New England Journal of Medicine*, Vol. 342, 2002, pp. 1301–1308.

[19] Sun, Y., I. Kohane, and A. R. Stark, "Fuzzy Logic Assisted Control of Inspired Oxygen in Ventilated Newborn Infants," *Proceedings of the Annual Symposium of Computer Applications in Medical Care*, 1994, pp. 756–761.

[20] Adlassnig, K., "The Section on Medical Expert and Knowledge-Based Systems at the Department of Medical Computer Sciences of the University of Vienna Medical School," *Artificial Intelligence in Medicine*, Vol. 21, 2001, pp. 139–146.

[21] Nemoto, T., et al., "Automatic Control of Pressure Support Mechanical Ventilation Using Fuzzy Logic," *American Journal of Respiratory and Critical Care Medicine*, Vol. 160, 1999, pp. 550–556.

[22] Riley, R. L., and A. Cournard, "Ideal Alveolar Air and the Analysis of Ventilation-Perfusion Relationships in the Lungs," *Journal of Applied Physiology*, Vol. 1, 1949, pp. 825–847.

[23] Kety, S. S., "Exchange of Inert Gas at Lungs and Tissues," *Pharmacological Reviews*, Vol. 3, 1951, pp. 1–41.

[24] West, J. B., *Respiratory Physiology: The Essentials*, 6th ed., Baltimore, MD: Lippincott Williams and Wilkins, 2000.

[25] Sharanm, J., M. P. Singh, and A. Aminataei, "A Mathematical Model for the Computation of the Oxygen Dissociation Curve in Human Blood," *Biosystems*, Vol. 22, 1989, pp. 249–260.

[26] Kelman, G. R., "Digital Computer Subroutine for the Conversion of Oxygen Tension intro Saturation," *Journal of Applied Physiology*, Vol. 21, 1996, pp. 1375–1376.

[27] Evans, C. W., *Engineering Mathematics: A Programmed Approach*, London, U.K.: Chapman & Hall, 1993.

[28] Zetterstrom, H., "Assessment of the Efficiency of Pulmonary Oxygenation. The Choice of Oxygen Index," *Acta. Anaesthesiol. Scand.*, Vol. 32, 1988, pp. 579–584.

[29] Jang, J. S. R., "ANFIS: An Adaptive-Network-Based Fuzzy Inference System," *IEEE Trans. on Systems, Man and Cybernetics*, Vol. 23, No. 3, 1993, pp. 665–685.

[30] Hagan, M. T., and M. B. Menhadj, "Training Feedforward Networks with Marquardt Algorithm," *IEEE Trans. on Neural Networks*, Vol. 5, No. 6, 1994, pp. 989–993.

[31] Goode, K. M., et al., "Development of a Fuzzy Rule-Based Advisor for the Maintenance of Mechanically Ventilated Patients in ICU," *Biomed. Eng. Appl. Basis Comm.*, Vol. 10, No. 4, 1998, pp. 60–70.

[32] Lumb, A. D., *Nunn's Applied Respiratory Physiology*, 5th ed., Oxford, U.K.: Butterworth Heinemann, 2000.

[33] Rutledge, G. W., et al., "The Design and Implementation of a Ventilator-Management Advisor," *Artificial Intel. Med.*, Vol. 5, 1993, pp. 67–82.

[34] Barber, D. C., and B. H. Brown, "Applied Potential Tomography," *J. Phys. E: Sci. Instrum.*, Vol. 17, 1984, pp. 723–733.

[35] Brown, B. H., "Electrical Impedance Tomography (EIT): A Review," *Journal of Medical Engineering & Technology*, Vol. 27, No. 3, 2003, pp. 97–108.

CHAPTER 8

Hybrid Modeling of Healthy Subjects Experiencing Physical Workload

This chapter promises first to look at the effect of physical (workload) stress on the cardiovascular system and proposes a hybrid closed-loop physiological model that describes the dynamics behind heart rate, blood pressure, and respiration in response to physical stress.

8.1 Introduction

The cardiovascular system consists of a complex set of interacting subsystems, and a complicated control system with several inputs, which regulates the whole system. As McDonald [1] remarked, the many mechanisms that affect the heart can be loosely grouped into the following three main types:

1. The intrinsic cardiac mechanisms that are the source of cardiac rhythmicity;
2. The cardiovascular reflexes that are the basis of circulatory control;
3. Integrated patterns of cardiovascular response to a variety of functional demands such as exercise, thermal regulation, and emotional state.

It is widely agreed that the complexity of the cardiovascular system makes it difficult to study, as it is almost impossible to keep all the influencing factors constant while varying one input. For example, if the subject has eaten recently, the digestion process will affect the cardiovascular system, and the extent of this influence is very difficult to ascertain. Figure 8.1 is a diagram showing the complexity of the physiological interactions in the cardiovascular system [2].

Exercise has a great influence on the cardiovascular system, most noticeable in the effect on the heart rate (HR). At the start of exercise, there is a sharp increase in heart rate, which then increases steadily with increased workload. There is an almost linear relationship between workload and heart rate, ignoring the initial abrupt increase, and when close to maximal working levels. As the exercise continues, more blood is pumped through the circulatory system (affecting the skin) to enable heat dissipation. This results in less blood being pumped to the muscles, and eventually the muscle will be unable to sustain further effort. The relationship between exercise and blood pressure (BP) is more difficult to identify. However, there is an inverse relationship between blood pressure and heart rate. If the blood pressure level drops, the heart rate is increased to restore the blood pressure level to

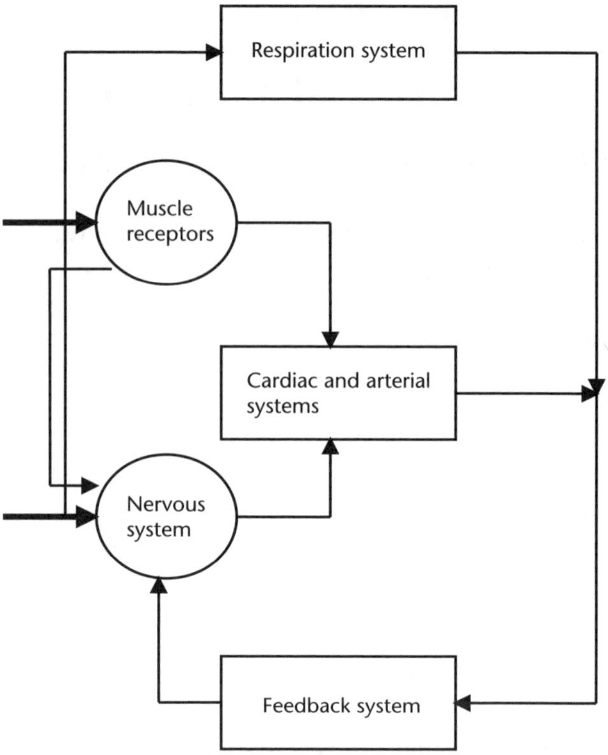

Figure 8.1 A simplified block diagram of physiological interactions in the cardiovascular system.

normal. This relationship is known as Marey's Law of the Heart. This is one example of the many control systems included in the cardiovascular system.

There are two main oscillatory control systems associated with the heart; these are the blood pressure control system and the temperature control system. They both have their own oscillation frequency that can vary from day to day. A graph of a typical power spectrum of a subject's resting heart rate is shown in Figure 8.2 [3]. The regions of interest are marked with arrows. The region marked T is due to thermoregulatory control system oscillations; the peak B is due to blood pressure control system oscillations; and the peak R is due to respiration (RESP).

The frequency of oscillation of each of the control systems is not unique; it will differ between individuals and will vary from day to day. There are many factors that affect the frequency of oscillation and the magnitude of the spectral peak. These include the digestive system, exercise, and health.

An adequate functioning of the cardiovascular system is very important for homeostasis. Homeostasis is responsible for the control of blood pressure and heart rate through the corresponding brain centers to supply the body cells with their needs for oxygen and nutrients under various conditions. In the absence of external perturbations, it might be expected that blood pressure and heart rate will remain constant, but nevertheless, compensation for even the smallest internal perturbations does take place.

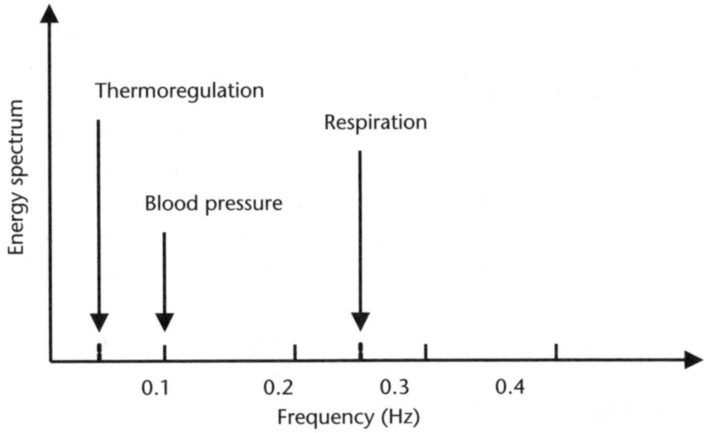

Figure 8.2 The three main frequencies involved in the heart rate spectrum after entrainment.

Blood pressure changes in response to the contraction and relaxation of the heart muscle as well as the instantaneous heart rate (beats/minute) (beat to beat). This is usually referred to as the *quick control loop* of blood pressure [2]. Furthermore, the vasodilation and vasoconstriction of vessels alter the blood pressure, which is controlled by the vasomotor center in the brain [4], this being referred to as the *slow control loop* of blood pressure [2].

Heart rate is controlled by the autonomic nervous system through two branches of effectors: sympathetic and parasympathetic (vagus) groups of fibers. These two fibers are themselves controlled through the cardiac center in the brain [4]. Stimulation of the sympathetic branch increases the HR as well as the strength of the cardiac muscle, which results in an increase of blood pressure, while the stimulation of the parasympathetic branch decreases the HR as well as the strength of the cardiac muscle, which results in a decrease of blood pressure.

It has been found that the respiration signal, which is controlled by the respiratory center, introduces certain disturbances to blood pressure [4, 5]. Furthermore, body temperature affects blood pressure either through vasodilation to release the excess of heat, or by vasoconstriction to reserve heat, since preserving a constant body internal temperature is essential for certain metabolic reactions [6]. Thermoregulation is controlled by the hypothalamus, which is a part of the diencephalon. The hypothalamus serves as an integrator for the temperature and it contributes to the function of the vasomotor center, to control the contraction state of the blood vessels [4].

In addition to the model proposed by Luczak et al. [2, 5], deBoer [7] introduced a closed-loop model that describes the interactions in the cardiovascular control system on a beat-to-beat basis. Later, Whittam [8] modified this model by extending its ability to match the actual HR and BP signals. Seydnejad and Kitney [9] proposed a closed-loop representation for the cardiovascular system, which is able to explain the generation of the high frequency (HF), low frequency (LF), and very low frequency (VLF) peaks existing in the heart rate variability (HRV) spectrum. This model is based on the incorporation of the individual models previously presented in the literature.

In the study presented in this chapter, a hybrid model, which is based on the model structure first proposed by Luczak and coworkers, will be investigated. This hybrid model will include a combination of quantitative and qualitative information relating to blood pressure, heart rate, and respiration. At the center of this modeling architecture will be real-time data relating physical workload (physical stress) to changes in certain physiological variables; this data will be collected from experiments using human subject volunteers. These experiments will consist of imposing a sinusoidal workload, with a period of around 9 to 11 seconds, on healthy subject volunteers. If a periodic workload with a frequency close to the oscillation frequency of the blood pressure control system is imposed upon the subject, then the internal control system can become *entrained*. This means that its natural oscillating frequency changes to that of the exercise workload frequency. The extent of this entrainment will be studied for different subjects under various conditions as detailed in the following sections. The overarching aim of this model, apart from being able to predict accurately and in a transparent way the changes in human physiological behavior, is more ambitiously to exploit the model in a reverse engineering fashion to regulate stress so that *critical* physiological markers are not violated.

8.2 Experimental Setup

8.2.1 The Logistics

The experimental rig used in this research study includes the following components

A Cateye Ergociser Exercise Bicycle

The model EC-3700 is a recumbent-type high-performance exercise bicycle with a built-in computerized training system, as shown in Figure 8.3. The Ergociser features a heart rate sensor, which is clipped to the subject's ear to give measurements of the average heart rate in beats per minute (bpm) and not the beat-to-beat heart rate[1] (BTB). In the case of this particular machine, it is reckoned that the heart rate is averaged using ($n = 8$) samples. The bicycle is interfaced with the computer using a serial communication Port RS232-C. A previously written program, written in visual C^{++}, allows the user to interact with the bicycle by selecting the workload profile, frequency, magnitude, as well as the duration of the exercise. The program allows also the user to log all relevant data into files for later consultation and analyses. The fastest time interval that the user can select to sample this process is 1 second, but this is considered to be fast enough to capture any change of dynamics in the system as a result of stress. See Figure 8.4 for a screenshot of the corresponding GUI.

An Ohmeda Finapres Heart Rate and Blood Pressure Monitor

This equipment was used to acquire measurements of the systolic, diastolic, and mean blood pressure (mmHg) and also of the BTB heart rate (bpm). Finapres stands

1. This is also known as "instantaneous" heart rate.

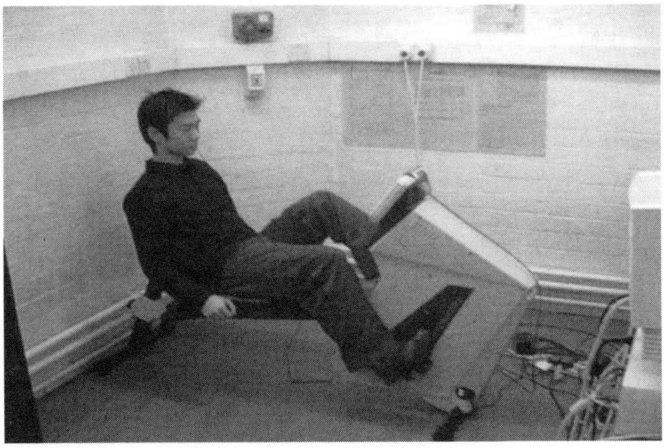

Figure 8.3 Picture showing the Cateye Ergociser EC-3700 fitness bicycle.

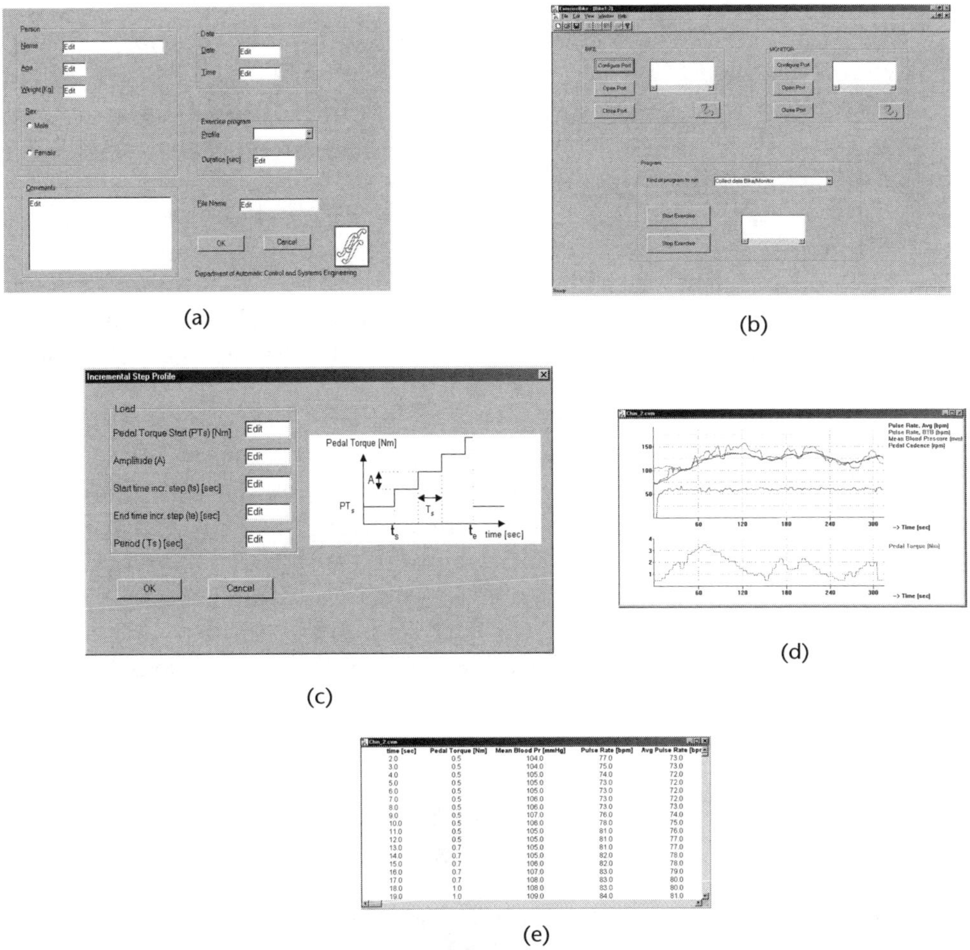

Figure 8.4 The GUI associated with the experimental setup of Figure 8.3: (a) the information form; (b) the communication form; (c) the incremental step profile dialog; (d) the graphic view; and (e) the table view.

for "finger arterial pressure," denoting that this is exactly what the device measures. The blood pressure monitor can also measure AVG heart rate. The BTB heart rate is averaged over 2 seconds, while the AVG heart rate is averaged over 6 seconds. The blood pressure monitor was set so that BTB heart rate is stored, since the average can be obtained via the sensor located in the Ergociser bicycle.

A ProComp+ Advanced Technology System for Measurement and Data Acquisition for Acquisition of Respiration, Temperature, and EEG Signals

The BioGraph/ProComp+ is an advanced technology measurement and data acquisition system. There are eight input channels labeled A to H (see Figure 8.5). The default channel settings are as follows:

A: EEG1 (Brain activity/ECG)	B: EEG2 (Brain activity/ECG)
C: EMG1 (Electromyogram)	D: EMG2 (Electromyogram)
E: SC (Skin conductance)	F: TEMP (Body temperature)
G: BVP (Blood volume pulse)	H: RESP (Resperation)

The inputs A and B are limited to EEG or ECG because they are fast channels with a sampling frequency of 256 Hz, while inputs C to H have a sampling frequency of 32 Hz. The sensors may be processed independently to produce up to 32 user-defined data channels. Up to 12 data channels can be displayed in the software simultaneously.

The EEG Signal

The ProComp+ has the ability to measure the EEG signal through two channels at the same time with a sampling rate of 256 Hz. Using these two channels, two types of connections can be applied to the EEG electrodes: the Bipolar Placement (2-EEG Sensor) and the Referential Placement (1-EEG Sensor). The latter connection, shown in Figure 8.6, was adopted throughout this research study. The ProComp+ measures the EEG signal between the positive and negative poles of the sensor with the positive pole was placed over the Cz location of the 1,020 montage, while the negative pole was placed over the other earlobe. This placement introduces large amplitudes for the recorded EEG signal because the common point (earlobe) is uncorrelated with this EEG signal. This helps to ensure good signal-to-noise ratios for recording the generalized activity changes in the cerebrum that have been associated with thalamic and brainstem arousal systems, which are distinct from local cerebral activation. These recordings are highly correlated to the activities related to

Figure 8.5 The Procomp+ measurement device.

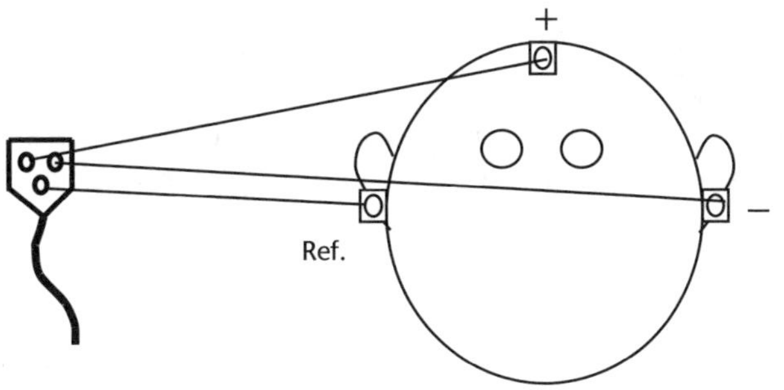

Figure 8.6 Referential EEG electrodes placement.

the cardiovascular, respiratory, and thermoregulatory systems, as the centers that control these systems are located in the brainstem and hypothalamus. The EEG signal is acquired at a sampling rate of 256 Hz.

The Body Temperature Signal (TEMP)

The temperature sensor detects the temperature of the underlying tissue of the test site. Temperature changes as a function of the amount of blood perfusing the tissue. The arterioles supplying blood to the peripheral tissues are surrounded by smooth muscles that are innervated by the sympathetic nervous system. When sympathetically aroused, the muscles constrict, causing vasoconstriction and reducing blood flow to peripheral tissues. The temperature signal is acquired at a sampling rate of 32 Hz.

The Respiration Signal (RESP)

The respiration sensor can be placed over either the sternum (above the breasts) for thoracic monitoring or over the diaphragm for diaphragmatic monitoring. The respiration signal is measured as a function of the change in the thoracic volume. In our case, the respiration sensor will be placed over the sternum and the respiration signal is acquired at a sampling rate of 32 Hz.

The Blood Volume Pulse Signal (BVP)

Photoplethysmography is the process of applying a light source and a light sensor to an appendage such as a finger or a toe and measuring the light reflected by the skin. At each contraction of the heart, blood is forced through the peripheral vessels, producing engorgement of the vessels under the light source and thereby modifying the amount of reflected light to the photo-sensor. The resulting pressure waveform can be monitored and displayed. The BVP signal is acquired at a sampling rate of 32 Hz and can be used to derive other signals such as the heart rate and the blood pressure.

Two IBM Compatible Personal Computers[2]

The first PC is used to control the workload on the Ergociser exercise bicycle and acquire the different monitoring signals from both the Finapres and the bicycle,

2. At the time when this research was conducted, restrictions on the use of two PCs rather than just one were imposed by the computer hardware with respect to the maximum number of serial interfaces one can have on each machine without creating software interrupt conflicts.

while the second PC is used for acquiring those signals sent by the ProComp+ system.

Figure 8.7 depicts the overall configuration of the system as used during the experiments, which will be described in the later sections.

8.2.2 Experimental Design

Although there are various types of excitation signals that can be used for identification purposes as being truly reflective of physical stress—such as sinusoidal, PRBS, and multifrequency—only the sinusoidal signal was retained for this research study as it was found to lead to entrainment and significant changes in all of the physiological variables considered, such as heart rate, blood pressure, respiration, body temperature, and EEG.

In this study, the applied physical workload was chosen to be a sinusoidal wave having the following form:

$$\text{Workload} = 0.75 \cdot \sin(2\pi 0.091 \cdot t) + 1.25 \qquad (8.1)$$

where t is the sampling time and Workload is the applied exercise profile.

These values were chosen to reflect a minimum of 0.5 kg · m, and a maximum of 2 kg · m workloads. These limits were found to be sufficient to excite all the systems under test without exhausting the subject during the period of the experiment. This period was chosen to be 5 minutes, which is the reasonable period for which it is reckoned that the human body can maintain its full efficiency under physical stress. The workload frequency was chosen to be slightly different from the spontaneous oscillation frequency of the blood pressure control system, which is approximately 0.1 Hz. The existence of such frequency (11 seconds or 0.091 Hz) in the blood pressure spectrum was used to check for the blood pressure entrainment to ensure the full transmission of the workload effect to the cardiovascular system [see Figure 8.8 for the workload profile and its spectral analysis using the fast Fourier transform (FFT)]. Each volunteer was asked to maintain the pedaling speed in the range 60 to

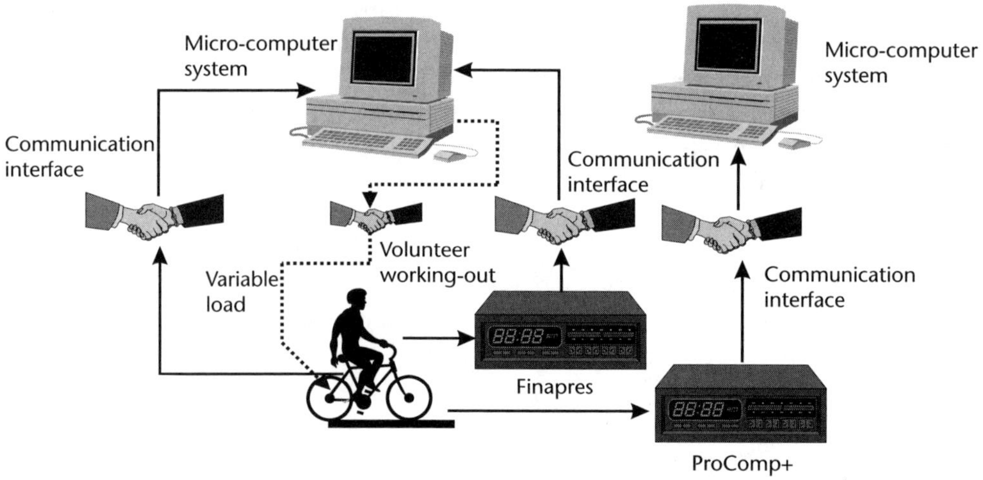

Figure 8.7 Diagram showing the experimental setup.

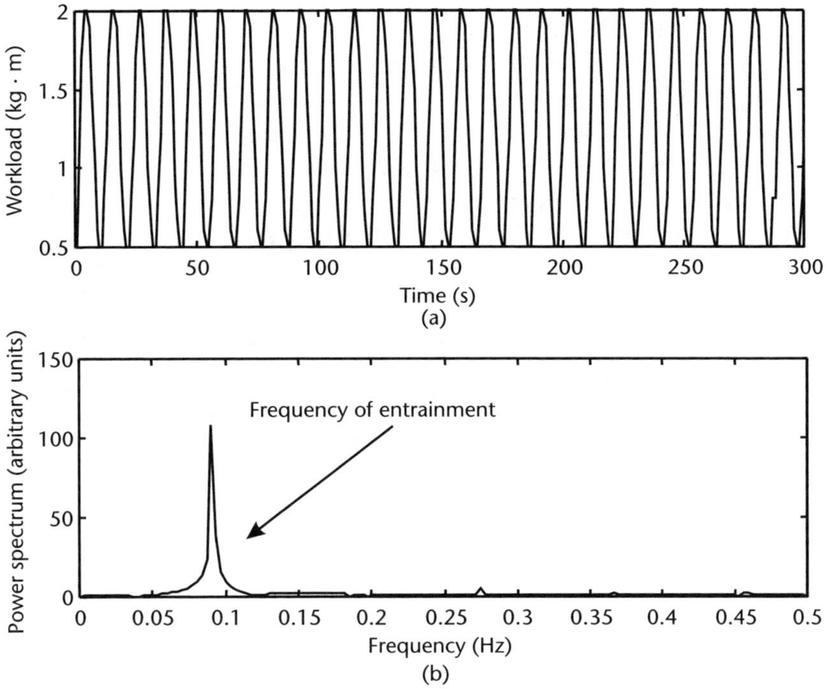

Figure 8.8 A sinusoidal workload in (a) the time domain and (b) the frequency domain.

70 rpm, which ensures that the effect of the workload was indeed being induced. The sampling frequencies were chosen to be 1 Hz for the HR and BP signals; 32 Hz for RESP and TEMP signals; and 256 Hz for the EEG signal. Our data sets include two main age groups: the first one contains 12 subjects with an average age of 34 years (first group), while the second one includes 4 subjects with an average of 29.5 years (second group).

8.3 Modification of the Original Luczak/Raschke Physiological Model

The main objective of this research is to study and model the physiological interactions of the cardiovascular system (CVS) of the human body, with possible extensions to include other functions such as respiration, body temperature, and brain activity in response to physical stress. At the start of this challenging and yet exciting task, there was a common feeling within my research team that it was important to start from a solid base with a model structure that would reflect faithfully causal relationships between *stress* and *effect* on bodily functions but also with a structure that allowed one to expand on other forms of stress (psychological) and to include other meaningful variables. An extensive literature search indeed revealed a vast wealth of submodel structures that tackle certain body substructures with one variable or two as the effect, but such a search did not return many comprehensive[3] models that describe physiological interactions in a "white-box" fashion and yet

3. I have to be careful here not to make the model sound too promising!

retain Einstein's idea of parsimony.[4] Hence, our search revealed that the model proposed by Luczak and Raschke [5], which describes the behavior of heart rate, blood pressure, and respiration rate in response to stress, represents an ideal model structure for this study. However, before analyzing this model as a whole and exploring how to quantify its components, it is useful to analyze sample data sets that were collected using the experimental rig described in Section 8.2 and assess the difficulties and pitfalls relating to such processes.

8.3.1 Direct Model Identification for Heart Rate and Blood Pressure Under Stress Conditions

As a case study for this section, previous data files collected as part of an M.Sc. final project in my department were used to identify the models relating to heart rate and blood pressure by employing various data-driven model identification tools. These files include heart rate and blood pressure data for several subject volunteers who were subjected to sinusoidal workloads with different frequencies of periods 9 to 11 seconds. The modeling tools that will be investigated here[5] will focus on autoregressive with exogenous input (ARX), neural networks (NN), and fuzzy logic based models.

The Least Square ARX (LSARX) Model

The general form of an nth-order ARX model with offset is as follows:

$$y(k+1) = -\sum_{i=1}^{i=n} a_i \cdot y(k+1-i) + \sum_{j=1}^{j=m} b_j \cdot u(k+1-j-d) + e(k) + offset \tag{8.2}$$

where:

k represents the sampling time.

d represents the integer value of the time delay.

y, u, e represent the output, input, and residual errors, respectively.

offset represents the fact that some signals start from a significant baseline value.

A detailed analysis, based on the MSE between the actual and estimated signals, was carried out, for both HR and BP signals, to select the best delay (in samples) and also the best order of the model. It was found that the best delays are 2 samples for HR and 3 samples for BP. The best order was found to be 5, for both HR and BP, but comparing the MSE of the fifth-order model with that of the first order led to little significant difference. Therefore, in order to alleviate the computational burden and conserve simplicity, the optimum order of the model was chosen to be equal to 1.

Hence, the specific forms of the model used for heart rate and blood pressure identifications are

4. Einstein's idea of model parsimony suggests that "a model should be as simple as possible but no simpler."
5. These have been chosen to reflect the old and classical and the new and emerging.

$$HR(k+1) = -\hat{a}_{1HR} \cdot HR(k) + \hat{b}_{1HR} \cdot PS(k-d) + \hat{b}_{2HR} \cdot PS(k-1-d) + offset_{HR} \tag{8.3}$$

$$BP(k+1) = -\hat{a}_{1BP} \cdot BP(k) + \hat{b}_{1BP} \cdot PS(k-d) + \hat{b}_{2BP} \cdot PS(k-1-d) + offset_{BP} \tag{8.4}$$

where *PS* stands for physical stress.

The parameters $\hat{a}_1$, $\hat{b}_1$, $\hat{c}_1$, and *offset* in the models of (8.3) and (8.4) are calculated using the least squares algorithm. In a matrix form one can write

$$\mathbf{Y} = \mathbf{\Phi} \cdot \mathbf{B} \tag{8.5}$$

Y is the vector containing all the output samples obtained during a measurement; that is,

$$\begin{bmatrix} HR(1) \\ HR(3) \\ \dots \\ \dots \\ HR(N+1) \end{bmatrix}$$

The vector **B** contains the model parameters, for example, for first order,

$$\begin{bmatrix} \hat{a}_1 \\ \hat{b}_1 \\ \hat{b}_2 \\ offset \end{bmatrix}$$

The matrix **Φ** has the following form:

$$\mathbf{\Phi} = \begin{bmatrix} -HR(1) & PS(1-d) & 0 & 1 \\ -HR(2) & PS(2-d) & PS(1-d) & 1 \\ -HR(3) & PS(3-d) & PS(2-d) & 1 \\ \dots & \dots & \dots & \dots \\ \dots & \dots & \dots & \dots \\ -HR(N) & PS(N-d) & PS(N-1-d) & 1 \end{bmatrix}$$

The least squares estimate is obtained by minimizing the sum of the squares of the model errors. The MSE is calculated as follows:

$$MSE = \frac{1}{N} \sum_{k=1}^{N} \left(y_k - y_k^p \right)^2 \tag{8.6}$$

where y_k is the actual output at sample instant k, y^p_k is the predicted output at the same sample instant, calculated by multiplying the row of matrix $\mathbf{\Phi}$ corresponding to this sampling instant with the column vector $\hat{\mathbf{B}}$ containing the model parameters.

In a matrix form, the mean square error can be written as follows:

$$MSE = \frac{1}{N}\left(\mathrm{Y} - \mathbf{\Phi}\hat{\mathrm{B}}\right)^T\left(\mathrm{Y} - \mathbf{\Phi}\hat{\mathrm{B}}\right) \tag{8.7}$$

Solving with respect to $\hat{B}$ leads to the following expression:

$$\hat{\mathrm{B}} = \left(\mathbf{\Phi}^T\mathbf{\Phi}\right)^{-1}\mathbf{\Phi}^T\mathrm{Y} \tag{8.8}$$

Then vector $\hat{\mathbf{B}}$ contains the estimates of the model parameters.

A sample of the estimated HR and BP using the LSARX model is shown in Figure 8.9. From the figure, a good fit is apparent for the HR signal and an acceptable fit is apparent for the BP signal.[6]

The Neural Network Model

Both linear and nonlinear neural network structures can be used to identify the relation between either HR or BP and the physical stress. However, the relationship

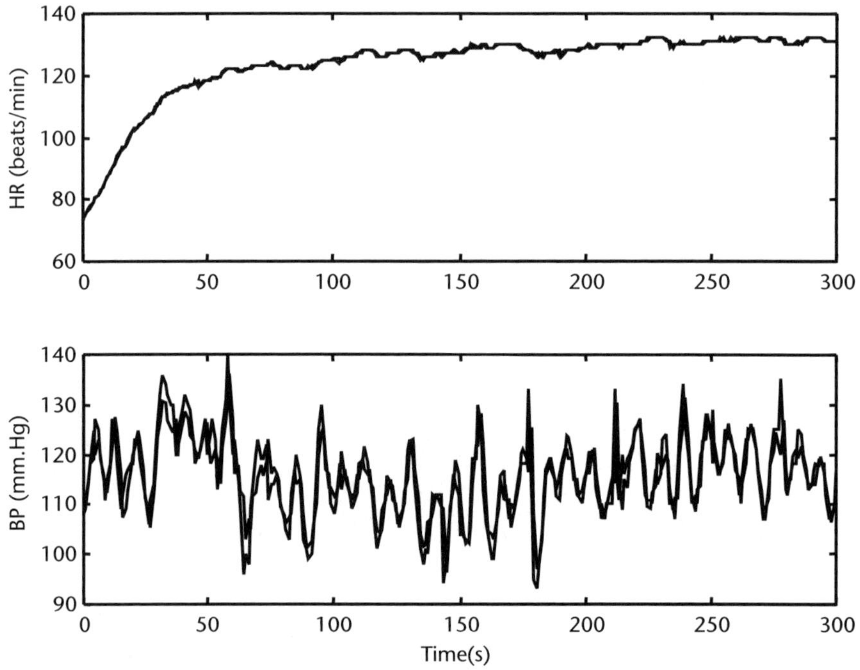

Figure 8.9 Actual and predicted HR and BP signals using the least-square linear model.

6. From the author's own experience, blood pressure is always the most problematic signal to identify due to its "tricky" dynamics!

between physical stress and either the HR or BP signals is nonlinear, hence a nonlinear neural network was used. The structure of the nonlinear neural network is shown in Figure 8.10.

The use of past samples from the output signals (HR or BP) introduced some dynamics into the constructed model. By testing all the combinations, between the delays for both the input and output in terms of the MSE, it was found that the best structure is to use a 6-sample delay for the output with 2 samples for the input. The network configuration was chosen to be as follows: 10 neurons in the input layer, 5 neurons in the hidden layer with linear activation function, and 1 neuron in the output layer with sigmoid activation function to predict either HR or BP signals. The selection of activation functions, as well as the number of neurons in the hidden layer was done based on the MSE after trying various structures for the neural network.

A sample of the estimated HR and BP using the nonlinear neural network-based model is shown in Figure 8.11. From this figure, one can visually notice better results than the LSARX model. This is expected, because the neural network is a nonlinear identification method, which has the ability to efficiently capture all the nonlinear dynamics existing in HR and BP control systems.

The Fuzzy Logic Based Model

The ANFIS architecture introduced in Chapter 6 (Appendix 6C), which allows one to elicit fuzzy logic based rules of a Takagi-Sugeno Kang (TSK) type of the following form, was used on the same data files as previously:

Rule i: IF x_1 is A_{1l} and x_2 is A_{i2} and ... and x_m is A_{im} THEN $= f(x_1, x_2, \ldots, x_m)$

where:

$(x_1, x_2, \ldots, x_m)$ are the inputs to the system.

y is the output.

A_{ij} are the linguistic labels such as zero (ZE), negative small (NS), or positive big (PB).

$f(x_1, x_2, \ldots, x_m) = c_0 + c_1x_1 + \ldots + c_mx_m$ is usually taken as a linear[7] mapping between the input and output spaces.

A sample of the estimated HR and BP signals using the fuzzy-based model is shown in Figure 8.12. This figure shows that the fuzzy logic based model reached a

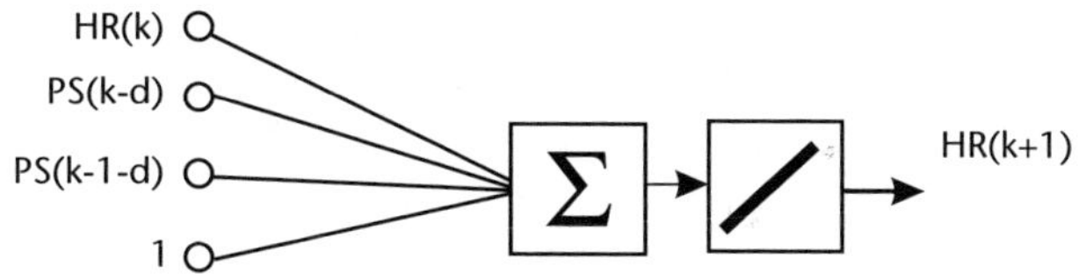

Figure 8.10 The structure of the neural network model.

7. The overall fuzzy mapping is still a nonlinear relationship despite the fact that the function f is linear!

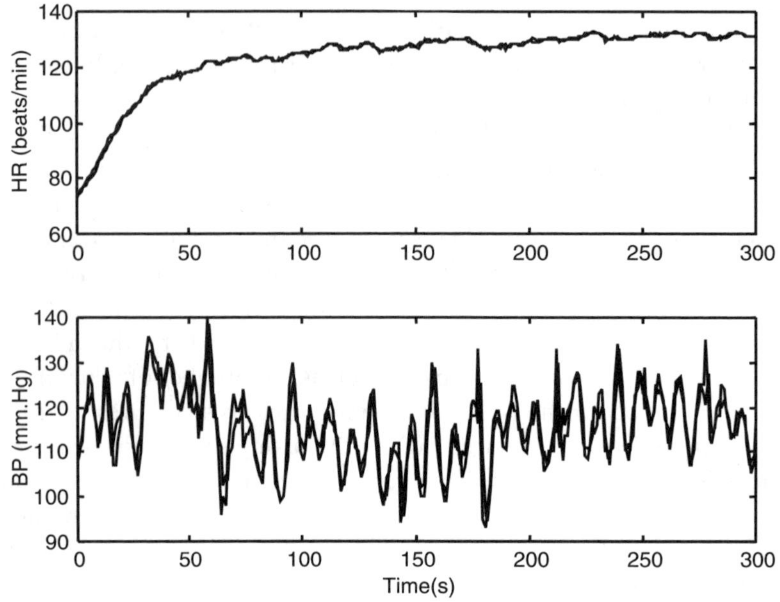

Figure 8.11 Actual and predicted HR and BP signals using the neural network model.

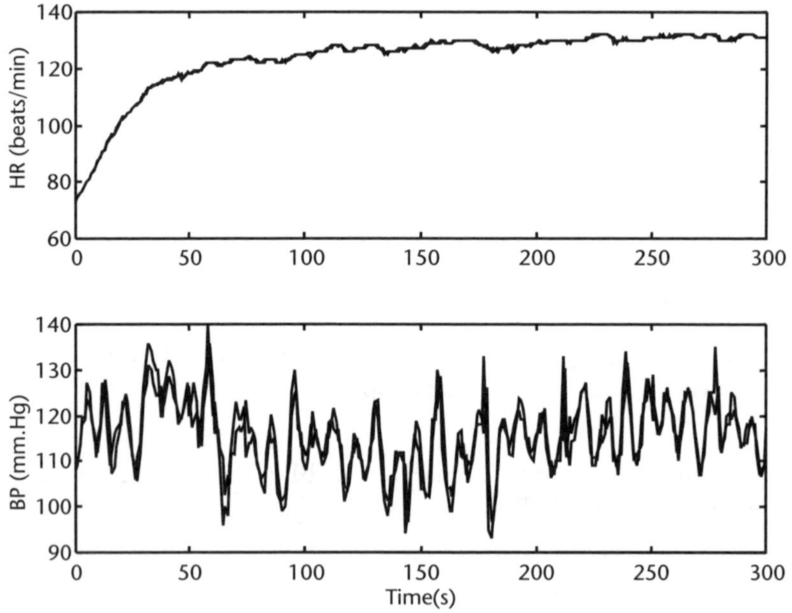

Figure 8.12 Actual and predicted HR and BP signals using the neuro-fuzzy model versus ANFIS.

good fit and the performance suggests a similar performance as the nonlinear neural network model.

When comparing the three model performances against this data, Table 8.1 would suggest that the fuzzy logic–based model is the model that led to the lowest

Table 8.1 Comparing the Three Modeling Techniques Using HR and BP Signals

Signal Type	*Heart rate*				*Blood Pressure*			
Method Used	*Linear*		*Nonlinear*		*Linear*		*Nonlinear*	
	LSARX	*NN*	*NN*	*Fuzzy*	*LSARX*	*NN*	*NN*	*Fuzzy*
MSE	0.3607	0.539	0.3483	0.1889	19.9741	21.34	18.66	11.33

MSE index for heart rate as well as blood pressure. Furthermore, when the three types of models were validated against the remaining sets of available subject volunteers' data, the comparative analysis led to Table 8.2, which allowed one to conclude the following:

1. A nonlinear model structure is recommended.
2. For heart rate modeling a fuzzy logic based model is best.
3. For blood pressure, the neural network based model performs better.

In the light of the above, one can conclude that the fuzzy logic based modeling paradigm, which uses a neural network architecture to elicit the fuzzy logic based rules inherently, will be employed as the modeling tool for the remaining investigations into the physiological closed-loop model proposed by Luczak and his coworkers [2, 5].

8.3.2 A New Gray-Box[8] Physiological Closed-Loop Model Describing Stress

In the previous section it transpired that it was possible to elicit models that describe the effect of physical stress on variables such as heart rate and blood pressure, but it is also clear that such model structures, however accurate and transparent they can prove to be, cannot be classed as "implicitly" closed-loop.[9] Figure 8.13 shows the structure of a more detailed model, which was first proposed by Luczak et al. [4, 5]. It includes the most significant interactions in the cardiovascular system. Within this "generic" model representation, the authors summarized the interactions within the cardiovascular system as follows:

Table 8.2 Comparing the Three Modeling Techniques Using HR and BP Signals Against All Remaining Sets of Data

Signal Type	*Heart rate*				*Blood Pressure*			
Method Used	*Linear*		*Nonlinear*		*Linear*		*Nonlinear*	
	LSARX	*NN*	*NN*	*Fuzzy*	*LSARX*	*NN*	*NN*	*Fuzzy*
MSE	3	0	27	32	1	0	54	28

8. A gray-box model is a combination of white-box (detailed information) and black-box (not so detailed) components.
9. By "closed-loop" model, I mean a recirculatory model, where one component in this model can act as a cause as well as an effect; all these interactions should be explicitly described in such a model.

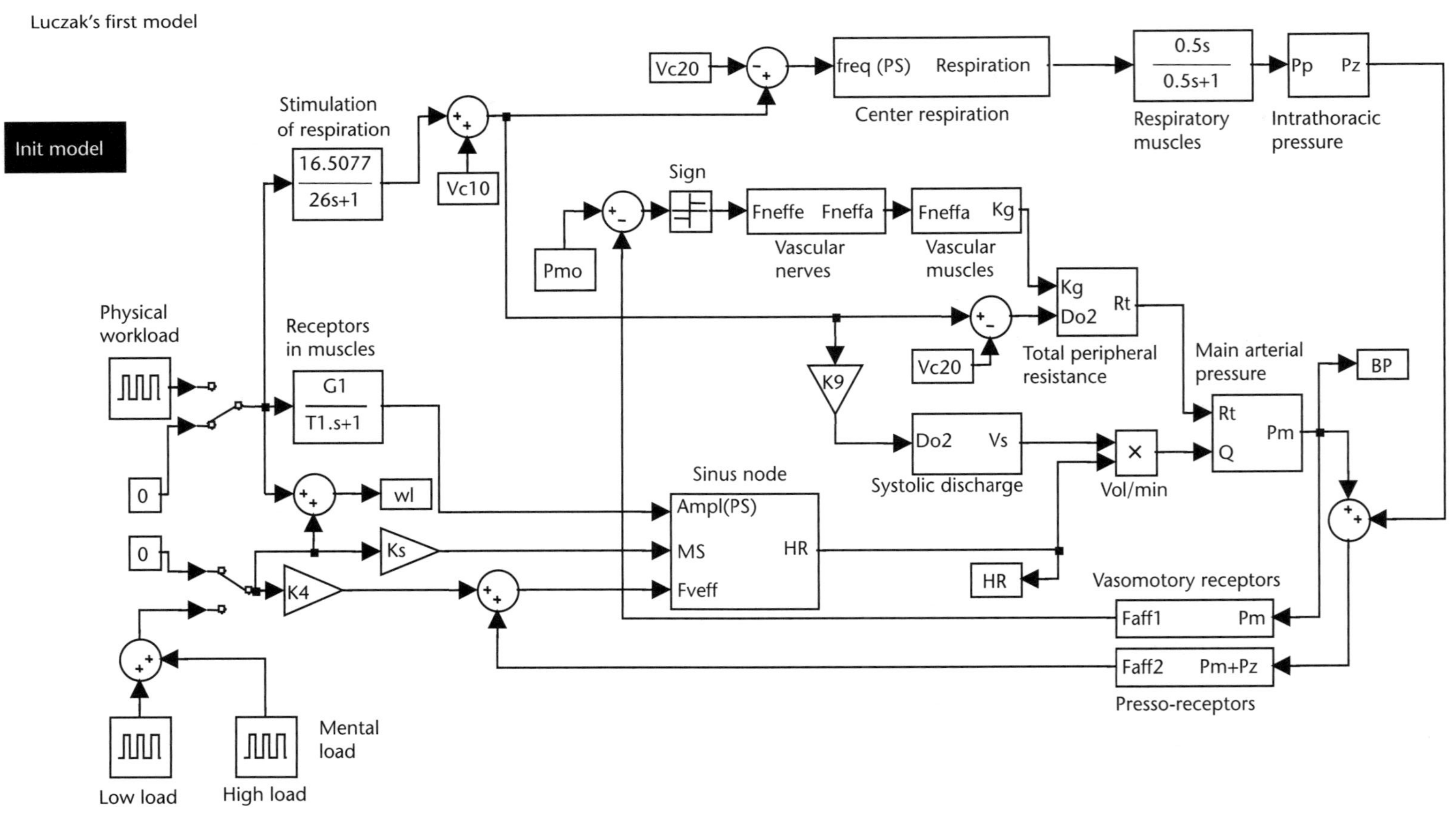

Figure 8.13 The detailed model structure corresponding to the diagram of Figure 8.1.

- There is a direct relationship between the RESP and the BP signals, such that RESP acts as a disturbance to BP, which produces blood pressure variations. These variations are detected by the presso-receptors (baro-receptors).
- There is an indirect relation between the RESP and the HR signals, such that the disturbance of BP by the RESP signal is detected by the presso-receptors which in turn generate efferents that stimulate the cardiac center to alter the HR signal accordingly.
- Any variation in the HR signal affects the BP signal through the cardiac output variation, as the BP signal is a result of the multiplication of the cardiac output by the peripheral resistance of blood vessels. This relation is called the "quick loop" and controls the BP signal.
- The variation of blood pressure is detected by the vasomotor receptors, which affect the vasomotor center, which, in turn, controls the blood pressure through either vasoconstriction or vasodilation of blood vessels. This relation is called the "slow loop" and controls the BP signal.
- The existence of physical stress, in the form of exercise, affects all of the three signals (HR, BP, and RESP) in the following ways: it increases the respiration frequency by increasing the respiratory impulses from the respiratory center in the brain; it increases the HR by increasing the sympathetic efferents to the SA node; and it alters the blood pressure through the variation in both the cardiac output and total peripheral resistance. These variations result in providing the muscles, which are involved in the exercise, with their requirements of oxygen.

Because it is possible to acquire real-time data using the experimental rig—which has already been described in Section 8.2—it was decided to modify this closed-loop model to (1) reflect this newly collected data, (2) generalize the model to several categories of subjects' dynamics, and (3) avoid the curse of finding the optimal values of "gains" and "time constants," which were missing in Luczak and coworkers published model data [4, 5]. Hence, all the applied modifications are based on the use of the fuzzy-based models, in the form of ANFIS structure, which can be summarized as follows:

- The calculation of the HR signal was carried out by the multiplication of the sympathetic and vagal branches, which are responsible for increasing and decreasing the HR respectively. The modification is related to the use of an ANFIS submodel to calculate HR using the output from the receptors in muscles block, which represents the effect of workload on varying the HR signal, and the output from the presso-receptors, which detect any variations in the blood pressure, as shown in Figure 8.14.
- The calculation of the BP signal was carried out by multiplying the total peripheral resistance (TPR) and the blood volume per minute, which is the blood flow (Q). In this case, the modification lies in the use of an ANFIS submodel to calculate the BP from the two previously mentioned inputs, as shown in Figure 8.15.
- The calculation of the respiration signal was carried out by generating a signal with a frequency of 0.25 Hz, then adding the effect of physical stress through

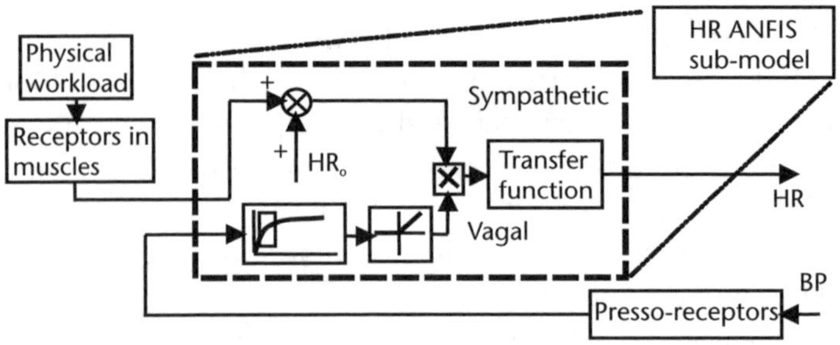

Figure 8.14 The modification that was applied to the blocks affecting HR in the original Luczak model.

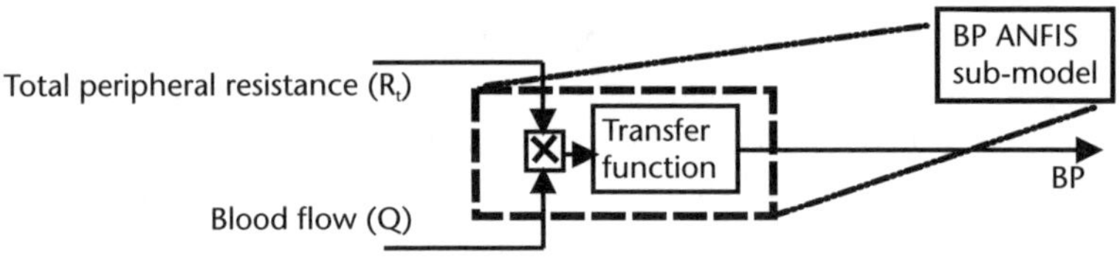

Figure 8.15 The modification that was applied to the blocks affecting BP in the original Luczak model.

the stimulation of respiration. The modification lies in the use of an ANFIS submodel to calculate the respiration signal using the physical stress as an input, as shown in Figure 8.16.

The cardiovascular model was built by importing the three proposed ANFIS submodels into Luczak's first model. Figure 8.17 shows a block diagram of the proposed cardiovascular model. Using this model representation, real data collected from subject volunteers will be used to train and validate such neuro-fuzzy based models.

The Training and Testing Data Sets

In order to elicit the three ANFIS-based models included in Figure 8.17, five data sets emanating from five subject volunteers were presented to the previously described three ANFIS structures. The various inputs used for each network are summarized in Table 8.3. Furthermore, because each one of these models needs to reflect the *dynamic* behavior of the associated variables, "time" was included in each network to show that behavior is *time-varying*.

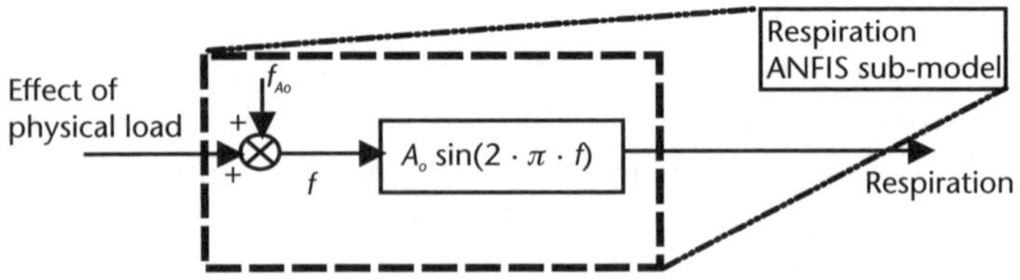

Figure 8.16 The modification that was applied to the blocks affecting RESP in the original Luczak model.

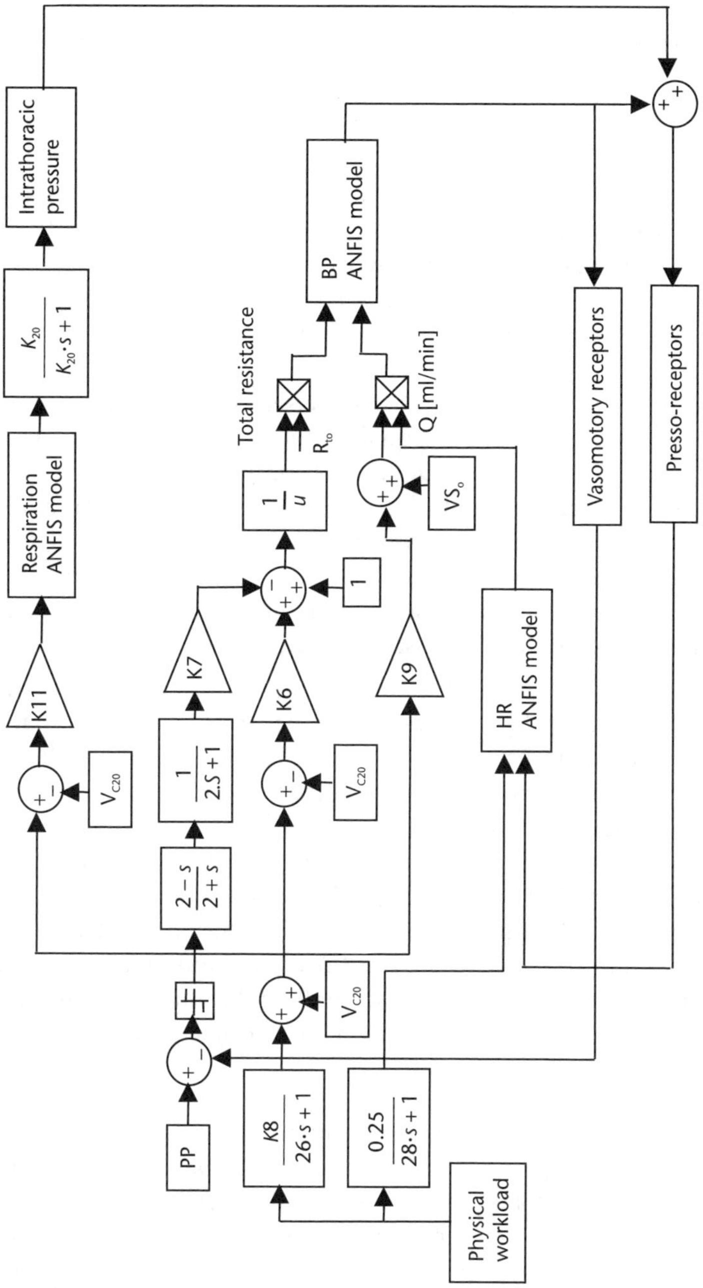

Figure 8.17 The overall physiological model including the ANFIS-based models.

Table 8.3 The Various Input/Output Mappings Used for HR, BP, and RESP Submodels

Submodel	*Inputs*		*Output*
	First input	*Second input*	
HR ANFIS submodel	Sympathetic efferents	Vagal efferents	HR
BP ANFIS submodel	Peripheral resistance	Blood flow	BP
RESP ANFIS submodel	Effect of physical load	Delayed effect of physical load by one sample	RESP

Hence, the training and testing sets were constructed by dividing the whole data record into two sets with a ratio of 2:1. Since the length of the BP signal is 300 samples (corresponding to 300 seconds worth of data), the training set will be chosen as 200 samples and 100 samples for the testing set. Two methods were used to construct the ANFIS structure: grid partitioning and subtractive clustering.

- *Grid partitioning.* This method is based on dividing the input space for each variable into a number of equally spaced fuzzy membership functions (MFs). Then the rules are constructed as a combination for all the MFs of the input variables. Since the number of rules is the multiplication of all the numbers of MFs for the inputs, this number will dramatically increase as the number of MFs in the input space is increased. This method is not preferred due to its computational overhead and difficulty in the fine-tuning of the ANFIS submodel, which will be discussed later.
- *Subtractive clustering.* This method is based on a clustering technique proposed by Chiu [10] to calculate the parameters of the submodel's MFs. This technique is implemented as a subroutine ready to be used within the MATLAB environment. The selected radii for the clusters in the inputs and output spaces affect the resultant number of clusters, which correspond to the number of rules. These rules are constructed by assigning each MF from each input to one MF in the output, which makes the number of rules equal to the number of MFs (number of clusters). This method is characterized by its reduced number of MFs over the grid partitioning method, which reduces the computational overhead. Additionally, it has the ability to capture the existing dynamics in the input-output data set through clustering of data as well as ease in fine-tuning.

Table 8.4 shows the results of a comparative study of the mean square error (MSE) indices relating to the blood pressure signal over the training and testing sets.[10] From these results it is apparent that the method that applies clustering leads to a good compromise between accuracy and efficiency of the model.

It is worth noting perhaps at this stage that the study has concentrated exclusively on the blood pressure signal. This is because its dynamics are the trickiest to identify when compared to heart rate or respiration. Hence, the analyses below will continue to focus on this signal.

10. All data were normalized in the range [0, 1] prior to modeling.

Table 8.4 Comparative MSE Criteria Relating to BP Between Grid Partitioning and Subtractive Clustering

ANFIS Type	*Grid Partitioning (3 MFs, 27 fuzzy rules)*		*Subtractive Clustering (6 MFs, 6 fuzzy rules)*	
Process Type	Training set	Testing set	Training set	Testing set
Normal Data	18.29	20.33	28.01	27.45
Normalized Data	10.64	12.30	15.9	15.97

Analysis and Further Fine-Tuning of the Fuzzy Logic–Based Models

Blood pressure ANFIS submodels for Subject 5 were constructed using the subtractive clustering algorithm previously described. Hence, to analyze the stability of these submodels, the corresponding three-dimensional surfaces relating to the inputs/output mappings were constructed. Figure 8.18 shows the resulting surfaces.

It is clear from Figure 8.18, that although the output was guided so that it lies within the normalized range [0, 1] (normalized output), there exist certain regions in each of the graphs in the figure that include out-of-range values for the output. This is mainly caused by the attempt to extrapolate beyond the remit of the data, therefore making the estimates less reliable. As a result the model ended up with a relatively large number of fuzzy rules (13 in this case, corresponding to 13 MFs for each variable). Figure 8.19 shows the actual versus estimated blood pressure signals for the blood pressure ANFIS submodel over the testing set.

To improve on the above prediction performance and at the same time improve on the model transparency by limiting the model's ability to extrapolate too much beyond the data it trained for, the automatic tuning technique suggested by Chao [11] and Castellano and Fanelli [12] was used; this consists of simply removing the

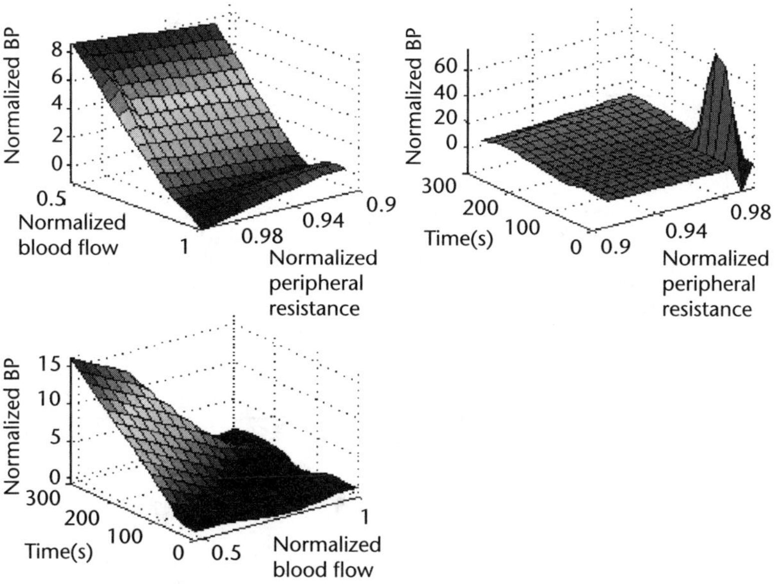

Figure 8.18 The three-dimensional surfaces relating to the fuzzy logic–based mappings between the normalized inputs and BP.

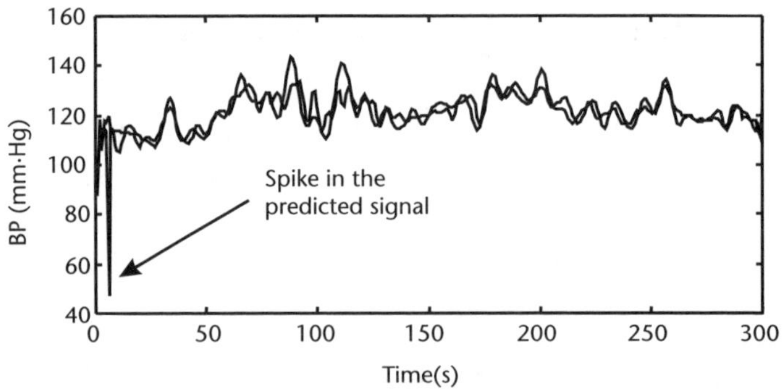

Figure 8.19 Actual versus predicted BP using the ANFIS-based submodel via subtractive clustering technique.

redundant fuzzy rules or those rules that do not contribute significantly[11] to the model predictions. The algorithm calculates the NORM for the actual output of the rules. Then it selects the rule that has the minimum NORM value which corresponds to the rule with the minimum effect on the model. Finally, it updates the rules' weights accordingly. Using this method, the performance of Figure 8.20 was obtained, which led to an MSE index of 16.22 with a total of seven rules compared to an MSE criterion of 34.78 and 13 fuzzy rules prior to this fine-tuning operation. In turn, Figure 8.21 shows the three-dimensional surface mappings between inputs and outputs, which, despite an improved performance, still includes regions where the normalized range has been violated.

In these circumstances a further tuning can be carried out, which consists of intervening directly onto the fuzzy MFs to modify the positions of the peaks and the widths, but this was not found to be to affect the model dynamics too much, leading to the performance of Figure 8.22 which does not allow one to track the signals' fluctuations.

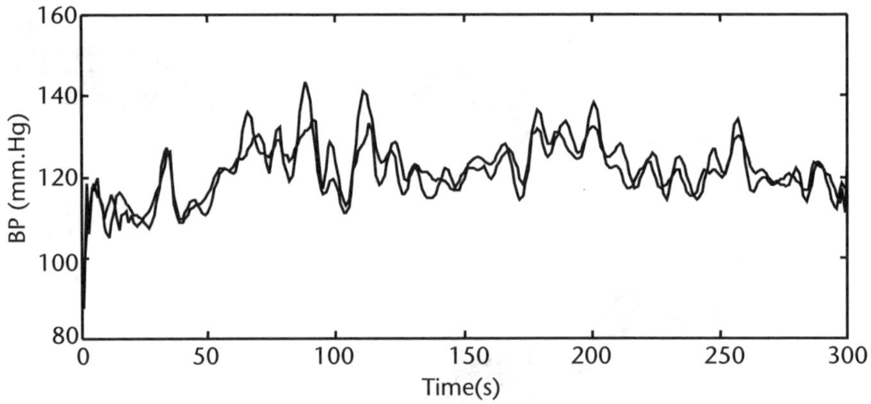

Figure 8.20 Actual versus predicted BP using the ANFIS-based submodel via subtractive clustering technique after tuning.

11. I refer to the positive contribution.

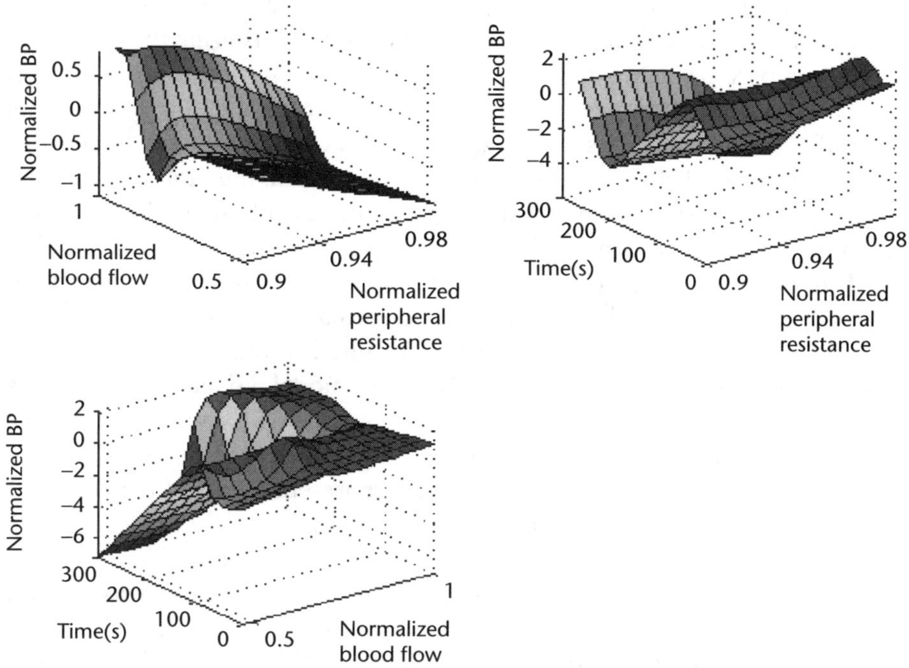

Figure 8.21 The three-dimensional surfaces relating to the fuzzy logic based mappings between the normalized inputs and BP with subtractive clustering and post-tuning.

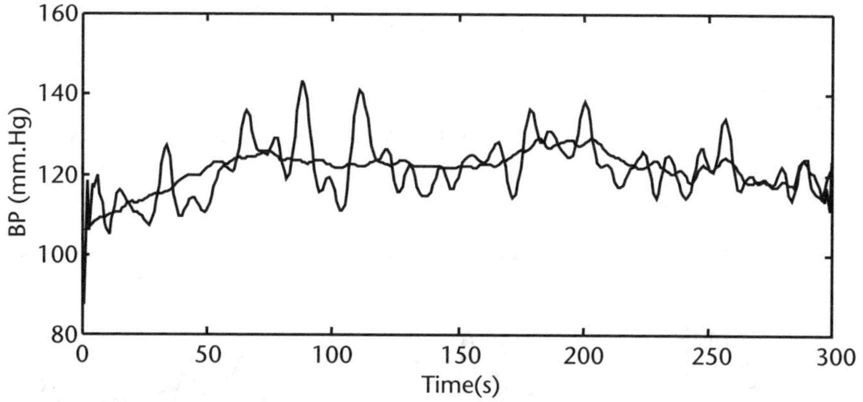

Figure 8.22 Actual versus predicted BP using the ANFIS-based submodel via subtractive clustering technique after fine-tuning.

Calculating the optimum number of clusters existing within an input-output data set is an essential issue that affects the resultant fuzzy model. The primary controller for this number is the selected cluster validity measure, such as the one presented by Xie and Beni [13] [see also Chapter 6, (6.2)]. However, another validity measure along with a new algorithm based on the competitive neural networks for clustering data were presented by Chen and Linkens [14, 15]. The proposed validity measure achieved the best results when compared with other measures. The proposed algorithm is based on estimating initial clusters' centers and radii using competitive neural networks (CNN). Moreover, the optimum number of clusters is calculated using the FCM algorithm (see Chapter 6, Appendix 6A) for clustering the resultant centers from the first part according to the new fuzzy partition measure above for cluster validity. The radius (δ) of the neurons in the CNN controls the initial number of clusters. A small value of (δ) results in a large number of clusters, while a large value results in a small number of clusters. The value of (δ) lies within the range [0, 1], as the input-output data set must be normalized as a preprocessing phase. Following several trials, the optimal value of (δ) was found to be 0.2, which results in a large number of clusters. Moreover, choosing smaller values than 0.2 for (δ) results in a great increase in the number of clusters, which affects the performance of the algorithm.

The application of this algorithm resulted in only seven clusters, which correspond to seven fuzzy rules in the ANFIS submodel. Given the relatively small size of the rulebase, only minimal fine-tuning of the fuzzy MFs was applied to eliminate any regions outside the normalized range of [0, 1]. The MSE achieved for the blood pressure ANFIS submodel was 26.44, which represents the best performance yet. Figure 8.23 shows the actual versus the estimated BP signals with this ANFIS submodel, while Figure 8.24 shows its three-dimensional surfaces.

In the light of the above results, the modeling method was considered for the elicitation of the HR and RESP ANFIS submodels. Figure 8.25 shows the actual versus the estimated HR and RESP signals, with MSE of 0.67 and 0.0018, respectively. Figures 8.26 and 8.27 display the three-dimensional surface mappings between inputs and outputs, which lead to five and seven fuzzy rules for heart rate and respiration, respectively.

Validation of the Modified Integrated Closed-Loop Physiological Model

In order to validate the closed-loop physiological model, which now includes the fuzzy logic based models relating to heart rate, blood pressure, and respiration, "unseen" data from another subject was used for comparison with predicted time and frequency responses.[12]

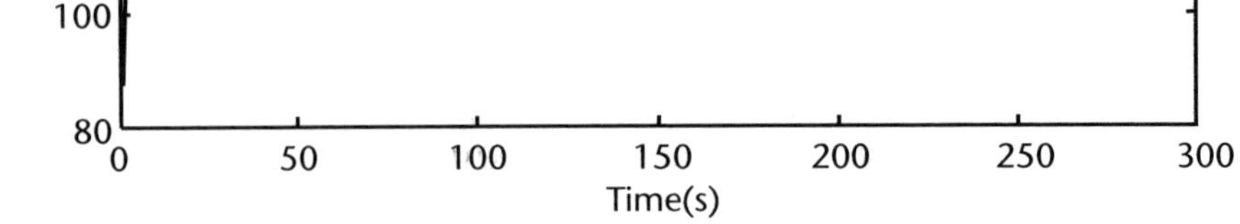

Figure 8.23 Actual versus predicted BP via a model built using a competitive neural network [14, 15].

12. It has always been my belief that a relatively low MSE index does not necessarily mean that the model is valid.

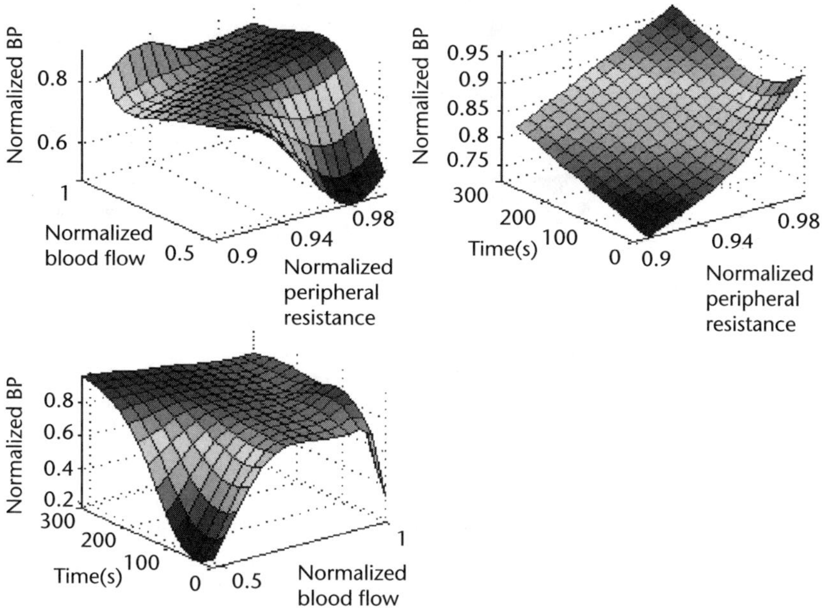

Figure 8.24 The three-dimensional surfaces relating to the fuzzy logic based mappings between the normalized inputs and BP for the model built using a competitive neural network [14, 15].

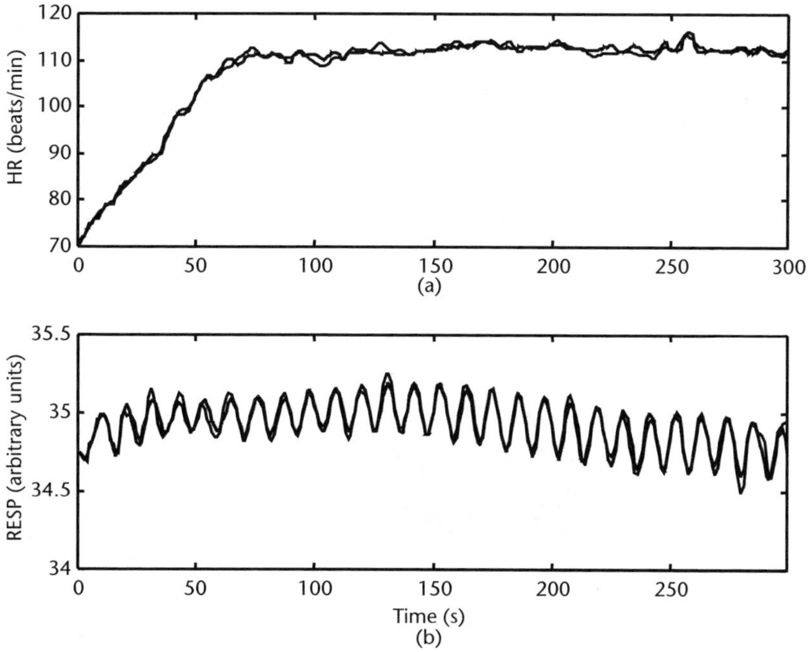

Figure 8.25 Actual versus predicted (a) HR and (b) RESP, obtained via their respective submodels using a competitive neural network.

Figures 8.28 and 8.29 show such responses. It is apparent from these figures that the model failed to capture the dynamics relating to respiration. Indeed, The respiration frequency spread disappears from both the actual and estimated signals

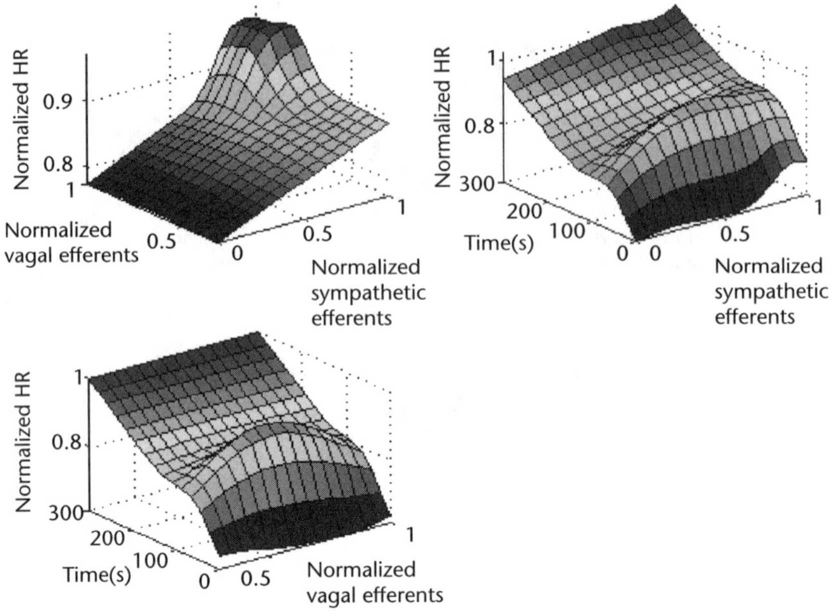

Figure 8.26 The three-dimensional surfaces relating to the fuzzy logic based mappings between the normalized inputs and HR for the model built using a competitive neural network [14, 15].

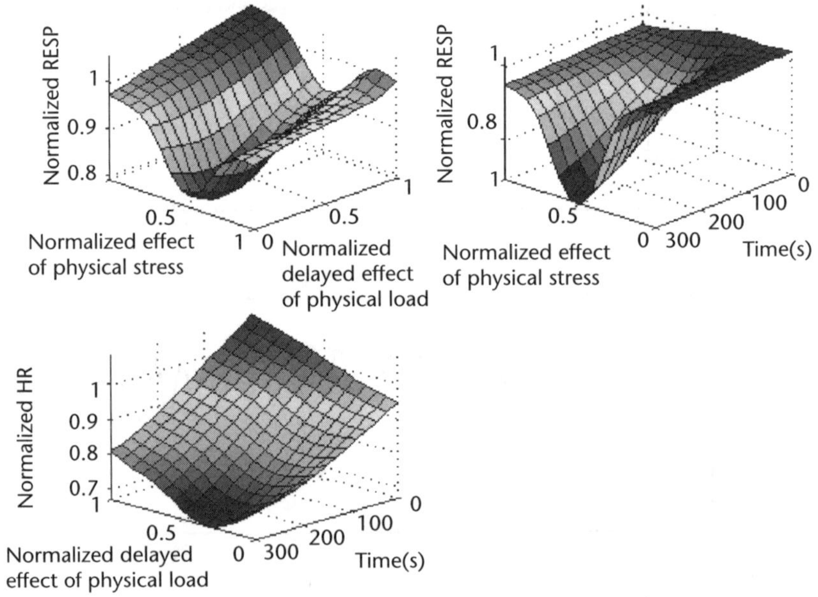

Figure 8.27 The three-dimensional surfaces relating to the fuzzy logic based mappings between the normalized inputs and RESP for the model built using a competitive neural network [14, 15].

due to the increase in this frequency, under physical stress, beyond the 0.5-Hz limit. This increase resulted in a signal distortion, because of the down-sampling applied to the respiration signal.

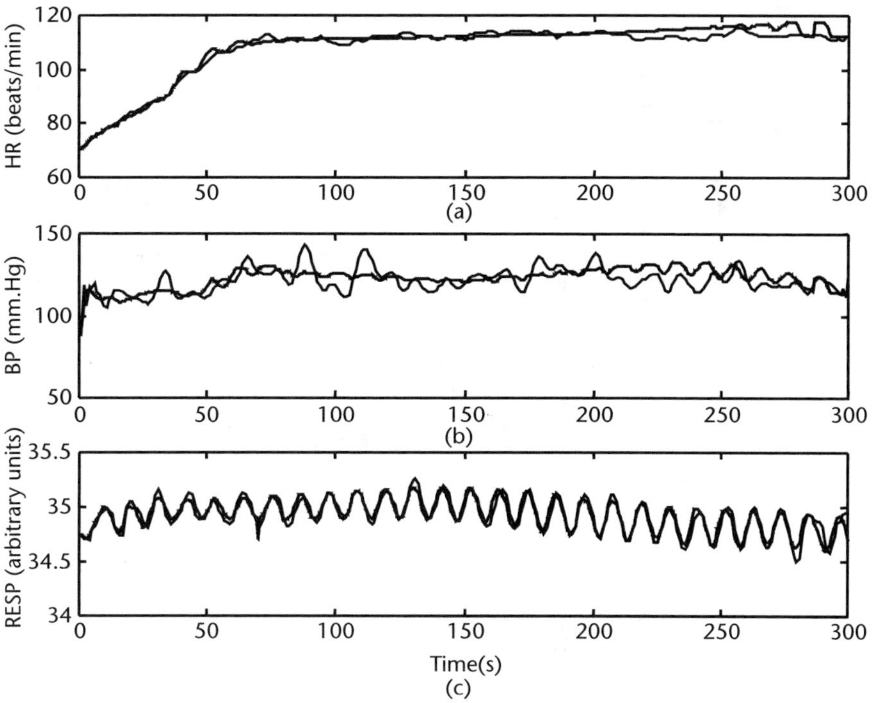

Figure 8.28 Actual versus predicted (a) HR, (b) BP, and (c) RESP signals when simulating the overall closed-loop model.

As a result, the respiration signal was down-sampled from 32 to 2 Hz, instead of 1 Hz, and a new ANFIS submodel was constructed for this signal. Furthermore, all the signals within the model were resampled at a frequency of 2 Hz. Figure 8.30 shows the actual versus the estimated signals as well as a comparison between their spectra for the RESP signal.

It can be seen that while the actual signal regained its spread at frequency 0.5 Hz, the predicted signal failed to capture such dynamics. Hence, the respiration submodel was modified according to the following.

The respiration signal was originally represented by Luczak et al. [4, 5] as a sinusoidal signal with a frequency of 0.25 Hz (15 breaths/minute), which increases under the effect of physical stress. However, according to the physiology of the respiratory system, there are two types of efferents (called *respiratory impulses*) emitted from the brain, which affect the respiration frequency. Hence, to ensure a correct representation of this system, the respiratory impulses will be used as an input to the RESP fuzzy-ARX model whose output will be the RESP signal. These impulses were modeled as the representation of the RESP signal in the original Luczak's model with certain modification to their frequency. This modification is achieved by using the gain, K_{freq}, with a value of 0.014. This value was chosen according to the average dominant respiration frequency of all subjects experiencing the sinusoidal workload (12 subjects in total). The proposed modification for the RESP submodel is shown in Figure 8.31. In this model structure, the fuzzy ARX submodel assumes the following structure:

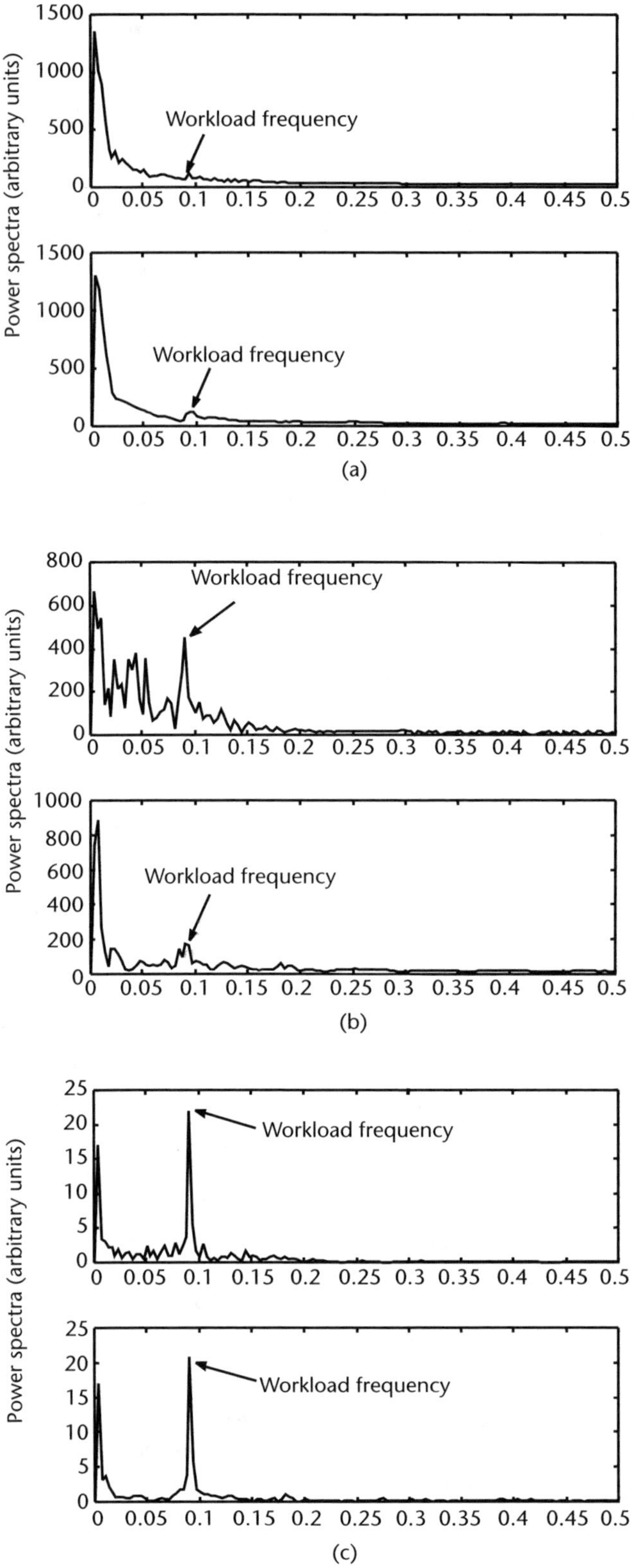

Figure 8.29 Actual (top) versus predicted (bottom) (a) HR, (b) BP, and (c) RESP using the overall closed-loop model.

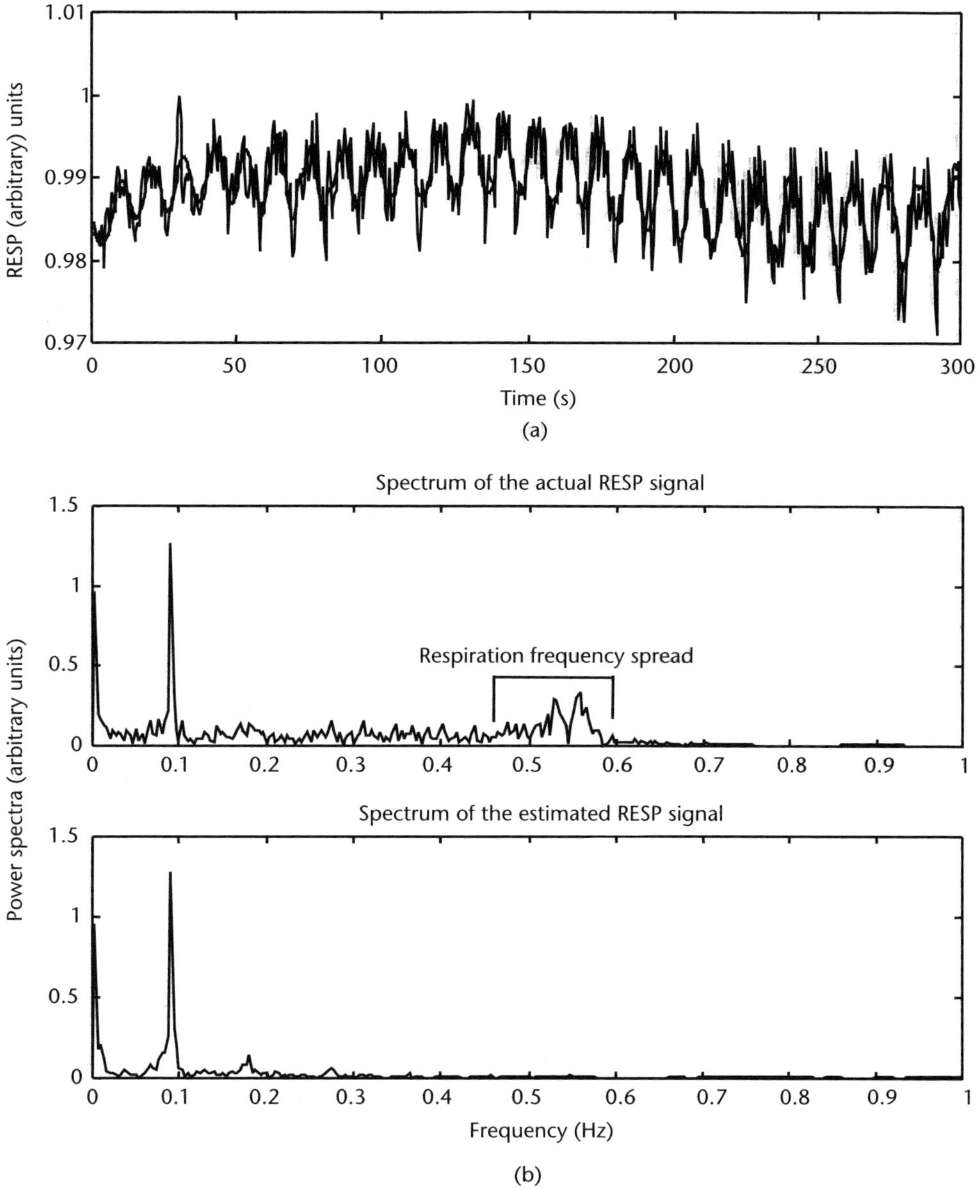

Figure 8.30 Actual versus predicted RESP signals in (a) the time domain and (b) the frequency domain.

$$R^{(i)}: IF\ u\ is\ A_i\ THEN\ y(t+1) = -\hat{a}_1^{(i)} y(t) - \hat{a}_2^{(i)} y(t-1) + \hat{b}_1^{(i)} u(t) + \hat{b}_2^{(i)} u(t-1) \tag{8.10}$$

where $R^{(i)}$ denotes the ith rule in the rule-base ($i = 1, 2, \ldots, M$), u is the input, y is the output, and A_i is a fuzzy linguistic label such as zero (ZE), negative small (NS), positive big (PB).

Figure 8.32 shows the actual versus the estimated RESP signals that result from the new submodel. This figure shows that the fuzzy-ARX submodel succeeded in

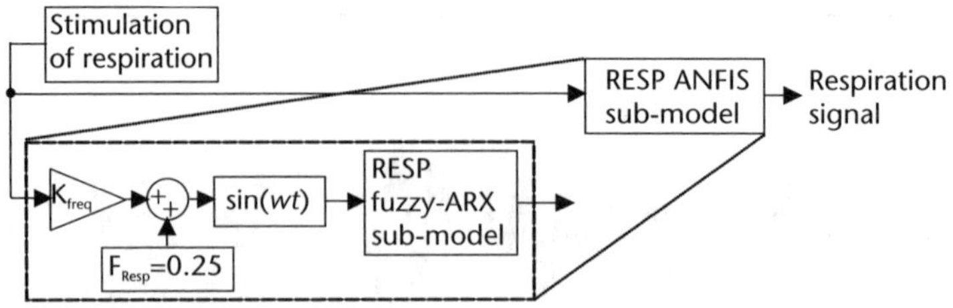

Figure 8.31 The proposed modification to the RESP model.

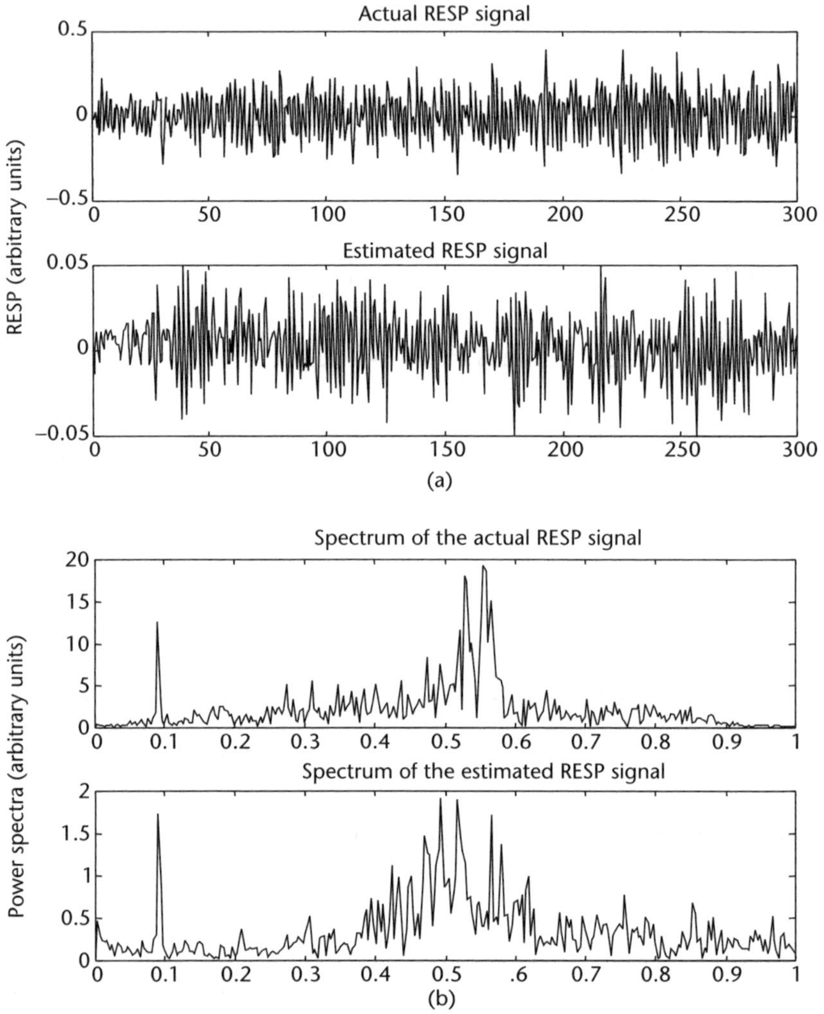

Figure 8.32 Actual versus predicted RESP in (a) the time domain and (b) the frequency domain, after including the modification of Figure 8.31.

estimating the RESP signal in the time domain as well as in the frequency domain with a small deviation. Furthermore, it succeeded in estimating the workload component, which exists in the actual spectrum of the RESP signal.

Using the model structure of Figure 8.31 for respiration, the submodels for blood pressure were reoptimized using the technique described by Chen and Linkens [14, 15] and which was described previously and was found to include six fuzzy rules. Fine-tuning was then applied to ensure its stability by visually inspecting the three-dimensional surface mappings between inputs and output. Figures 8.33 and 8.34 show the time and frequency domains of the actual and estimated signals resulting from the closed-loop simulations. As can be seen from these figures, the model fittings were generally good and point at a compromise between the

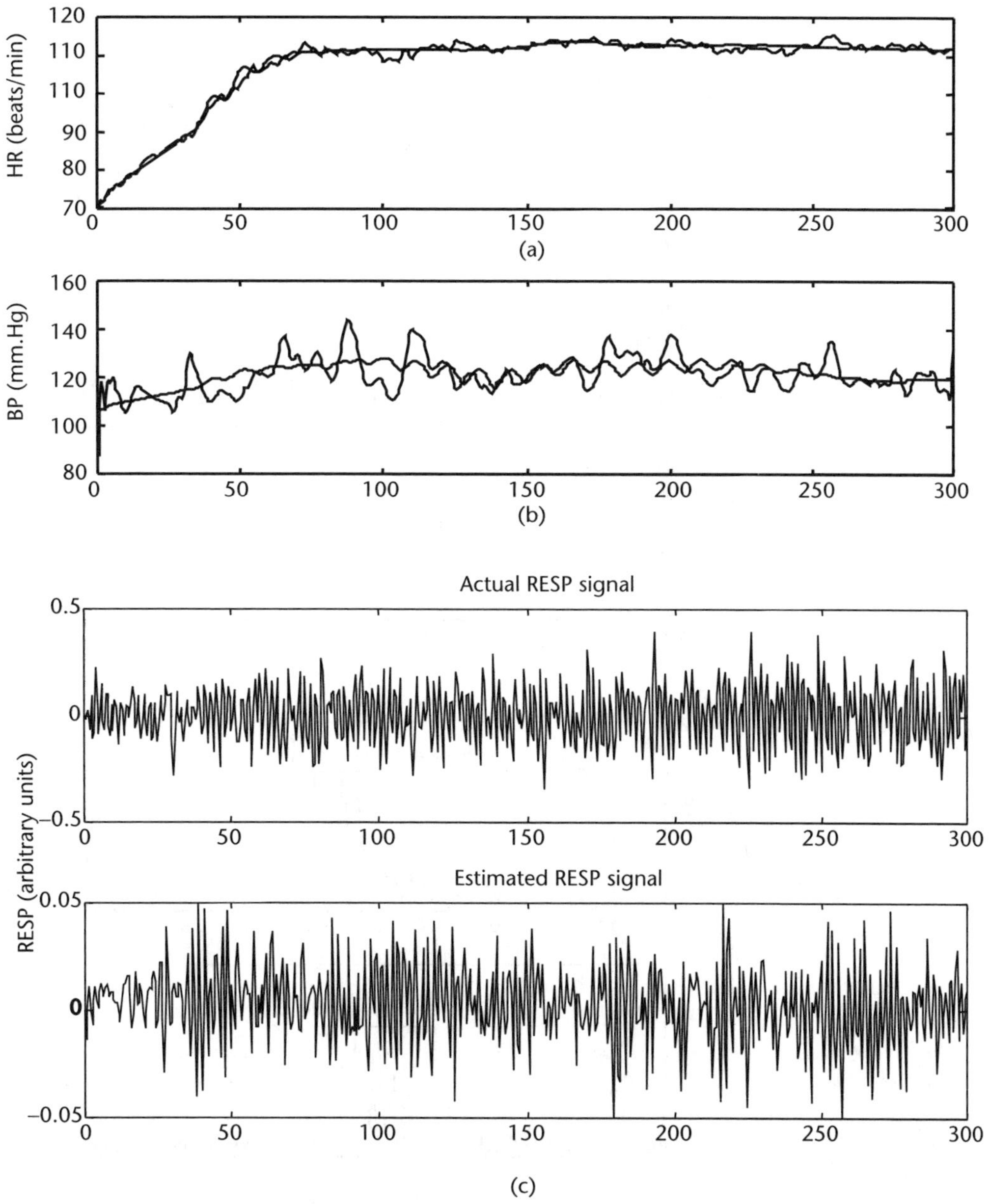

Figure 8.33 Actual versus predicted (a) HR, (b) BP, and (c) RESP, when using the closed-loop model with the modification of Figure 8.31.

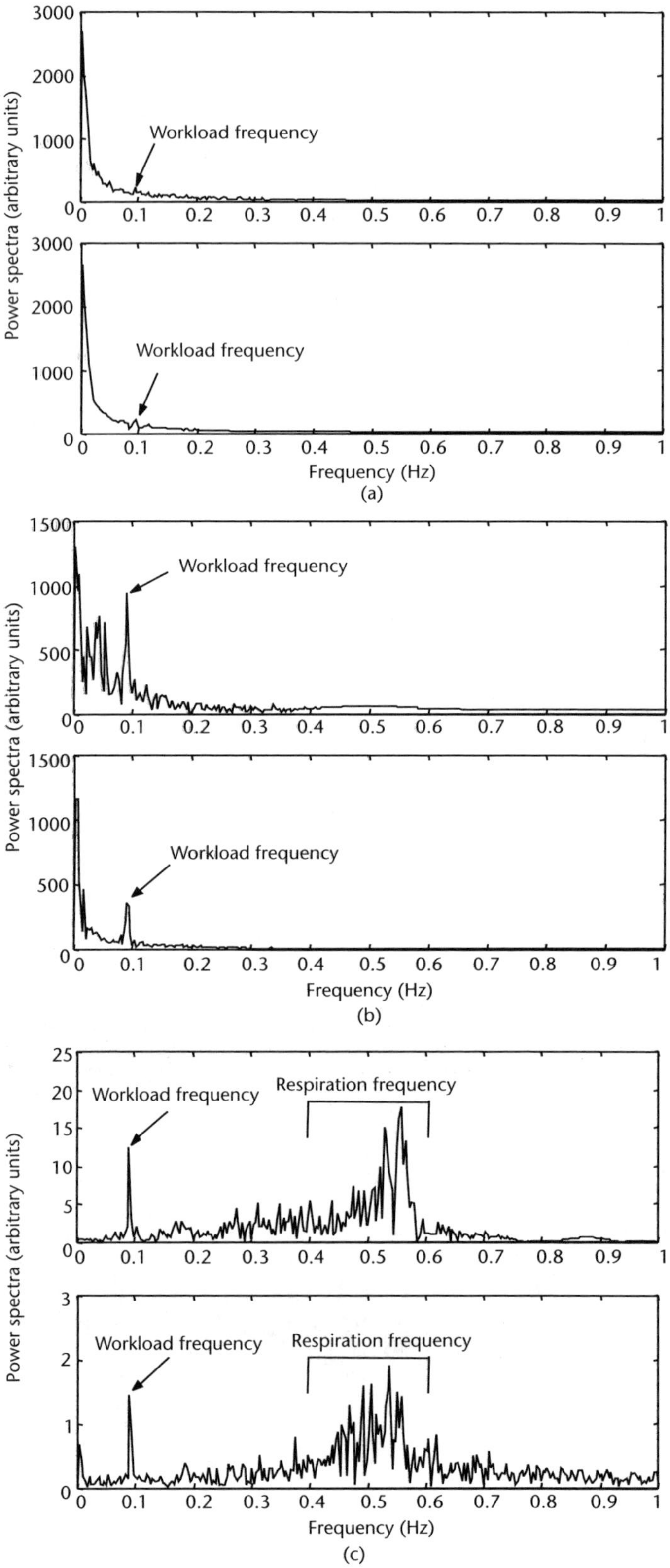

Figure 8.34 Actual versus predicted (a) HR, (b) BP, and (c) RESP, in the frequency domain when using the closed-loop model with the modification of Figure 8.31.

reduction of the MSE index and the capturing of the main frequency components (magnitude *versus* frequency).

8.4 Conclusions

In this chapter, a hybrid model, which describes the effect of physical workload (stress) in human subject volunteers on the cardiovascular system via heart rate, blood pressure, and respiration functions, has been proposed. The "hybrid" structure of the model refers to the combinations between mechanistic expressions and linguistic fuzzy rules.

The starting point for the elicitation for such a model was the structure that was proposed by Luczak et al. [4, 5], which allowed one to describe the effect of physical workload (stress) on the cardiovascular system. Early investigations using the original version of this model suggested the absence of several parameters, relating to gains and time constants, which were either not published in the original material or were simply missing. To obtain estimates of these parameters, actual measurements of heart rate, blood pressure, and respiration were used within an optimization routine. The parameters and the structure of the model were then adapted/modified via optimization using new real-time data, which was collected from healthy subject volunteers.

A series of model identification experiments were conducted using the ANFIS tool, which finally lead to model structures that reconciled the reduction of MSE between the measured and the predicted signals and the location of the fundamental frequency in the power spectrum relating to the heart rate, blood pressure, and respiration. The overall predictions, in a closed-loop fashion, were good, and the model structure, hence adopted, allows for model extensions to include thermoregulation as well as brain activity, which will be the subject of the next chapter.

References

[1] McDonald, A. H., "Mechanisms Affecting Heart Rate," Chapter 1 in R. I. Kitney and O. Rompleman, (eds.), *The Study of Heart-Rate Variability*, New York: Oxford University Press, Oxford Medical Engineering Series, 1980.

[2] Luczak, H., U. Philipp, and W. Rohmert, "Decomposition of Heart Rate Variability Under the Ergonomic Aspects of Stressor Analysis," Chapter 8 in R. I. Kitney and O. Rompleman, (eds.), *The Study of Heart-Rate Variability*, New York: Oxford University Press, Oxford Medical Engineering Series, 1980.

[3] Rompelman, O., "The Assessment of Fluctuations in Heart Rate," Chapter 4 in R. I. Kitney and O. Rompleman, (eds.), *The Study of Heart-Rate Variability*, New York: Oxford University Press, Oxford Medical Engineering Series, 1980.

[4] Thibodeau, G. A., and K. T. Patton, *Physiology of the Cardiovascular System*, 4th ed., London, U.K.: Mosby Books, 1999.

[5] Luczak, H., and F. Raschke, "A Model of the Structure and the Behavior of Human Heart Rate Control," *Biological Cybernetics*, Vol. 18, 1975, pp. 1–13.

[6] Huizenga, C., H. Zhang, and E. Arens, "A Model of Human Physiology and Comfort for Assessing Complex Thermal Environment," *Building and Environment*, Vol. 36, 2001, pp. 691–699.

[7] deBoer, R. W., J. M. Karemaker, and J. Strackee, "Hemodynamics Fluctuations and Baroreflex Sensitivity in Humans: A Beat-to-Beat Model," *American Journal of Physiology*, Vol. 253, 1987, p. H680-9.

[8] Whittam, A. M., et al., "Heart Rate and Blood Pressure Variability in Normal Subjects Compared with Data from Beat-to-Beat Models Developed from deBoer's Model of the Cardiovascular System," *Physiological Measurements*, Vol. 21, 2000, pp. 305–318.

[9] Seydnejad, S. R., and R. I. Kitney, "Modeling of Mayer Waves Generation Mechanisms," *IEEE Engineering in Medicine and Biology*, Vol. 20, No. 2, pp. 92–100.

[10] Chiu, S. L., "Fuzzy Model Identification Based on Cluster Estimation," *Journal of Intelligent and Fuzzy Systems*, Vol. 2, No. 3, 1994, pp. 267–278.

[11] Chao, C. T., Y. J. Chen, and C. C. Teng, "Simplification of Fuzzy-Neural Systems Using Similarity Analysis," *IEEE Trans. on Systems, Man and Cybernetics*, Vol. 26, No. 2, 1996, pp. 344–354.

[12] Castellano, G., and A. M. Fanelli, "An Approach to Structure Identification of Fuzzy Models," *Sixth IEEE International Conference on Fuzzy Systems*, Barcelona, Vol. 1, 1997, pp. 531–536.

[13] Xie, X. L., and G. Beni, "A Validity Measure for Fuzzy Clustering," *IEEE Trans. on Pattern Analysis and Machine Intelligence*, Vol. 13, 1991, pp. 841–847.

[14] Chen, M. Y., and D. A. Linkens, "A Fast Fuzzy Modeling Approach Using Clustering Neural Networks," *IEEE World Congress on Computational Intelligence*, New York, Vol. 2, 1998, pp. 1088–1093.

[15] Chen, M. Y., and D. A. Linkens, "A Fuzzy Modeling Approach Using Hierarchical Neural Networks," *Neural Computing & Applications*, Vol. 9, No. 1, 2000, pp. 44–49.

CHAPTER 9

Physiological Model Extension and Model Exploitation Via Real-Time Fuzzy Control

In this chapter, the closed-loop model that was described in Chapter 8 is extended to include a thermoregulatory path under physical stress conditions. Furthermore, a simplified model to describe the brain activity, via EEG signals, is also included. The proposed model is based on a gray-box modeling approach, utilizing several neural-fuzzy structures, which were used because of the sufficient level of details that they provide. A higher level, represented by a neural network, is hierarchically superimposed on the overall model to produce a "generic" cardiovascular model that is able to predict a wide range of dynamics associated with previously "unseen" (new) subjects.

9.1 Introduction

The model described in Chapter 8 allows one to predict the behavior of heart rate, blood pressure, and respiration in response to physical workload. Hence, the study was limited to these three variables only. Consequently, because of the wealth of data already generated in the author's own Research Laboratory and most importantly because of the structure of the current model, which can accept additional models, it was decided to increase the level of understanding of the current physiological model by including other effects such as body temperature (TEMP), brain centers, and brain activity via EEG.

Most experiments on body heat–related stress hitherto conducted lead to models that focus on the human response to the thermal environment, including skin temperature, core temperature, sweating, shivering, vasodilation, and vasoconstriction. Thermoregulation is understood to be achieved by regulating blood flow, such that vasoconstriction and vasodilation regulate blood distribution to control skin temperature and increase or decrease heat loss to the environment. Sweating reduces body temperature through the stimulation of the sweat glands, while shivering increases body temperature due to the heat emitted from the metabolism of foods in the muscles.

Research work that explored the interactions of the cardiovascular system with other control systems such as respiratory, thermoregulatory, and brain, is limited. In this chapter, the representation of thermoregulation is referred to here as "hybrid," in that the proposed model borrows ideas from the cardiovascular representation originally described by Luczak et al. [1, 2] and Stolwijk and Hardy's thermoregulation representation [3]. Consequently, the proposed model architecture is presented in the form of white-, black-, and gray-box components. The gray-box models were used in the form of neural fuzzy systems for generating the mappings relating to HR, BP, and TEMP, while a fuzzy- ARX model was used for generating the RESP signal. The model for EEG, which is based on the phase-locking phenomenon [4, 5], should be able to relate EEG to the other signals generated from the model (BP, RESP, and TEMP); a fuzzy-ARX model was used to describe EEG mapping. It is also worth noting that the proposed EEG model is only open-loop, since the inclusion of a closed-loop model would increase the complexity of the overall model to an unmanageable level as well as increase the difficulty for noninvasively acquiring the signals required for the identification (brain efferents and afferents).

In order to give the model even more flexibility in predicting the physiological behavior of several categories of subject volunteers, the extended model is supplemented with an "intelligent" supervisory layer that will allow one to tune the most sensitive[1] model parameters to optimal output predictions. Finally, the study will conclude by showing how such a comprehensive model can be exploited not only to predict the behavior of certain parts of the body but also to determine ways in which to influence these behaviors in a "controlled" manner via automatic feedback control strategies.

9.2 A Model to Describe Thermoregulation

One of earliest, most comprehensive model descriptions of the thermoregulatory process was presented by Stolwijk and Hardy [3] who divided the human body into six segments with four layers each. The model can provide accurate monitoring of the body temperature in humans when subjected to external perturbation such as physical workload. In addition to the authors' own validation of this model using a small limited number of experimental data, Konz et al [6] reinforced such validation with their own set of experimental data. Kitney [7] also proposed a model representing thermoregulation in the hand. This model simplified the hand into 15 blood vessels each having a rectangular cross section (5 arteries and 10 veins), hence reflecting human anatomy and thermal physiology. Huizenga et al. [8] proposed a modified version of the model by Stolwijk and Hardy [3] by increasing the number of segments and by enhancing the blood flow model. In the study presented in this chapter, the model by Stolwijk and Hardy [3] was chosen as the basis for the proposed model architecture for representing thermoregulation, which will be integrated within the closed-loop physiological model already presented in Chapter 8, as the next section explains.

9.2.1 Model Analysis

As stated above, the model architecture adopted for this research is similar to the one proposed by Stolwijk and Hardy [3], which was later extensively exploited by

1. Those model parameters that influence the predictions significantly.

Konz et al. [6]. The model should provide a continuous variation of the subject's mean temperature including all other activities related to it, such as shivering, sweating, and vasomotor activities. This model is based on a feedback control system comprising the *regulated system* and the *regulating system* (*regulator*). The difference between the feedback and the reference is a measure of the disturbance on the controlled variable, in this case the *temperature*. Such control actions represent means of modifying heat loss, heat production, or heat conservation by sweating, shivering, and vasomotor activities. Figure 9.1 is a diagram representing the modified thermoregulation model with respect to the overall closed-loop physiological model presented in Chapter 8.

It is worth noting that in order to estimate the dynamic model related to temperature, two inputs were used: the first input relates to the *total heat* (generated heat minus lost heat); the second input relates to *time* (this is to reflect the dynamic nature of temperature behavior). The ANFIS-based model was constructed using a similar approach to the one used to elicit the BP and HR fuzzy TSK models in Chapter 8 (including four fuzzy rules). Figure 9.2 shows the actual versus the estimated TEMP signals using the resultant ANFIS submodel, emphasizing the very good fit;

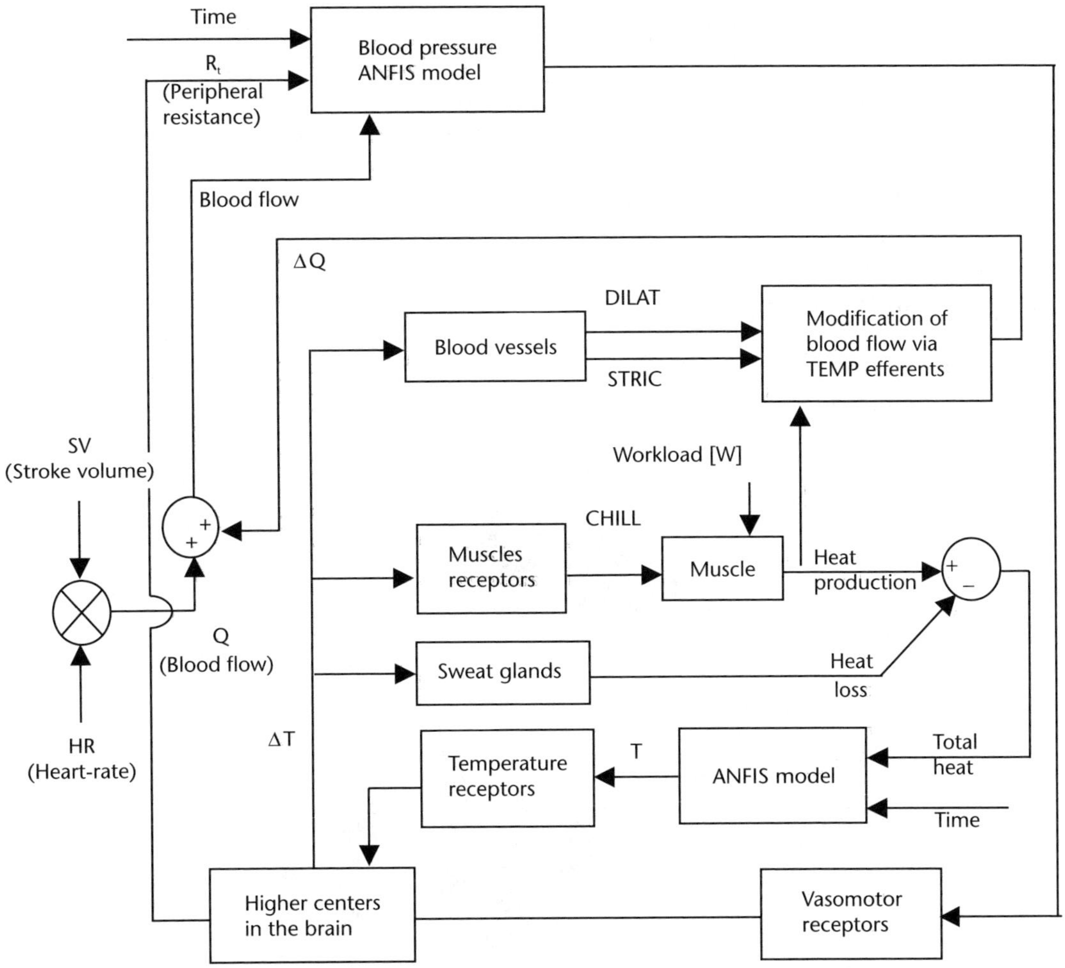

Figure 9.1 The proposed thermoregulation model.

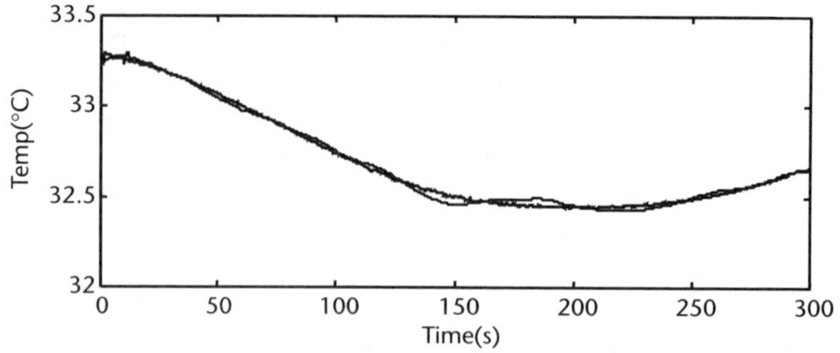

Figure 9.2 The actual (gray) versus estimated (black) TEMP signals for Subject 5 using the ANFIS submodel.

while Figure 9.3 shows its three-dimensional surfaces, underlying the nonlinear input-output mappings via such fuzzy relationships.

9.2.2 Model Validation

After having imported the TEMP ANFIS submodel into the model that was elicited in Chapter 8, simulation experiments were carried out using this overall model, the results of which are shown in Figure 9.4. From this figure, it can be seen that the performances for HR, RESP, and BP are similar to the ones obtained in Chapter 8. The TEMP fit can also be described as a very good fit. The MSE index figures between *actual* and *model predicted* traces are shown in Table 9.1 and indicate similar conclusions with the added remark that the MSE index for BP is better than in the case of the model that did not include the thermoregulation model (see Chapter 8). It seems that the inclusion of an extra input in the BP ANFIS model (blood flow in this case) improved the prediction ability of the model.

Because the author of this book does not believe that fitting models in terms of time-series outputs necessarily means that the most important dynamics have all been captured, the corresponding spectra relating to these signals need also to be

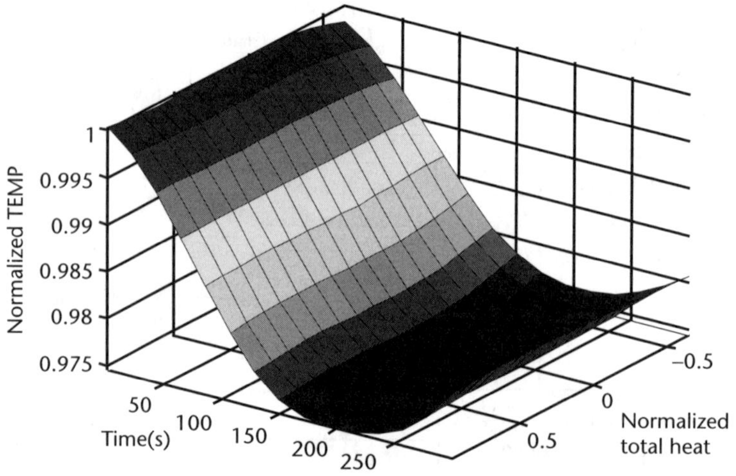

Figure 9.3 The three-dimensional surfaces of the TEMP ANFIS submodel for one subject.

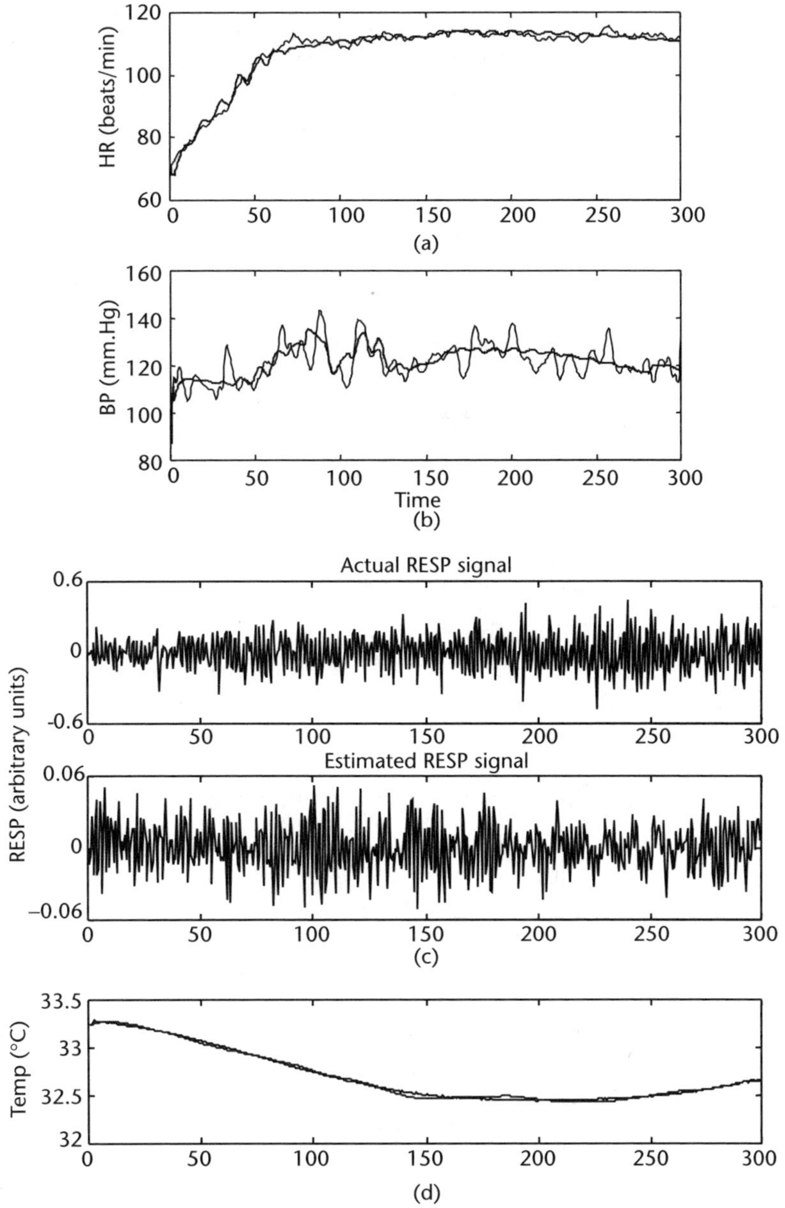

Figure 9.4 The actual versus estimated (a) HR, (b) BP, (c) RESP, and (d) TEMP signals of the closed loop simulation in time domain for the extended model.

analyzed. Hence, Figure 9.5 shows the power spectra relating to HR, BP, RESP, and TEMP signals. It can be seen that the model succeeded in capturing the workload frequency of 0.09 Hz in the HR, RESP, and BP spectra. In addition, the RESP spectrum also succeeded in capturing the respiration frequency that lies between 0.5 and 0.6 Hz.

Table 9.1 MSSE Values Between Actual and Predicted Signals Using the Model That Includes Temperature

Submodel	*HR*	*BP*	*RESP*	*TEMP*
MSE	2.3	30.04	0.0223	0.0004

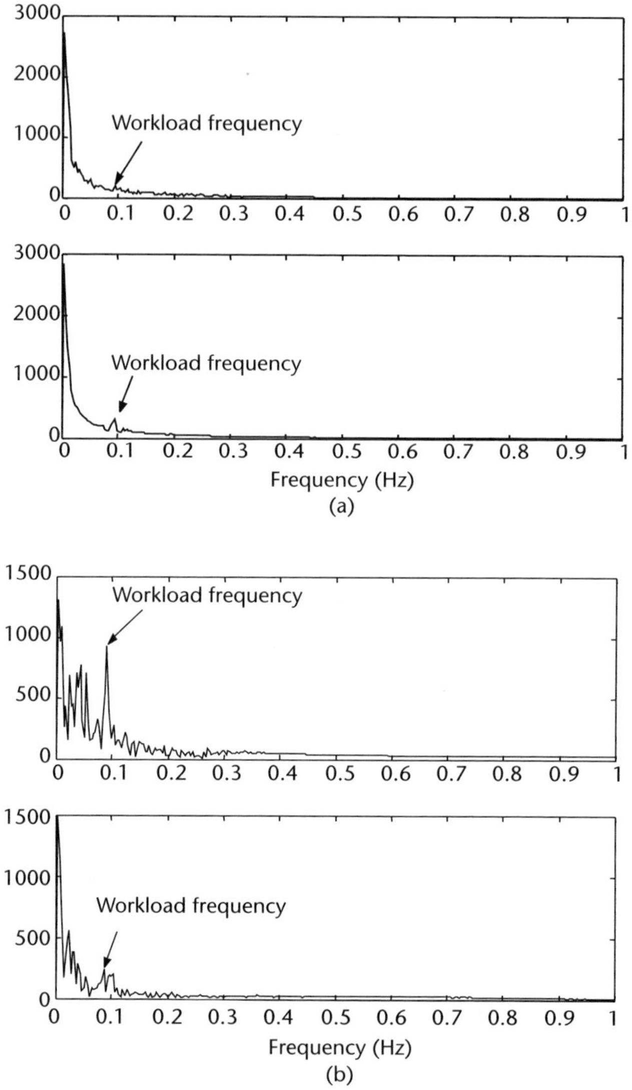

Figure 9.5 The actual (top) versus estimated (bottom) signals of the closed-loop simulation in frequency domain of (a) HR, (b) BP, (c) RESP, and (d) TEMP signals for the extended model.

In the light of the above considerations, this extended model can be deemed to be a good model that describes the behavior of HR, BP, RESP, and TEMP under external disturbances such as workload. The next section will focus on a second type of extension linking brain activity with the previous variables, indirectly via the brain centers, and directly via the analysis of EEG signals.

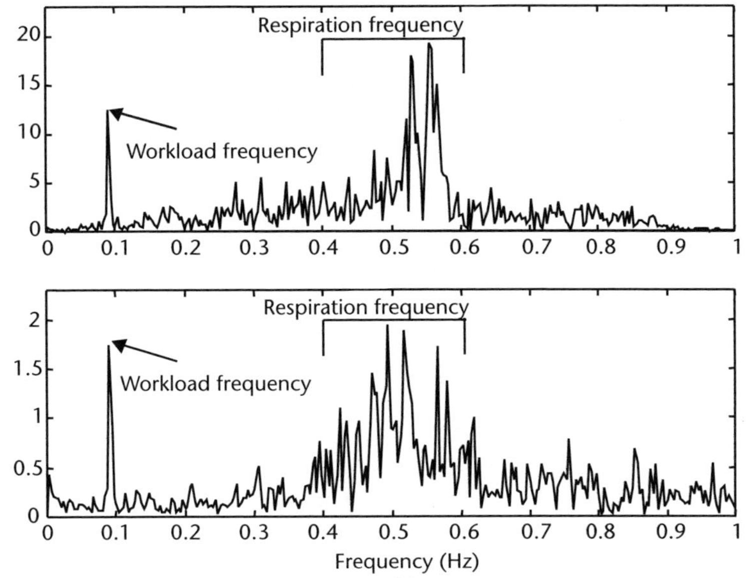

(c)

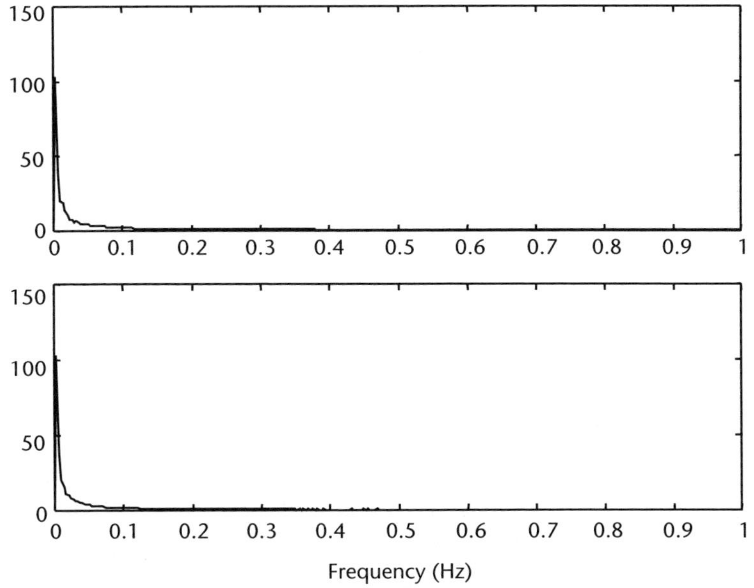

(d)

Figure 9.5 (continued.)

9.3 Representation of the Brain Centers

In this section, some enhancements are added to the previous model to include the role of the brain in controlling such systems (cardiovascular, respiratory, and thermoregulatory). The proposed modifications can be summarized as follows.

9.3.1 The Cardiac Center

It is reckoned that the cardiac center represents that part of the nervous system that controls HR via the sympathetic and parasympathetic (or vagal) fibers. Located in the brain stem, the cardiac center uses feedback from the baroreceptors[2] to control these two fibers. Any change in BP (increase or decrease) triggers the cardiac center to initiate stimulation of the sympathetic and parasympathetic branches to decrease or increase in HR and strength of the cardiac muscle. Rosenblueth and Simeone [9] expressed the sympathetic and vagal efferents as follows:

$$Symp = 1 + \frac{F_s}{HR_0\left(c + d \cdot F_s\right)} \tag{9.1}$$

$$Vagal = 1 - \frac{F_v}{HR_0\left(a + b \cdot F_v\right)} \tag{9.2}$$

where:

Symp is the sympathetic efferents to the heart.

Vagal is the vagal efferents to the heart.

HR_0 is the baseline heart rate with no influence of any efferents from the brain (sympathetic or parasympathetic).

a, *b*, *c*, and *d* are constants.

F_s, F_v is the frequency of the sympathetic and vagal activities, respectively.

Figure 9.6 shows the structure of the proposed model for the cardiac center where the combination of outputs from (9.1) and (9.2) is referred to as HR impulses [10].

9.3.2 The Vasomotor Center

All the blood vessels, except the capillaries, have smooth muscle fibers that are supplied by nerves from the autonomic nervous system. These nerves arise from the vasomotor center, which is located in the brain stem, specifically in the medulla oblongata. Their function is to change the diameter of the blood vessels, hence controlling the volume of blood they contain. Only sympathetic nerves supply the smooth muscle of the blood vessels and there is no parasympathetic nerve supply to most of these vessels. The tone of the smooth muscle is determined by the degree of the sympathetic nerve stimulation [11]. In other words, the decreased nerve stimulation causes the smooth muscle to relax; this process is called *vasodilation* and results in increased blood flow under less resistance. Conversely, when the nervous activity is increased the smooth muscle contracts; this process is called *vasoconstriction*. Hence, the smooth muscles can acquire three states: dilated, normal, and constricted. In addition, the regulation of the temperature affects the state of the smooth muscles such that the smooth muscles vasodilate to increase heat loss, while they

2. Baroreceptors act similarly to sensors to detect changes in blood pressure.

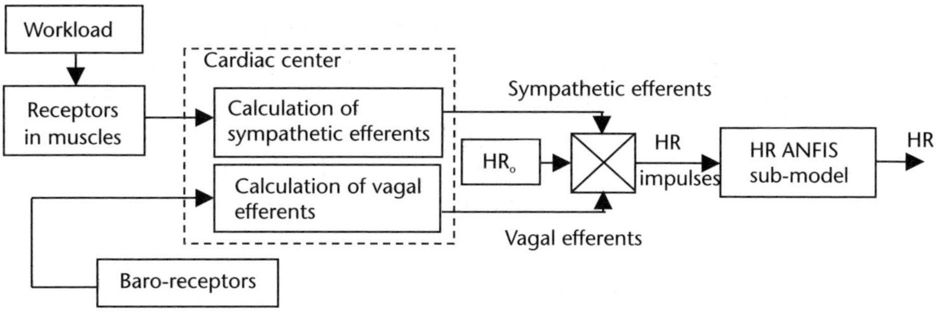

Figure 9.6 Representation of the cardiac center in the brain.

vasoconstrict to decrease heat loss and preserve the heat inside the body [11]. Consequently, the activity of the vasomotor center is a combined activity that compromises regulating both the temperature and blood pressure under external disturbances, making it a very complex control mechanism.

The smooth muscles in our original representation (see previous sections) have only two states: vasodilation (–1) and vasoconstriction (1). According to the previous discussion, however, this does not match the true control mechanism of the smooth muscles. Hence, a modification is included in this representation in order to model the three states of the smooth muscles, such that (1) corresponds to vasoconstriction, (0) corresponds to normal, and (–1) corresponds to vasodilation. Following this, the efferents from the hypothalamus due to temperature variations as well as that of the vasomotor receptors due to blood pressure variations are added together. The resultant signal represents the efferents to the blood vessels to control the peripheral resistance, which is an input to the BP ANFIS model. Figure 9.7 shows the proposed representation for the vasomotor center.

The previous figure shows that the function (Singe) is used to represent the three states of the blood vessels based on certain limits. These limits are as follows: for temperature, they were chosen to equal ±0.75°C, which is the maximum allowable variation in temperature [11]. For blood pressure, they were chosen to be ±10% of

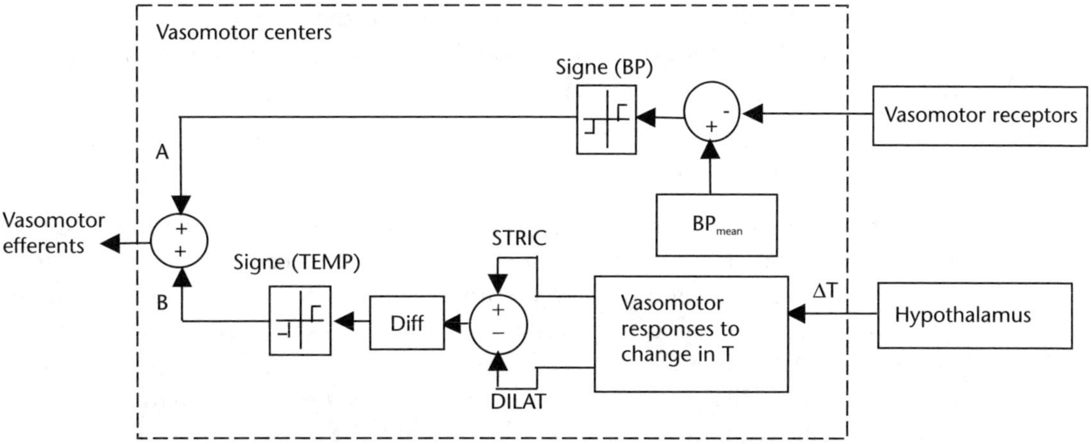

Figure 9.7 Representation of the vasomotor center in the brain.

the average value of the actual BP signal for each subject. This value was chosen to be the mean of the maximum variations of the actual BP signals from their average values over all subjects. In addition, the (Diff) function is used to remove any trends related to the TEMP signal. Finally, the overall vasomotor response is the summation of the vasomotor response due to the blood pressure variations (A) and the vasomotor response due to the temperature variations (B).

9.3.3 The Respiratory Center

The respiratory center is constructed from groups of nerve cells that control the rate and depth of respiration. They are situated in the brain stem, specifically in the medulla oblongata and the pons. The interrelationship between these groups of cells is very complex. In the medulla oblongata there exist the inspiratory and expiratory neurons, which are called the *medullary rhythmicity area*. Furthermore, neurons in the apneustic and pneumotaxic centers, situated in the pons, influence the inspiratory and expiratory neurons of the medulla [11]. The apneustic center stimulates the inspiratory center to increase the length and depth of inspiration. In addition, the pneumotaxic center normally inhibits both the apneustic and the inspiratory centers [12]. Before discussing the representation of the respiratory center, an important issue should be clarified, which is the role of both the pneumotaxic and apneustic centers in the model. Since the identification is carried out under the influence of the physical workload, the effect of the pneumotaxic center will dominate the apneustic center. Moreover, since the closed-loop model is very complex, modeling these pneumotaxic and apneustic centers will increase the level of complexity of the model, making its analysis very difficult. Thus, there is no need to include the apneustic center in the closed-loop model. The respiration impulses' frequency was modeled by generating a sinusoidal signal that has a frequency related to the applied level of workload. Figure 9.8 shows the model proposed for the respiratory centers.

The block called "stretch receptors in lungs" is added to represent the influence of these receptors, which existed in the thoracic cage, to match the actual physiological operation of the respiratory system. These receptors were introduced by Luczak [1] as a modification to enhance the performance of his closed-loop model that was previously introduced in 1975 [2]. The influence of the stretch receptors must be subtracted from the output of the sinus oscillator, which generates the respiratory impulses signal.

9.3.4 The Hypothalamus

The hypothalamus plays an important role as a temperature-regulating center in response to the temperature of the circulating blood. This center controls the body temperature through autonomic nerve stimulation of the sweat glands when the body temperature rises, or stimulation of the muscles to shiver when the body temperature falls. Additionally, it affects the vasomotor center through the nerve impulses, since vasodilation and vasoconstriction play a role in regulating the body temperature. This regulation involves either dilating the blood vessels to remove the excess of heat or constricting the blood vessels to preserve the heat in the body [11].

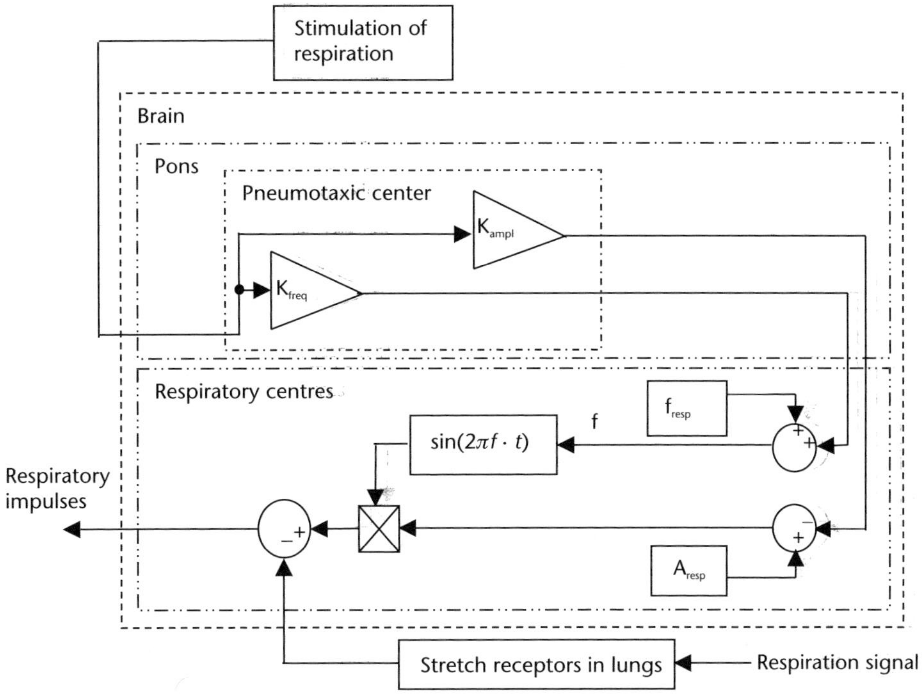

Figure 9.8 Representation of the respiratory center in the brain.

Hence, the representation of the hypothalamus can be implemented as two blocks that generate efferents responsible for regulating the temperature. These blocks generate efferents to sweat glands, to stimulate sweat secretion as well as efferents to muscles, which stimulate the metabolism of food. In addition, there is a link between the hypothalamus and the vasomotor center that is responsible for regulating the contraction state of the blood vessels according to the temperature variations. Figure 9.9 shows the representation of the hypothalamus located in the diencephalon.

The process of building the ANFIS models was carried out as follows. Given the constraints of the Finapres and ProComp+ monitors, the sampling frequencies were automatically set at 1 Hz for the HR and BP signals, 32 Hz for RESP and TEMP signals, and 256 Hz for the EEG signal. The data sets include two main age groups; the first one included 12 subjects with an average age of 34 years (first group), while the second one included 4 subjects with an average of 29.5 years (second group). The first group was used for training/testing the model (model development), and the second group (made up of totally unseen data by the model) was used for testing (model verification). This division on age was more coincidental than anything and depended on the availability of suitable subject volunteers. It is worth noting that the final model adopted a unified sampling rate of 2 Hz; hence, the HR and BP signals were resampled (via a simple interpolation) to reflect such a higher frequency, while the RESP, TEMP, and EEG signals were down-sampled (the original frequency is reduced). As a result, each data set used included a total of 600 data points per signal.

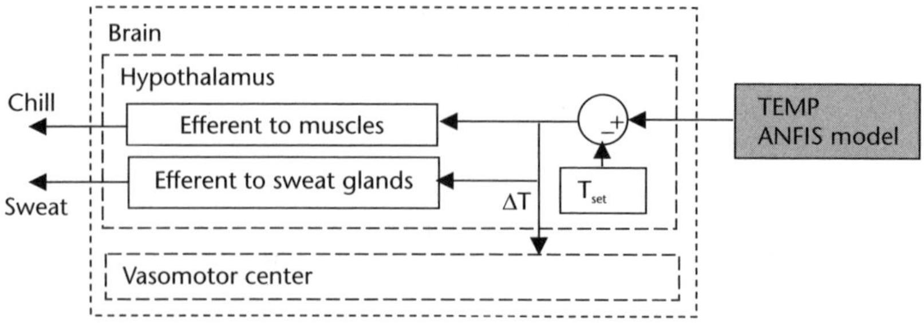

Figure 9.9 Representation of the hypothalamus in the brain.

Neural networks–based clustering [13] was selected as a preprocessing operation to estimate an optimal number of clusters that can efficiently represent the mapping between the input and output. The ANFIS model includes time as an extra input to characterize the dynamic nature of the models and to improve the predictions. On the other hand, the fuzzy-ARX model was constructed using triangular MFs equally spaced to partition the input space with the use of trapezoidal membership functions for the first and last one to enhance the prediction ability of the model. The least squares method was used to estimate the second-order ARX models.

Having imported all the submodel structures relating to the brain centers onto the extended closed-loop physiological model, which was last evaluated in Section 9.3, a simulation experiment was carried out to evaluate the new predictions relating to HR, BP, RESP, and TEMP. Figure 9.10 shows the various signals obtained. It can be seen that good fits were obtained generally despite the differences in peak-to-peak magnitudes in the RESP signals. Indeed, Table 9.2 shows slightly better MSE values when compared to those of Table 9.1. Here again, it seems that more transparency in the model (via the inclusion of the brain centers) seems to have contributed to a modest but nevertheless improvement in accuracy. In turn, from the power spectra of Figure 9.11 one can conclude that the workload frequency was captured by the BP frequency response and the RSP signal succeeded in capturing the workload frequency as well as the respiration frequency albeit with a small deviation from the one on the actual data. The model was later validated against the remaining 11 patients' data and led to the MSE values of Table 9.3 between predicted and actual outputs, which "expectedly"[3] vary within ranges but are reasonable generally. It was concluded that the existing extended closed-loop model will be retained as the basis for further model extensions, as the next section shows.

9.4 Modeling the Brain Via EEG Measurements

It is fair to say that the EEG signal, measured at the surface of the scalp, represents a *lumped* information source of what goes on in certain parts of the brain in terms of signals received and sent from and to other parts of the body. It is possible to model

3. It is not reasonable to expect the model to generalize perfectly given the intrasubject parameter variability in terms of steady-state gains and dynamics.

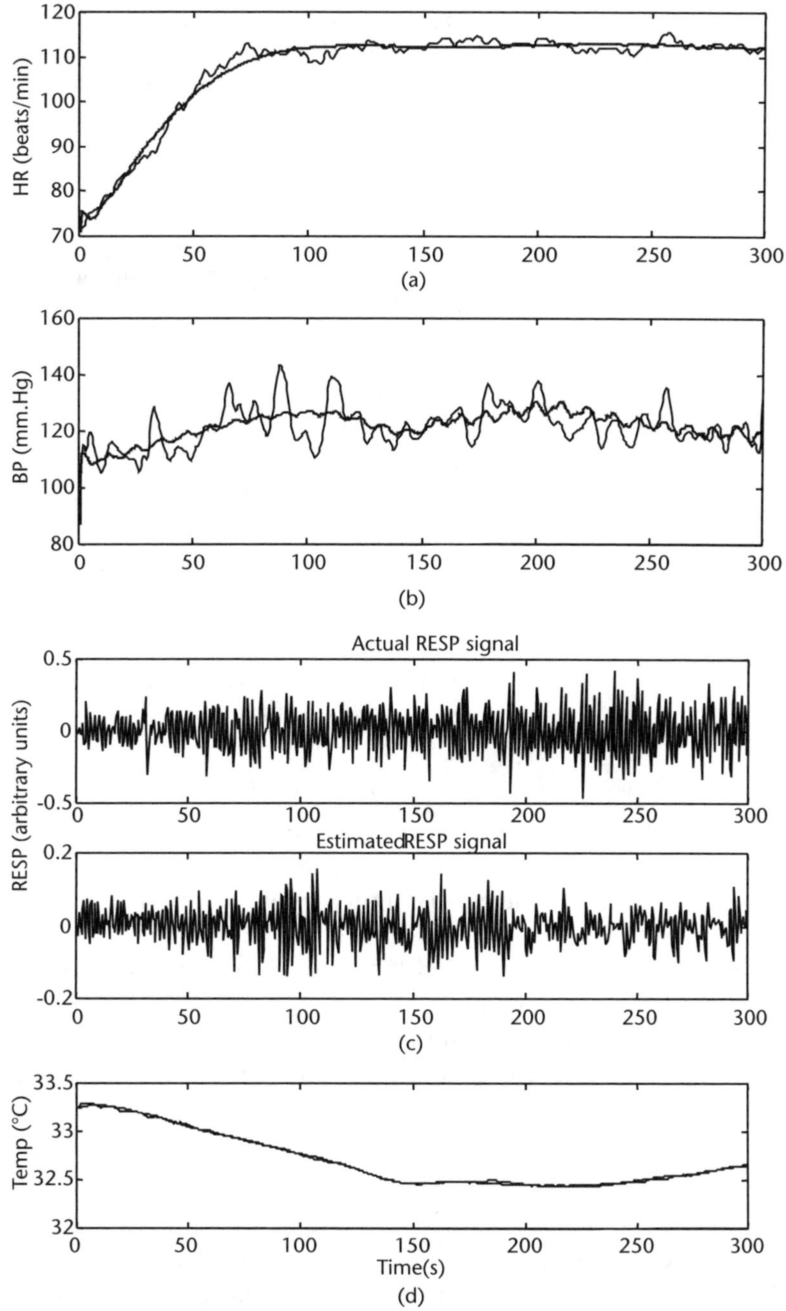

Figure 9.10 The actual versus the estimated (a) HR, (b) BP, (c) RESP, and (d) TEMP signals of the closed-loop simulation in time domain after importing the brain centers.

this information transfer *completely* using a combination of invasive[4] and noninvasive[5] methods, and *partially* through noninvasive methods. In this study the

4. For measuring the signal's transfer via the nerve fibers.
5. Electrical electrodes placed on the scalp.

Table 9.2 MSE Values Between Actual and Predicted Signals Using the Model That Includes Temperature of the Brain Centers

Submodel	*HR*	*BP*	*RESP*	*TEMP*
MSE	2.12	33.61	0.0223	0.00016

Table 9.3 MSE Values Between Actual and Predicted Signals Across the 12 Subject Volunteers

Subject No.	*MSE*			
	HR	*BP*	*RESP*	*TEMP* [$\times 10^{-5}$]
1	3.13	16.38	0.0344	3.58
2	2.45	25.45	0.0299	29.34
3	1.52	24.22	0.1835	0.918
4	2.54	20.39	0.0418	15.02
5	2.12	33.61	0.0233	16.1
6	1.94	38.82	0.0661	0.688
7	1.5	14.23	0.1988	27.15
8	2.18	24.37	0.1506	59.83
9	4.81	18.89	0.0744	6.22
10	1.86	15.87	0.3998	0.899
11	6.08	53.59	0.1451	35.73
12	2.15	24.66	0.0176	25.92

latter is proposed as a means of counteracting the problem of invasive interventions and all the complexities surrounding their corresponding processes. As a result, the link between the existing closed-loop model, which has so far been extended, and any derived EEG model will be open-loop.

In order to identify which variables in the closed-loop physiological model the EEG signal will relate to, the phenomenon of *phase locking* will be investigated.

9.4.1 Phase Locking

Phase locking stipulates that, "if two interacting oscillators coexisted then they are N: M phase locked, if marked events of one oscillator occur at fixed phases of the other oscillator" [4, 5]. This phase synchronization condition implies also that the frequency synchronization between the two rhythms (i.e., N periods of the first rhythm have exactly the same duration as M periods of the second one). Synchronization between two signals is often referred to as *entrainment* or *phase locking*, which can be detected by the existence of common peaks in their spectra. Therefore, the spectrum of the EEG signal was compared to the spectra of all the signals generated from the closed-loop model, which are HR, HP, RESP, and TEMP. A preprocessing phase was first applied to the EEG signal in order to remove the existing noise, which was carried out using the following steps:

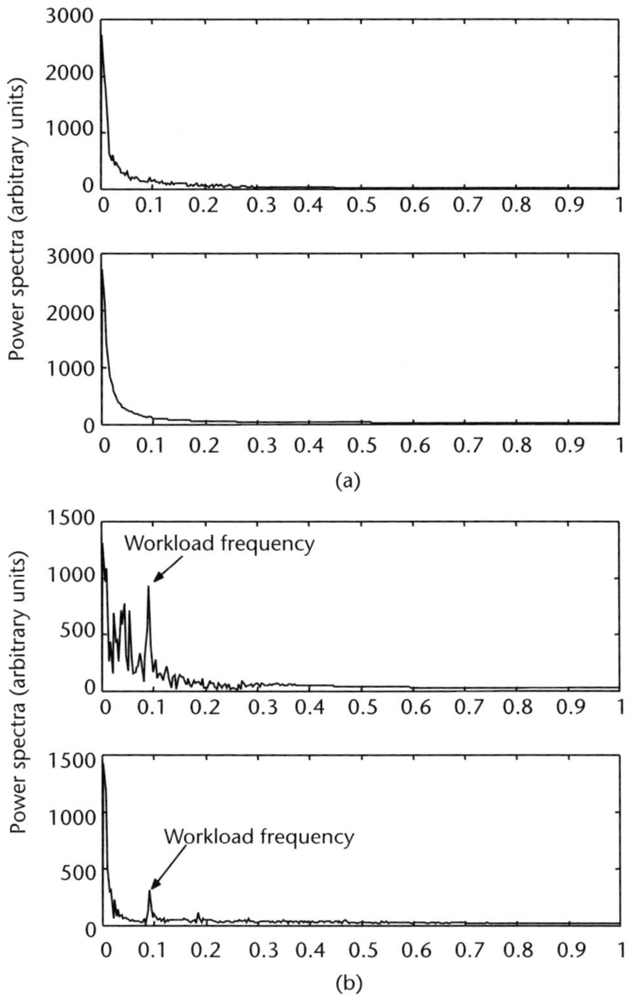

Figure 9.11 The actual versus the estimated (a) HR, (b) BP, (c) RESP, and (d) TEMP signals of the closed-loop simulation in frequency domain after importing the brain centers.

- A fifth-order lowpass Butterworth filter with a cutoff frequency of $F_{cutt} = 32$ Hz, was applied to the EEG signal to remove the 50-Hz noise corresponding to the electrical mains.
- The resultant EEG signal was downsampled from 256 Hz, which represents the original sampling frequency, to 64 Hz.

Wavelet transforms (see Chapter 6) were used to decompose the EEG signal into multilevels and then reconstruct the approximation part at the desired level, which was chosen to be the level corresponding to the frequency range 0 to 4 Hz. The mother wavelet was chosen to be the "Daubechies" [14] with order 24 to increase accuracy of the extracted EEG signal within the selected frequency range. As a final step for extracting the EEG signal, a downsampling from 64 to 8 Hz was applied.

The results of the spectra comparison revealed that there exists phase locking between the EEG spectrum and the RESP, BP, and TEMP spectra. On the other hand, there was no phase locking between the EEG and the HR spectra. Figure 9.12

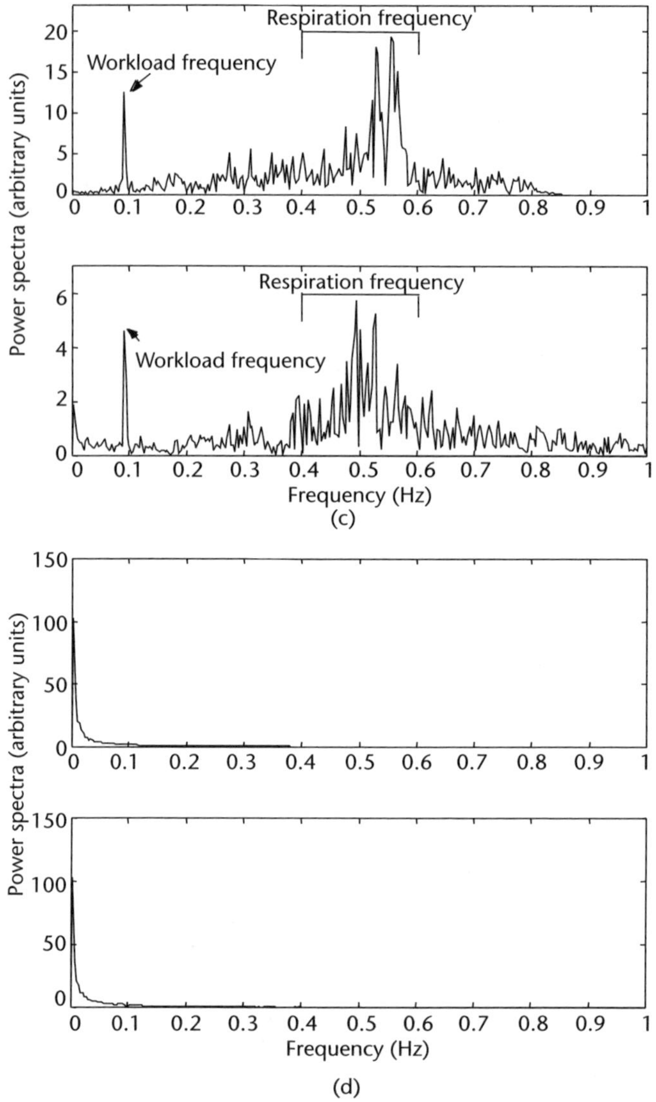

Figure 9.11 (continued.)

shows samples of the phase locking existing between the EEG spectrum and the RESP. It is worth noting that from the 12 subjects studied:

- All but one showed phase locking between EEG and RESP.
- All but five showed phase locking between EEG and BP.
- All but nine showed phase locking between EEG and TEMP.
- None showed phase locking between EEG and HR.

Now that the significant inputs that contribute to the EEG model have been identified, the strategy for constructing this model is to combine these inputs into one single additive input, which will comprise the relative contribution of each individual input. In order to identify the gains by which each input will be multiplied,

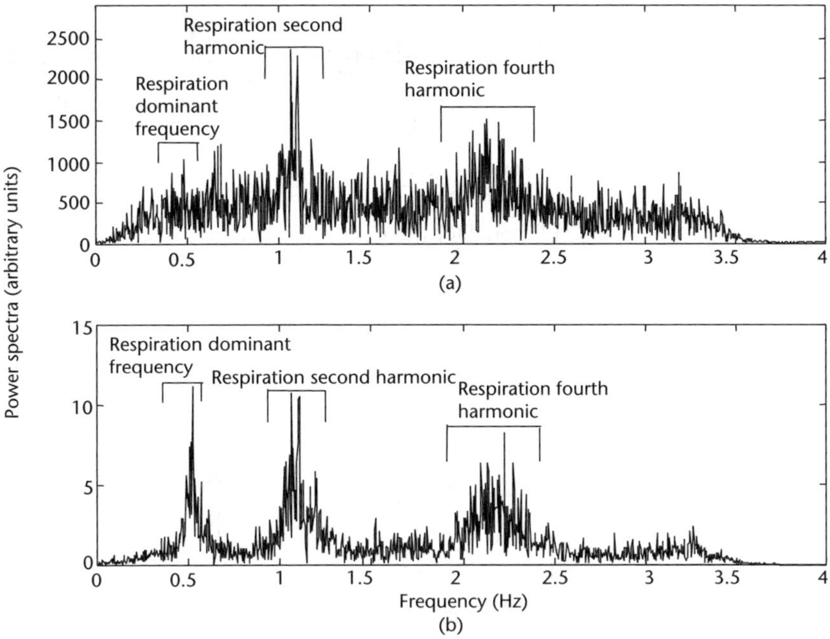

Figure 9.12 Phase locking between (a) EEG and (b) RESP spectra.

the energy existing in the EEG regions of the phase locking frequency ranges needs to be calculated for each subject. The selected regions are 0.001 to 0.005 Hz for TEMP, 0.08 to 0.13 for BP, and the respiration frequency range for RESP. Table 9.4 shows the normalized average energies found in the frequency ranges of TEMP, BP, and RESP using the 12 subjects' data. Such normalized values will represent the gains that will be applied to the three inputs before combining them into one single input as Figure 9.13 shows. Input "L" enters a fuzzy ARX submodel, with four fuzzy rules with a second-order consequent part, describing EEG behavior. Figure 9.14 shows the frequency-domain outputs using such a model. Although the magnitudes in the time domain are different, the frequency spectra seem to agree on the dominant modes.

9.4.2 Validation of the Overall Extended Closed-Loop Model

The closed-loop simulation using the proposed model, with the physical stress as the only input to the system, was carried out for all the subjects; a sample taken from 1 subject is shown in Figure 9.15.

Table 9.4 Average Energies Existing Within the TEMP, BP, and RESP Ranges over All Subjects

	Energy in TEMP Range	*Energy in BP Range*	*Energy in RESP Range*
Average	481.2	45889	312150
Normalized average	0.0015	0.147	1

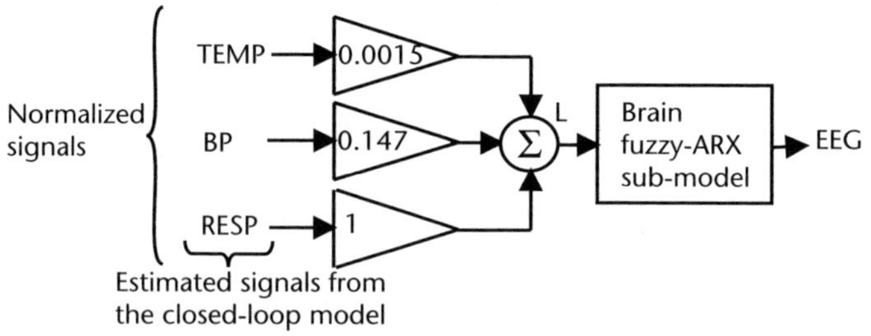

Figure 9.13 Fuzzy-ARX submodel for the EEG generation.

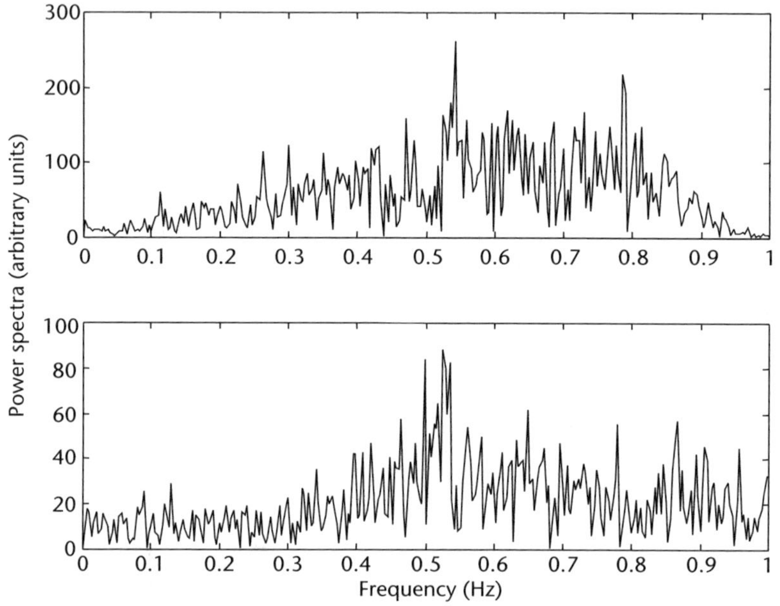

Figure 9.14 The actual (top) versus estimated (bottom) EEG signals with fuzzy-ARX model having four fuzzy rules in the frequency domain.

These results show generally a good fit with small MSEs for the HR, RESP, and TEMP signals, but the BP signal, more often reckoned to be a tricky signal, has led to a relatively high MSE. The MSE in the case of the EEG signal is also high, which is a result of the artifacts existing in the time domain actual EEG signal (see Figure 9.15). To complete the analysis of the simulation results for the proposed cardiovascular model, the spectra of the estimated signals were checked to ensure that they captured the most important dynamics that exist in the actual traces. These dynamics are the workload frequency (0.091 Hz), which is an indication for the blood pressure entrainment at that frequency, the respiration frequency spread, the temperature and heart rate dominant frequencies, and the frequency spread of the EEG signal. Figure 9.15 shows a comparison between the spectrum of the actual and estimated signals. The comparison revealed a good match for the workload, temperature, and respiration frequencies. The next section will discuss generalizing the model to a wide range of subjects' dynamics.

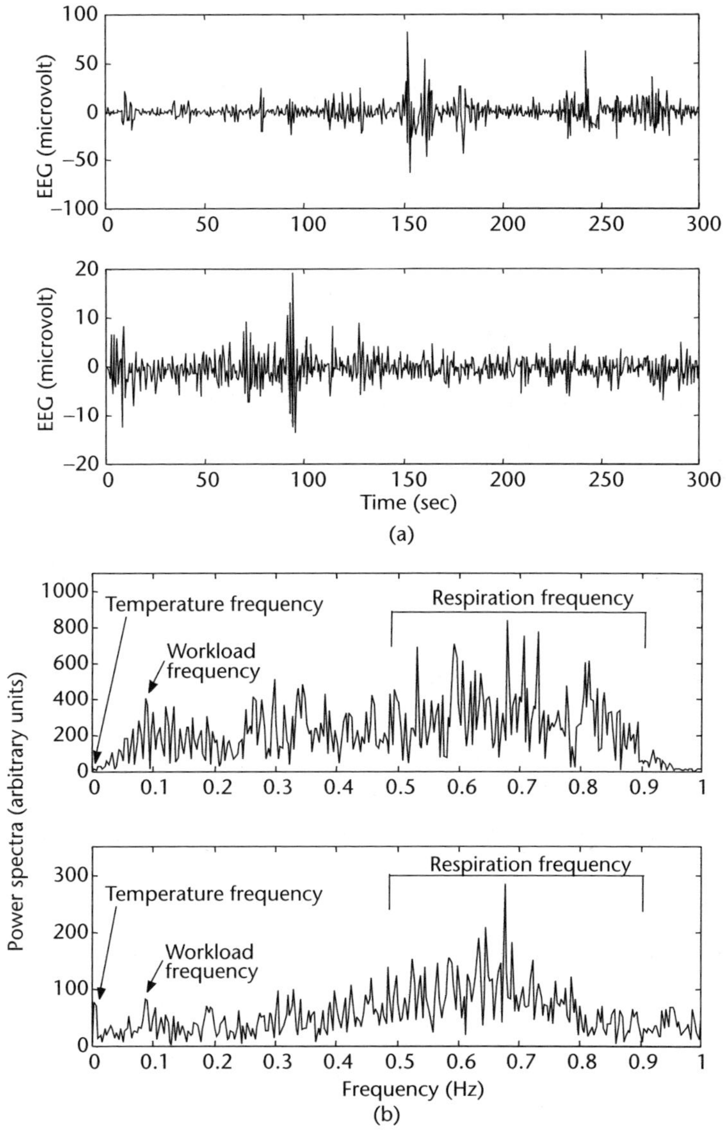

Figure 9.15 The actual (top) versus the estimated (bottom) EEG signals in (a) time and (b) frequency domains for Subject 9.

9.5 A Generic Model

In order to give a generic structure to the current model that will enable it to represent the behavior of a wide range of subjects' dynamics under physical stress, it is necessary to superimpose onto the existing model architecture an additional "intelligent" layer whose role is to map any incoming data to a category of average population dynamics to which they are closer, in order to enable the model to predict as accurately as possible the variables behavior.

It was decided to use neural networks for the construction of such a layer. The specific function of this layer is to map certain features, which will be extracted from the actual measurements, to the model parameters. The number of parameters, associated with each model, which should be estimated, is shown in Table 9.5 for the ANFIS models (HR, BP, and TEMP), and in Table 9.6 for the fuzzy-ARX models (RESP, and EEG). These parameters are usually associated with the centers and widths of the fuzzy MFs.

It is worth noting that in Table 9.6, the number of parameters associated with the membership functions is based on the use of triangular MFs with equal spacing, ensuring that the intersection between each successive MFs is at level 0.5 and that the first and last MFs are trapezoidal.

Moreover, in order to achieve the best performance for such an intelligent layer, the relationship between the extracted features and the required parameters should be causal. In other words, the extracted features should be descriptive and representative for the output parameters for the five models. Hence, all the available features from all the signals were noted, as shown in Table 9.7, and a correlation analysis was carried out for these features to reject the highly dependent ones. In the Table 9.7, note that the symbol (✓) indicates that the feature exists, while the symbol (–) indicates that the feature does not exist. Consequently, the results of the correlation analysis revealed that the best features are as follows:

- For the HR signal: minimum, rise time, and peak-to-peak;
- For the BP signal: minimum, maximum, and mean;
- For the RESP signal: respiration dominant frequency;
- For the TEMP signal: peak-to-peak;
- For the EEG signal: respiration dominant frequency, peak-to-peak.

Table 9.5 Number of Parameters Associated with HR, BP, and TEMP ANFIS Submodels

AFNIS structure	HR	BP	TEMP
Centers	4 MFs × 2 inputs=8	4 MFs × 3 inputs=12	4 MFs × 2 inputs=8
Widths	4 MFs × 2 inputs=8	4 MFs × 3 inputs=12	4 MFs × 2 inputs=8
Consequent parts	4 rules × 3 coefficients=12	4 rules × 4 coefficients=16	4 rules × 3 coefficients=12
Ranges			
Input	0.9927 : 1	0.9304 : 1	0.3322 : 1
	—	0.4439 : 1	—
	0.5 : 300	0.5 : 300	0.5 : 300
Output	0.5653 : 1	0.4742 : 1	0.9337 : 1
Number of normalization gains	1	2	1
Number of rules	4	4	4

Table 9.6 Number of Parameters Associated with RESP and EEG Submodels

Fuzzy-ARX Structure	*RESP*	*EEG*
Number of Rules	2	4
Number of Membership Functions' Parameters	4	6
Number of ARX Model Parameters	5 × 2 rules=10	5 × 4 rules=20

Table 9.7 A List of Available Features for All the Actual Signals

Signals / *Features*	*HR*	*BP*	*RESP*	*TEMP*	*EEG*
Minimum	✓	✓	—	✓	—
Maximum	✓	✓	—	✓	—
Mean	✓	✓	✓	✓	✓
Peak-to-peak	✓	✓	✓	✓	✓
Dominant frequency	—	—	✓	—	—
Rise time	✓	—	—	—	—

✓ = Existing feature — = Nonexisting feature

After deciding on the best features, three types of neural network intelligent layer were tested according to the MSE over the second group of measurements (four subjects) in order to select the best one for building this layer. The selected types are as follows:

1. Feedforward neural network with three layers: an input layer with a number of neurons corresponding to the number of input features, a hidden layer with 10 neurons having a logistic activation function, and an output layer with the number of neurons corresponding to the number of parameters to be estimated.
2. Generalized regression neural network (GRNN) with 1 input layer and two hidden layers having radial basis activation function for the first one and linear activation function for the second one, with only biases connected to the first hidden layer, and one output layer. This structure is a precompiled one within the environment of MATLAB.
3. Radial basis neural network with the same structure as the previous one but having biases connected to both the two hidden layers. This structure is also a precompiled one within the environment of MATLAB.

The results of the performance analysis revealed that the best neural network type that resulted in acceptable results over the second group of measurements is the GRNN. Figure 9.16 shows the structure of the intelligent layer, which is used for estimating the five models' parameters from unseen subjects' output data. In the figure, the two blocks, which are referred to as "correction of TEMP gain" and "correction of BP gain," are used to fine-tune the gains relating to the TEMP and BP ANFIS-based models.

In order to assess the model prediction performance, a validation test was carried out over the four subjects of the second group of measurements. The validation results showed good performance from the qualitative point of view. In other words, the selected models represent qualitatively to a high extent many of the dynamics associated with new (unseen) subjects. Figure 9.17 shows the estimated signal for one subject from the validation group. From the above figure, it can be seen that the selected models succeeded in estimating a qualitative representation for the behavior of the unseen subject for the HR, BP, and TEMP signals [see Figure

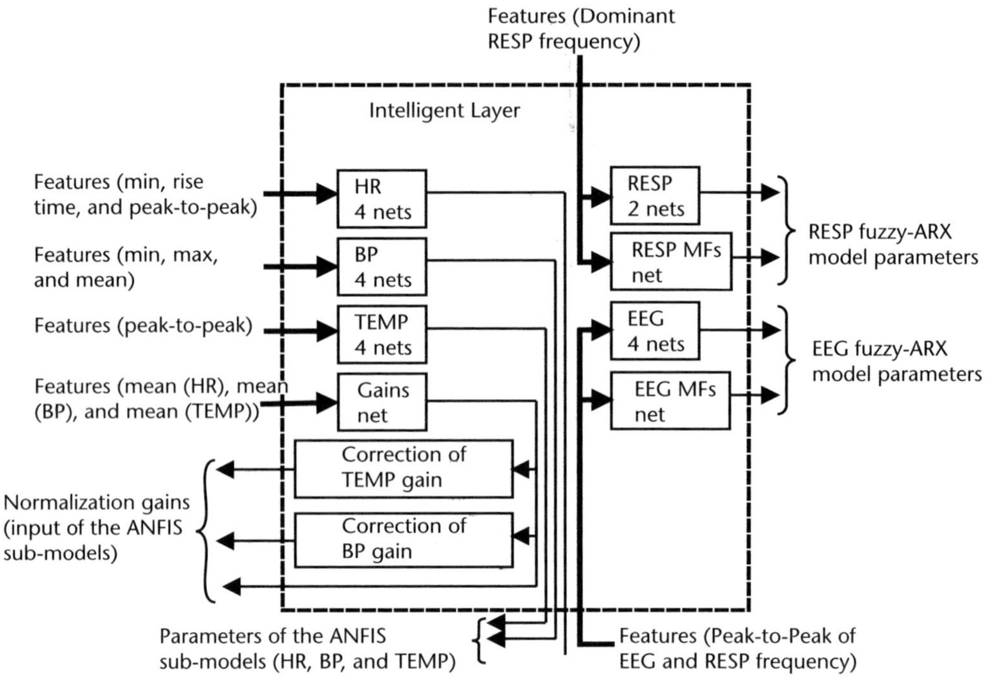

Figure 9.16 A schematic diagram of the intelligent layer that predicts the five models' parameters.

9.17(a, b, d)]. However, in spite of the difference in amplitude between the actual and estimated RESP signals, the estimated signal was still able to capture the respiration frequency with good accuracy [see Figure 9.17(c, f)]. In addition, in spite of the differences existing between the actual and estimated EEG signals due to the simplicity of the proposed EEG model and the complexity of the operation of the brain, the estimated EEG signal captured most of the dominant peaks that existed in the actual signal with a good accuracy.

9.6 Model Exploitation Via Feedback Control

There are two ways in which an elicited model can be exploited. One is to use it to *predict* the behavior of certain output variables given a certain input profile. The other is to *reverse-engineer* it in order to determine optimal recipes (routes) for ascertaining certain properties of the output (i.e., apply online feedback control). The idea of automatic control stems from the fact that one wishes to evaluate physical stress so that *safe* operating levels of physiological variables can be guaranteed. Several researchers considered such issues by monitoring pilots' cardiovascular behavior when subjected to stress [15–17]. While it is this author's opinion that controlling BP directly can be a tricky exercise, the control of HR, on the other hand, (which influences other functions within the cardiovascular system) can be proved to be a relatively easier operation. Although it is appreciated that the automatic control configuration that is proposed in this chapter will not be the exact one that will be used for monitoring subjects for stress in the workplace,

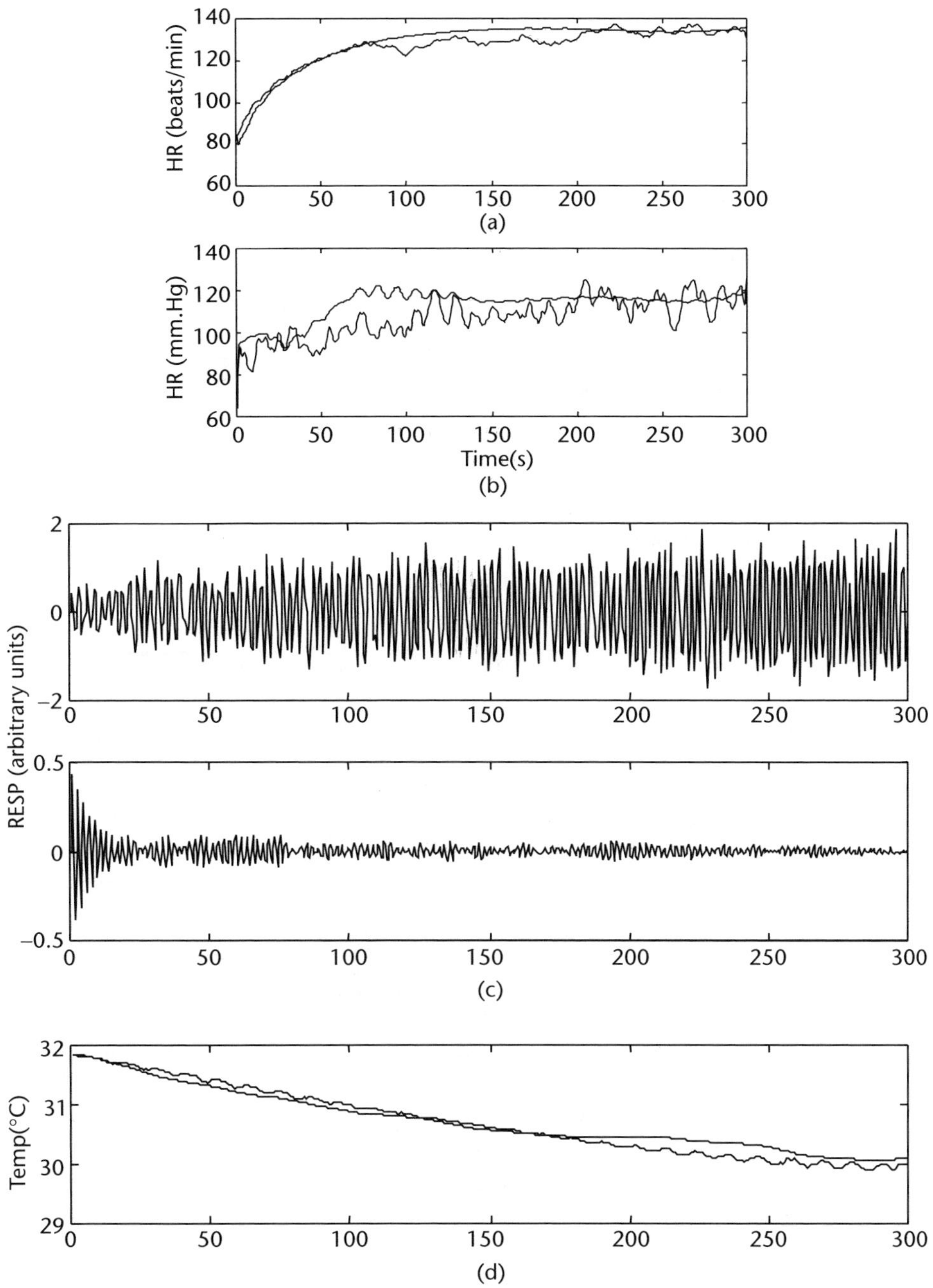

Figure 9.17 The actual versus the estimated signals with the intelligent layer (a) HR, (b) BP, (c) RESP, (d) TEMP, (e) spectra of the BP (f) spectra of the RESP, (g) EEG time series (actual and estimated), and (h) EEG spectra (actual and estimated).

the methodology will offer some useful lessons that should be learned regarding safe monitoring in general.

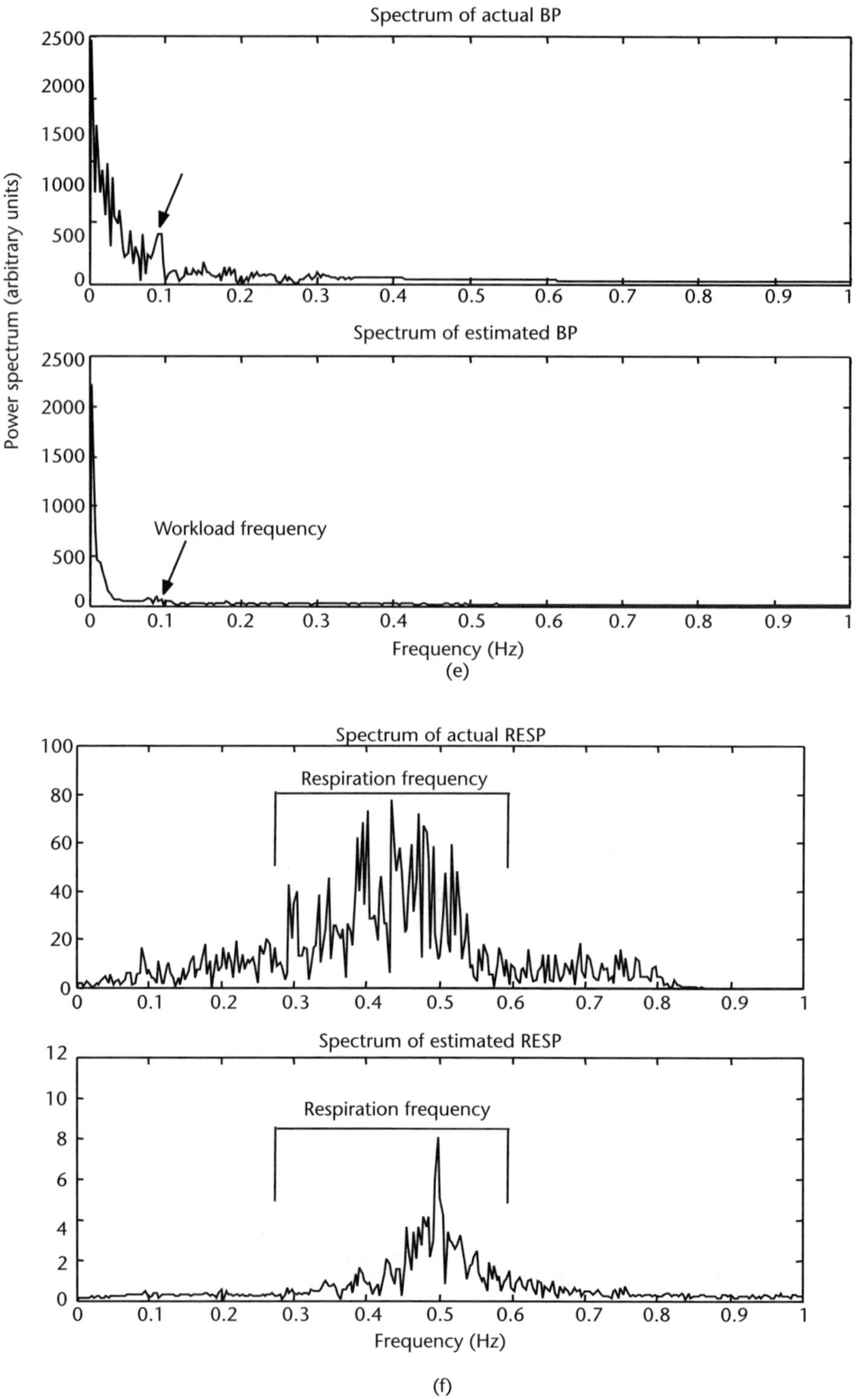

Figure 9.17 (continued.)

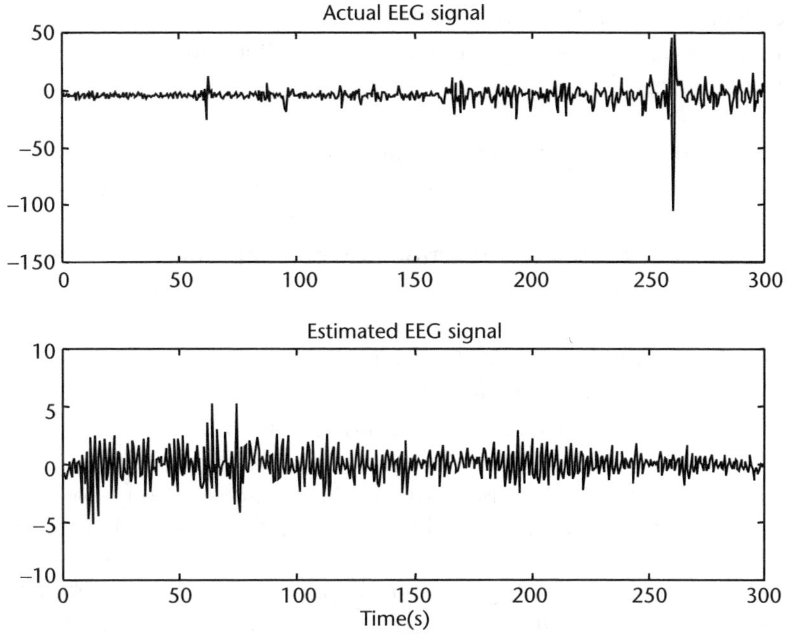

(g)

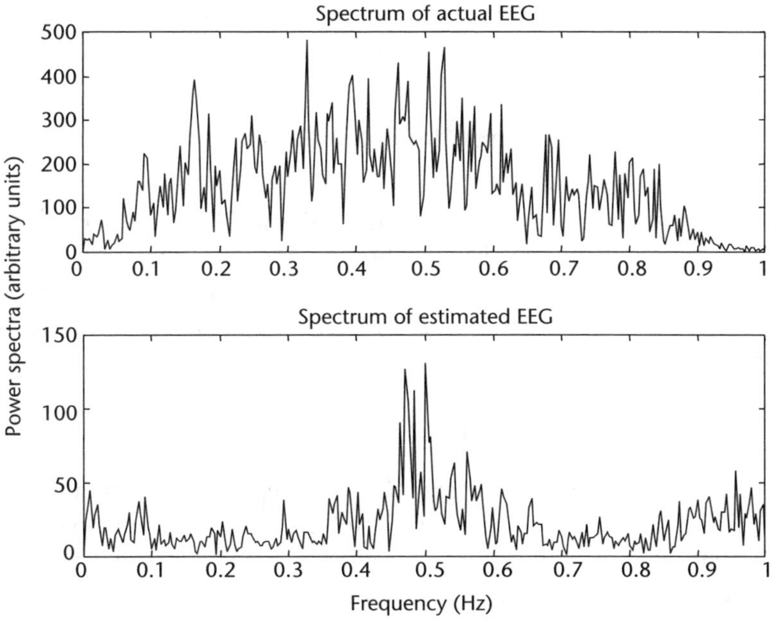

(h)

Figure 9.17 (continued.)

9.6.1 Control Structures

In this study, the classical proportional integral derivative control strategy was adopted as a common architecture for automatic feedback control. However, in this chapter the fuzzy version of such controllers will be investigated instead.

Fuzzy Proportional Integral Controller

The fuzzy equivalent of a classical PI controller consists of a set of fuzzy rules of the following form:

IF Error $e(t)$ is Positive Big (PB) and Change of Error $\dot{e}(t)$ is Negative Small (NS)
THEN the Change in Controller Output $\Delta u(t)$ is Positive Medium (PM)

Note that due to its integral action (lowpass filtering properties), the control actions of a PI tend to be smooth.

Fuzzy Proportional Derivative Controller

Similarly to the PI controller, the fuzzy equivalent of a classical PD controller consists of a set of fuzzy rules of the following form:

IF Error $e(t)$ is Positive Big (PB) and Change of Error $\dot{e}(t)$ is Negative Small (NS)
THEN the Change in Controller Output $u(t)$ is Positive Medium (PM)

Note that, conversely, due to its derivative action (highpass filtering properties), the control actions of a PD tend to be more abrupt than those of a PI controller.

These two controllers will form the basis for a closed-loop control experimental study (described in Section 9.6.2) in which HR and BP variables will be controlled using physical workload as the controlling variables.

9.6.2 Closed-Loop Control Simulation Results

Before initiating the simulation as well as the real-time experiments, it is vital to select the appropriate target values for HR and BP. These target values, also known as set points (targets), need to be chosen so that (1) they reflect the unique physiology of each subject, and (2) they can possibly be tracked by the controller.

Hence, a suitable set point for HR was found to respond to the following expression:

$$HR_{set-point} = round\left(\max\left(HR\right)/\alpha\right)\cdot\alpha \tag{9.3}$$

where α is a constant. The max(HR) value would be obtained from an off-line experiment with the volunteer subjected to minimum workload of 0.5 kg · m.

Equation (9.3) will ensure that whenever the controller wishes to revert to the minimum workload value (0.5 kg · m in this case), the HR signal will respond by decreasing to at least the HR baseline value.

Since BP is characterized by its spontaneous oscillations, which are located around 0.1 Hz under normal conditions, as well as its high frequency dynamics, the task of controlling such a signal is not easy. Indeed, the BP signal could only be controlled within a range that corresponds to two set points, comprising lower and upper limits. This means that the BP control becomes a process of keeping this signal between the upper and lower limits. This range[6] was chosen to be ±5 mmHg from the selected midpoint. This midpoint is halfway on the range between the lower and

6. Perhaps this range should have been slightly higher to reflect practices in clinical environments.

upper limits. The range width was selected to equal 10 mmHg, which is the maximum peak-to-peak fluctuation over the subjects of the first group. Additionally, since the midpoint should be selected such that there is no offset in the BP signal, the calculation of the BP set point was as follows:

$$BP_{mid-point} = \beta \cdot (\max(BP) - \min(BP) + \min(BP)) \tag{9.4}$$

where β is a constant. Similarly to (9.4), the max(BP) and min(BP) values would be obtained from an off-line experiment with the volunteer subjected to minimum workload of 0.5 kg · m.

In addition to (9.4), an algorithm that allows the fuzzy controller to control BP within a range rather than a set-point value was implemented according to the following rationales:

- The algorithm checks the value of the BP compared to the upper and lower limits. If this value is found to be larger than the upper limit, the algorithm assigns the lower limit to be the set point and vice versa. This is used to highly penalize the BP signal when outside the specified range.
- If the BP signal is found to be within the range, the algorithm checks for the difference of the BP from the midpoint. If this difference is found to be less than 2, which is almost half the range from the midpoint to either the lower or upper limits, the set point is set to the midpoint. Conversely, if the difference is greater than or equal 2, it is either subtracted from or added to the midpoint according to the sign of this difference. These commands are added to apply a smooth transition within the range of control, in order to reduce the high control effort applied by the controller;

Subjects 1, 5, and 6, from our original group of 12 subjects underwent the experiments of real-time control. The selected set points and ranges for HR and BP control for the three subjects are shown in Table 9.8. These were calculated according to (9.3) and (9.4). It is worth noting that the fuzzy PI control strategy was used for HR control, whereas the fuzzy PD control strategy was selected for BP control. A sample simulation result is shown in Figure 9.18, while the results of the corresponding real-time control experiment is shown in Figure 9.19, which all show acceptable performances.

Table 9.8 Set Points for the HR and BP Controls

Subject No.	*HR set point (beats/min)*	*BP set point (mmHg)*	
		Lower limit	*Upper limit*
1	120	91	101
5	115	121	131
6	115	117	127

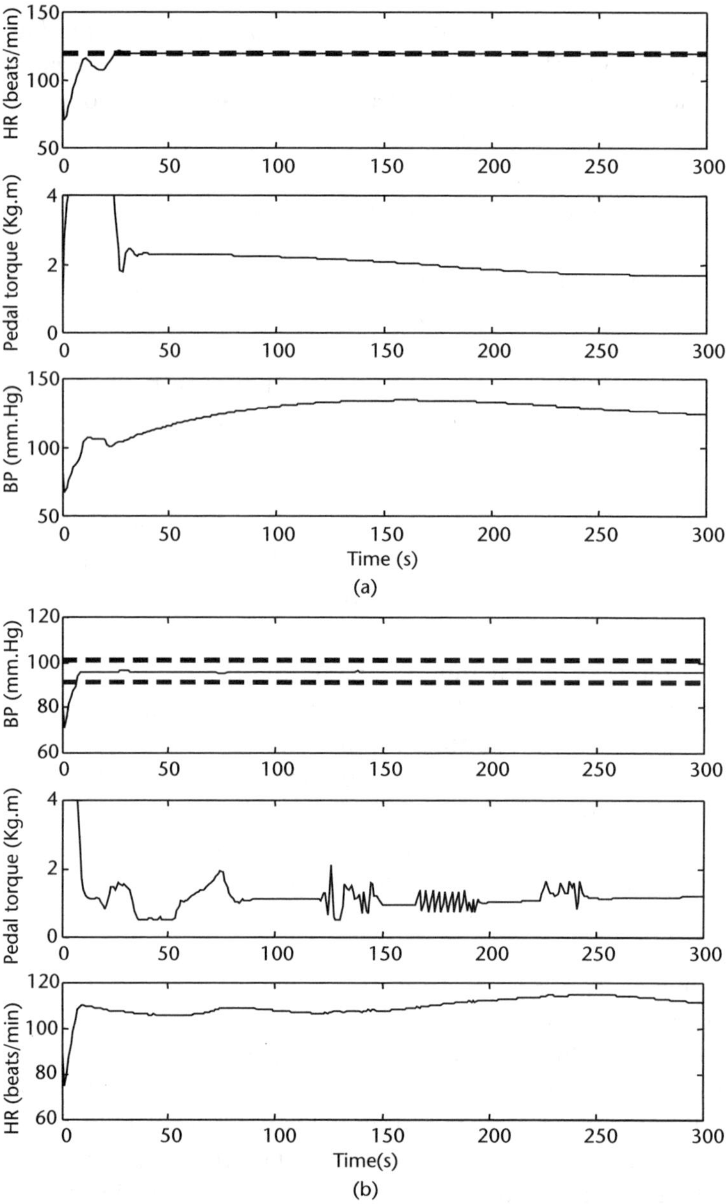

Figure 9.18 Results of the closed-loop simulation for (a) HR and (b) BP control for Subject 1. Dashed line: set point.

9.7 Conclusions

This chapter represents the extended work of the study conducted in Chapter 8, which relates to the elicitation of a closed-loop physiological model describing the behavior of the cardiovascular system (HR and BP), including respiration, in response to physical stress or workload. The model is hybrid in that it included a combination of physical equations and linguistic fuzzy rules that were derived using input/output data obtained from real-time experiments conducted in the author's own laboratory. In the study conducted in this chapter, the model was first extended

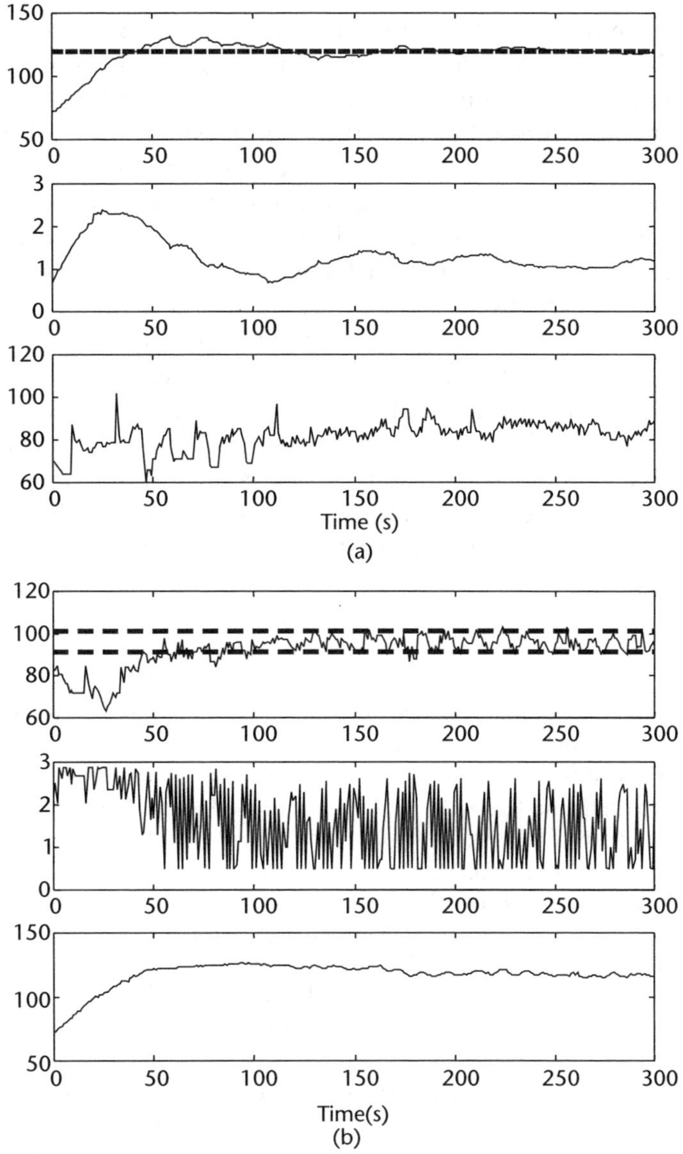

Figure 9.19 Results of the real-time closed-loop trials for (a) HR and (b) BP control for Subject 1. Dashed line: set point.

to include the effect of stress on body temperature. A fuzzy logic based model was embedded in a modeling architecture that took into account heat generation as well as heat losses. The model predictions were good compared to the actual measurements. The next modifications brought to the model concerned the inclusion of the brain centers, which influence directly or indirectly the cardiac, respiratory, and vasomotor activities. The extended model was validated using real-time data. Perhaps the bigger challenge of this study was the modeling of brain activity via EEG. It was recognized from earlier on that if any model was to be elicited for this it would have to take the form of an open-loop structure—in other words, it would not be possible to model the information transfer to *and* from the brain. What was proposed and modeled successfully, however, was a structure representing EEG and TEMP, BP, and RESP relationships identified using the phenomenon of phase locking. It is surprising that intersubject parameter variability and model generalization represent controversial but nevertheless interesting themes. It is in this spirit that a model that covers a wide range of subjects' dynamics was elicited by superimposing a supervisory layer to the existing model to categorize subjects according to certain features and tune the model accordingly to allow it to predict as best as possible.

Finally, it is this author's belief that a model is just as good as what it is intended to be used for. Perhaps the most interesting aspect of this research study is the model exploitation. Indeed, as already stated in Chapter 8, determining the levels of stress that can (or are likely to) cause "exhaustion" is the ultimate aim of the research. In other words, the aim is to identify physiological (and eventually psychological) markers that will allow one to determine thresholds, which, if reached, can result in undesirable consequences. The feedback control results that were obtained in the last sections of this chapter showed how can one control HR and BP via variations in stress levels given certain target values and constraints.

References

[1] Luczak, H., U. Philipp, and W. Rohmert, "Decomposition of Heart Rate Variability Under the Ergonomic Aspects of Stressor Analysis," Chapter 8 in R. I. Kitney and O. Rompleman, (eds.), *The Study of Heart Rate Variability*, New York: Oxford University Press, 1980.

[2] Luczak, H., and F. Raschke, "A Model of the Structure and the Behavior of Human Heart Rate Control," *Biological Cybernetics*, Vol. 18, 1975, pp. 1–13.

[3] Stolwijk, J., and J. D. Hardy, "Temperature Regulation in Man—A Theoretical Study," *Pflugers Archiv.*, Vol. 291, 1966, pp. 129–162.

[4] Censi, F., et al., "Transient Phase Locking Patterns Among Respiration, Heart Rate and Blood Pressure During Cardiorespiratory Synchronization in Humans," *Medical* & *Biological Engineering & Computing*, Vol. 38, 2000, pp. 416–426.

[5] Palus, M., "Detecting Phase Synchronization in Noisy Systems," *Physics Letters A.*, Vol. 235, 1997, pp. 341–351a.

[6] Konz, S., et al., "An Experimental Validation of Mathematical Simulation of Human Thermoregulation," *Computers in Biology and Medicine*, Vol. 7, 1977, pp. 71–82.

[7] Kitney, R. I., "The Analysis and Simulation of the Human Thermoregulatory Control System," *Medical and Biological Engineering*, Vol. 1, 1974, pp. 57-65.

[8] Huizenga, C., H. Zhang, and E. Arens, "A Model of Human Physiology and Comfort for Assessing Complex Thermal Environment," *Building and Environment*, Vol. 36, 2001, pp. 691–699.

[9] Rosenblueth, A., and F. A. Simeone, "The Interrelations of Vagal and Accelerator Effects on the Cardiac Rate," *American Journal of Physiology,* Vol. 110, 1934, pp. 43–55.

[10] Katona, P. G., et al., "Sympathetic and Parasympathetic Cardiac Control in Athletes and Nonathletes at Rest," *Journal of Applied Physiology*, Vol. 52, No. 6, 1982, pp. 1652–1657.

[11] Waugh, A., and A. Grant, *Anatomy and Physiology in Health and Illness*, 9th ed., London: Churchill Livingstone, 2001.

[12] Thibodeau, G. A., and K. T. Patton, *Physiology of the Cardiovascular System*, 4th ed., London: Mosby Books, 1999.

[13] Chen, M. Y., and D. A. Linkens, "A Fast Fuzzy Modeling Approach Using Clustering Neural Networks," *IEEE World Congress on Computational Intelligence*, New York, Vol. 2, 1998, pp. 1088–1093.

[14] Daubechies, I., "Orthonormal Bases of Compactly Supported Wavelets," *Communications on Pure and Applied Mathematics*, Vol. 41, 1988, pp. 909–996.

[15] Hunn, B. P., and M. J. Camacho, "Pilot Performance and Anxiety in a High Risk Flight Test Environment," *The Human Factors and Ergonomics Society 43rd Annual Meeting*, Santa Monica, CA, 1999, pp. 26–30.

[16] Watson, D. W., "Physiological Correlates of Heart Rate Variability (HRV) and the Subjective Assessment of Workload and Fatigue In-Flight Crew: A Practical Study," *People in Control 2nd International Conference on Human Interfaces in Control Rooms*, London, U.K., 2001, pp. 159–163.

[17] Bonner, M. A., and G. F. Wilson, "Heart Rate Measures of Flight Test and Evaluation," *The International Journal of Aviation Psychology*, Vol. 12, No. 1, 2002, pp. 63–77.

CHAPTER 10

Conclusion

This chapter concludes the book by reviewing its main contributions and outlining some future research trends in systems engineering that are directly applicable in medicine and healthcare.

10.1 Introduction

The concomitant desire of not time-stamping any research work so that it stands the so-called test of time and highlighting the potential of such a work to represent a strong platform for the future puts me somewhat in a difficult dilemma as an author and researcher when it comes to writing this concluding chapter. Because my research in this area, and others areas that share similar challenges, is still ongoing, I feel compelled to break the rule and do both.[1] In what follows, each chapter's contributions will be discussed and then new techniques that fall under some of the paradigms mentioned in this book will be highlighted.

10.2 Summary of the Book's Main Contributions

It is clear from the various chapters of this book that a computer-based system for use in medicine and healthcare may require the integration of several components for a successful outcome [i.e., to carry out the prespecified objective(s)]. These include measuring variables via sensor technology (which can represent a drawback due to costs and reliability), processing the acquired signals (which is feasible given the abundance of tools for dealing with complex trends), analyzing these signals (which may require experienced staff, the use of clinical judgment, human intuition and a certain level of subjectivity), decision making or control (which may be risky and dependant on reliable observations), and finally system implementation and possibly commercial exploitation (which is flawed with pitfalls relating to interface realization, safety, and customer satisfaction in general). Real-time implementations, data acquisition, signal processing/interpretation, system identification and modeling, and decision-making and control have all been featured as major themes in this book, whose contributions can be summarized as follows.

Fuzzy logic has been a major theme across most of the chapters in this book, and a comprehensive literature survey on the utilization of this theory in medicine and healthcare has been conducted. The study established the widespread use of

1. I see this as a privileged position to which not too many authors can claim to belong!

fuzzy logic technology, whether in modeling, signal processing, or decision support or control, by both the medical and the engineering communities. The study also revealed that the popularity of fuzzy logic in medicine and healthcare is mostly due to the fact that its users and designers relate to it better since it relies mostly on the expert's knowledge at the design stage and/or the produced design looks similar to the expert's knowledge. In other words, fuzzy logic does what other conventional methods failed to do, which is to capture systems' uncertainties and deal with the so-called curse of dimensionality whereby the mapping between the input and output spaces is highly dimensional and interactive. Indeed, it has long been argued that fuzzy logic emerged at the time when researchers felt disillusioned with conventional methods that often promised a lot and delivered so little. Furthermore, an analysis of the data trends provided a few vectors with reference to the future. First, application success of fuzzy logic in one area seems to inspire its use in neighboring areas. Next, while the number of medical journals that include fuzzy logic continues to grow, due perhaps to the emergence of promising areas where the systems engineering idea looks appealing, the number of computer science based publications has fallen for 2 years in succession. This can perhaps be explained by the fact that researchers are weighting their options via-à-vis these new emerging medical and healthcare areas such as genomics to decide whether to invest their research time investigating the feasibility of fuzzy logic. What is certain, though, is the encouraging news that fuzzy logic continues to be the focus of more research into more synergies within medicine and healthcare, and because of its ability to reconcile transparency (interpretability) with precision (accuracy), it will reach out to even more facets of this discipline.

Another significant contribution of this book was the development of two feedback control systems for administering muscle relaxant and anesthetic drugs to human patients in the operating theater during surgery. This has been carried out in a series of successful clinical trials at two U.K. hospitals. Prior to both systems being transferred to theater, they underwent extensive simulation studies, ranging from pharmacokinetics/pharmacodynamics[2] modeling to control design to simulation experiments, to establish that this form of control is effective but above all safe and robust. While muscle relaxation control proved easier to implement due to the availability of the measurement, anesthesia (unconsciousness) was more challenging for there is no single (direct) measurement that would indicate the level of depth of anesthesia. Instead, one single inferential variable, in this case mean arterial blood pressure, is used instead as a reliable measure when no emergency conditions occur such as blood loss. What the study also highlighted is that the reliance on one measurement only, which can be the subject of noise and disturbances, can give rise to safety issues: closed-loop control can almost be rendered obsolete unless a higher supervisory layer is superimposed on the original lower level control level which would look at the integrity of all input/output signals (jacketing). Finally, such closed-loop control systems for single or multidrug administration have proved very useful in controlling physiological variables to predefined target values, ensuring adequate drug dosage, and relieving tedium for anesthetists by allowing them to concentrate on other tasks in the operating theater.

2. This form of modeling assumes that the human body can be lumped into a finite number of compartments. Drug and fluid exchanges are assumed to occur between these compartments in accordance with dynamic physical equations (mostly linear).

This book has also shown that compartmental modeling of anesthetic/analgesic drugs can be extended to physiological modeling that mimics the various interactions of these drugs with the body organs. As a result, this technique follows the journey of the drug concerned from the moment it is administered to describe how it is exchanged/metabolized within the various structures of an organ and recirculated through the system. The model is generally closed-loop and relies on various assumptions that relate to:

- The number of body organs that need to be included;
- The structure of each organ with respect to how blood will distribute the drug (i.e., in a flow-limited or diffusion-limited fashion).

Most models are built using the flow-limited assumption with linear equations because they are simpler, they can be easily quantified, and there are analytical solutions available. A MATLAB-SIMULINK version of the Mapleson-Higgins physiological model for the drug fentanyl as a platform for successfully identifying auto regressive exogenous submodel structures for each organ using input/output data from this original model was described. The advantages of the proposed model lie in the fact that one can still monitor drug behavior within each organ in terms of drug amount or concentration (pharmacokinetics), extend the predictions to include drug effect (pharmacodynamics), and use tools which are at the heart of systems engineering. Model parsimony was also investigated via the *balanced realization* technique, which can also prove useful for closed-loop control design purposes.

In light of the above contributions, closed-loop control of anesthesia (unconsciousness) via blood pressure measurements was revisited. The anesthesia model used for control design and testing, prior to transferring the control protocol to operating theater for evaluation, relied on a closed-loop physiological architecture. The constrained version of generalized predictive control with *jacketing* procedures was successfully tested on 10 patients with the following achievements:

1. Steady levels of mean arterial pressure were obtained, which also led to adequate anesthesia via the monitoring of only one variable.
2. Low isoflurane drug consumptions (lower than manual control) by all patients were achieved, which entail lower risks of morbidity and possibly significant financial savings.
3. An efficient man-machine interface was established, which will help the system with being accepted by clinicians as an autonomous scheme for simpler surgical operations, as an advisor for more complicated surgical manipulations, and as a simulator for training medical staff, including junior doctors and nurses.

Another area where this book has made a significant contribution is in the measurement of depth of anesthesia. The correlation between brain activity and depth of anesthesia, which has long been the focus of research in this area, has been revisited, albeit from a different perspective. One anesthetic (propofol) and one analgesic (remifentanil), two modern drugs, were considered. Indeed, the case for using audio evoked potentials as reliable measures for depth of anesthesia was first made on the basis of overwhelming evidence from extensive research that linked activity in this signal with levels of unconsciousness without the usual drawbacks witnessed when raw EEG was considered. The signal being nonstationary, it was necessary to process it using wavelet analysis, rather than the usual fast Fourier transform analysis, to capture its most relevant features, which were then successfully classified using a fuzzy logic–based algorithm. The classifier combined knowledge of these features with the hemodynamic responses, including changes in heart rate and systolic arterial pressure. Second, the effect of the drugs on the cardiovascular system (heart rate and systolic arterial pressure) was also modeled indirectly via the surgical stimuli using Mamdani-type fuzzy rules in the form of IF-THEN statements. Such models are seen as filters for the surgical stimuli that affect HR and SAP. A comprehensive patient model was obtained via the integration of the previous models by linking the doses of the two drugs with AEP, ΔHR, and ΔSAP, using a neuro-fuzzy system. Finally, this model was exploited to successfully design a hybrid closed-loop control strategy via a combination of fuzzy logic based statements and a fuzzy logic based version of the well-known proportional plus integral controller. The controller's parameters were optimized using the new paradigm of Genetic Algorithm, which is an evolutionary computing technique relying on elitism (survival of the fittest).

This book has also contributed significantly to the idea of hybrid architectures. There is no doubt that hybrid architectures for modeling, signal processing, and control are set to dominate much of the research in the next decade or so, including the application areas of medicine and healthcare, since the new generation intelligent systems lend themselves naturally to some very interesting and very effective synergies and even to other seemingly totally different paradigms. Furthermore, it is believed that the integration of powerful techniques into one single paradigm tends to emphasize their respective advantages and attenuate any weaknesses these techniques may include. It is in this spirit that the Sheffield Intelligent Ventilator Advisor (SIVA) was designed for critically ill patients in the intensive care unit, a sensitive environment where physical equations have less of a role to play than do experts' knowledge. First, using the architecture of SOPAVENT as a supporting base, a new noninvasive estimation method of the shunt using the Respiratory Index via a neuro-fuzzy model was proposed and successfully carried out. Second, a hierarchical model architecture including two knowledge-based levels with Mamdani-type fuzzy rules was adopted for the model, which was then integrated with the SOPAVENT model structure to automatically set target values and to control the observed variables to these targets. SIVA was assessed by an independent clinician and it was deemed to have mostly confirmed the clinician's decisions. SIVA was also implemented using a friendly GUI and validated using realistic patient scenarios. It is worth noting that SIVA's integrated architecture is flexible enough to allow it to be extended further by including other supplementary measurements, which can then integrated for a more robust decision support.

This book has considered the role of the human operator, which has become increasingly complex and tedious with more operational demands, stress and fatigue, followed by an underlying threat to safety, efficiency, and reliability. If it were at all possible to detect the development of this state, we would be in a much better position to predict periods of increased operational risk and prevent serious human-machine system failure. It was proposed to examine the possibility of eliciting a comprehensive model on the basis of measured changes in the underlying states of the human operator that predict high operational risk. Such states can be physical or psychological, although the study in this book focused on physical stress only. Perhaps it is worth noting that such a study pertains to a research strand that drew a huge interest in the 1970s and 1980s from the biomedical and systems engineering research community. Since then, this interest has dwindled, at least as far as these communities are concerned. It seems that Research in Work Psychology have taken the baton in assuming the responsibility of making advances in understanding the cause-effect relationships between stress and physiological/psychological behaviors of humans. While this is taking place, it was seen as an opportunity to build a rapprochement between systems engineering tools and methods of signal processing, modeling, and control of such processes, which are by no means trivial. The modeling work evolved around what is considered to be a solid model architecture that was first proposed by Luczak et al. [1, 2]. Initially, an experimental platform for real-time data acquisition was established in the newly established Human Performance Laboratory. Cardiovascular system data allowed the elicitation of a gray-box[3] closed-loop physiological model that predicts the behavior of heart rate, blood pressure, and respiration rate in responses to physical workload (stress). The model was later extended to include EEG and body temperature predictions. A further extension of the model included a supervisory layer that allowed it to generalize its predictive capabilities to a wider categories of subject volunteers' dynamics. This layer, which is in the form of a neural network classifier, maps the input/output data of the subject volunteer who would be under investigation to the closest model characteristics so far elicited and applies such a level of mapping to predict the corresponding subject's behavior. Finally, the elicited model was successfully reverse-engineered to determine the level of workload (stress) necessary to achieve a predetermined set of physiological variables' behavior via feedback control. This series of results was very encouraging indeed in so far as determining the levels of stress, which may cause heart attacks or physical exhaustion in general.

10.3 Future Trends

There is no doubt that the need and desire to introduce safe and effective automation in the medical environment will increase for the foreseeable future. At the heart of such prediction is the increasing awareness of the benefits of using, in particular, intelligent systems-based paradigms to tackle the too-often difficult tasks of modeling, control, and optimization relating to the current complex and highly integrated systems. The term "intelligent systems" is usually taken to cover the application of machine intelligence. This attempts to understand and replicate, by using computers, phenomena that we humans associate with "intelligence" (i.e., the generalized, flexible learning and adaptive behavior that we see in the human brain). An intelligent system is designed so that it can achieve an *optimal* goal, while its internal components are not completely defined. This is hardly surprising because of the prospect of being able to mimic the operations of experts in the field, which is seen

3. This refers to a model structure that combines black-box models with white-box models.

by many as a significant achievement, and this can only be concomitantly beneficial to patients and the healthcare system as a whole. This is the good news!

The not-so-good news is that there are still many significant challenges ahead that will face the research community in the future, including:

1. Measuring everything measurable in a system and ensuring reliability of such measurements;
2. Reconciling interpretability (transparency) with accuracy (precision) in a system in order to provide it with a best chance of being taken beyond the simulation stage and into the real-time application arena.

It is perhaps worth noting that experience has taught us that parallel developments in both areas are necessary for systems to be effective. Indeed, a reliable and robust measurement system may prove useless if it is not exploited adequately by an efficient monitoring system, for instance; and conversely, the most sophisticated monitoring system may not be as effective if it did not use a reliable measurement system. The ultimate question, which seems to come back to haunt us time and time again, is: Are we actually measuring what we think we are measuring? In a general ICU, for instance, Electrical Impedance Tomography is currently being developed to provide robust, effective monitoring of the distribution of inflation in the lung. Indeed, ventilatory pressures, including peak and plateau airway pressure and PEEP, inflate different areas of the lung to different degrees. The distribution of this inflation can be monitored using EIT, which is a novel imaging technique with applications in medicine (see Figure 10.1). It is a thousand times cheaper than X-ray tomography and it requires no ionizing radiation. Developed in the 1980s [3], the technique generates images of impedance distribution within electrically conductive organ tissues. Because different biological tissues exhibit different electrical properties (e.g., impedance or resistivity), it is therefore possible to use such property to form anatomical images. From a set of measurements, an image reconstruction technique generates the so-called tomographic image. It was in 1985 that Professor Brown listed the imaging of the lungs and ventilation as a possible application of EIT [4], and since then several versions of the system have been produced, the latest including absolute impedance readings. Several research groups have used the technique for measuring impedance changes during ventilation [5, 6], but to my knowledge no group has yet attempted the use of EIT in an online integrated approach, similarly to SIVA.

Quoting Dr. Gary H. Mills, who is a senior ICU specialist at the Sheffield Royal Hallamshire Hospital [private communication, 2005]:

> The technique is non-invasive and close to being continuous. As a result it has the potential to identify regional, over, or under expansion in either lung. In this way it will be possible for one to modify the inflation or end-expiratory pressures that are applied to the lung in order to achieve the best compromise between pressure and potential barotraumas and re-expansion of the collapsed lung(s). Similarly, during weaning one can detect progressive lung collapse if respiratory support is withdrawn too quickly. Hence, not only has EIT potential as a routine monitor, it will also allow one to model lung expansion and behavior as respiratory support is altered in critically ill patients. Previously it has not been possible to look at this

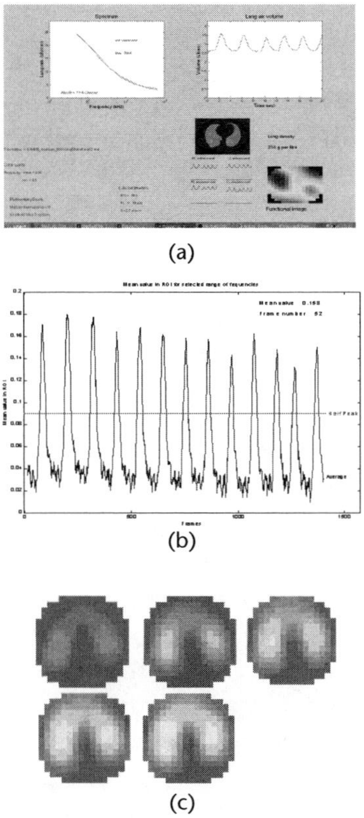

Figure 10.1 Information explosion relating to EIT: (a) regional lung ventilation during quiet breathing using the Sheffield EIT machine; (b) breathing: impedance against time (ROI: region of interest); and (c) a gradual inspiration in an adult.

behavior without the aid of radiological techniques such as CT scanning, which cannot be applied other than for very brief periods.

The signal can be considered in two forms: (1) an image of change of expansion in the lung over time; and (2) most importantly one can look at absolute impedance against time for regions within the lungs. These values are available to the observer as data/graphs of impedance against time for each of the four quadrants of the lungs. The point of the graph between each breath at the end of expiration will indicate how inflated or collapsed that region is. The changes corresponding to each inflation will inform one about the compliance of that region, as well as being related to the size of each supported breath. Respiratory rate will also be available. Online, this will enable an "intelligent" ventilator to learn about lung behavior and then modify the rate and inspiratory and expiratory pressures to optimize lung expansion, and even to identify some pathologies. Off-line analysis will enable us to analyze time constants and patterns of change in lung behavior with changes in ventilatory support.

In order to address the second issue, the refinement of techniques for modeling, signal processing, and control will continue to receive due attention, and one area that is currently the focus of interest is self-organizing models. Because it is important to emphasize that transparency and precision attributes should also be "in the eyes of the beholder" first rather than "in the eyes of the designer," schemes such as the one of Figure 10.2 epitomizes the concept of autonomy from the stage of data collection to model optimization with the user intervening only at the *results analysis* step. In this structure, data granulation, which is still at its infancy, promises to solve some of the problems associated with standard data clustering techniques, including interpretability of the clusters in relation to the nature of the data (see Figure 10.3).

Information granules [7] are collections of entities originating at numeric level that are arranged together due to their similarity, functional adjacency, or coherence. Because there is always a need to split any complex problem into a sequence of smaller subtasks, granular computing serves as an efficient vehicle for such modularization. Unlike numeric computing, granular computing is knowledge oriented and hence the level of abstraction, depending on the system and its constraints, will either be high (low granularity–low specificity) or low (high granularity–high specificity) (see Figure 10.4). By considering that the merger of each set of granules will inevitably lead to some loss of information, it is possible to link the geometrical distance to the information loss and hence plot an information loss graph during the granulation process. Information loss can be monitored during the granulation process, and this information can be used online to decide on a termination point or can also be used and off-line medium in order to determine the granulation performance for various compatibility parameters (see Figure 10.5).

Finally, it must be said that the many challenges that medicine is throwing at us suggest that there is still a lot to accomplish before full penetration of the so-called intelligent systems into the various facets of healthcare is possible. The recognition, by researchers such as myself, that a reliable measurement is needed in addition to the fact that any design must reconcile performance with transparency, fusion, and resolution of conflict between different sources of data and forms of knowledge,[4] is

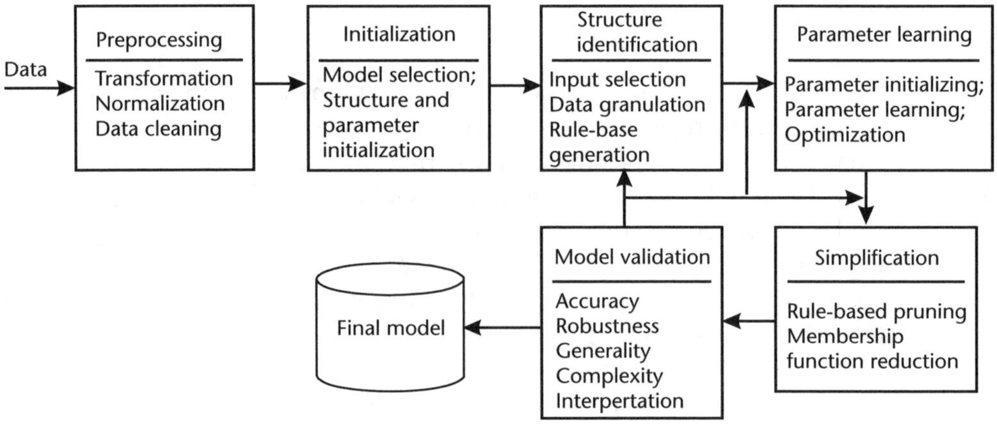

Figure 10.2 Methodology for constructing predictive and descriptive neural-fuzzy models.

4. Especially that U.K. plans are envisaged for a wider involvement of medical staff, including nurses and paramedics, into the diagnosis and treatment of patients.

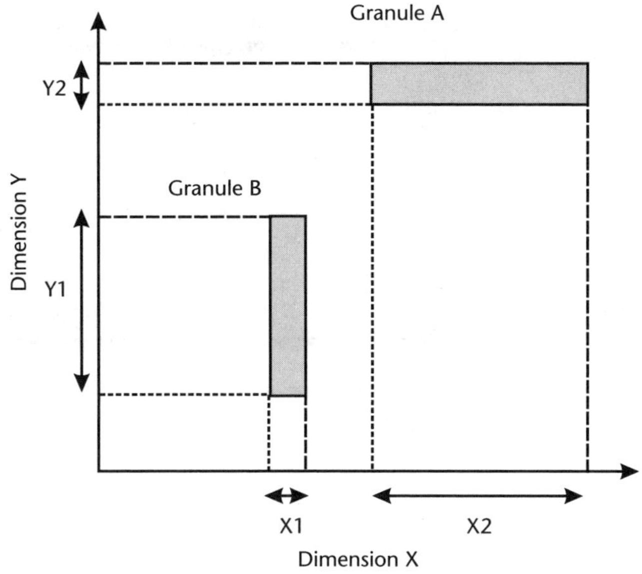

Figure 10.3 The following granulation example shows how the transparency and the additional information of the GrC process can assist the modeling process. Granules A and B have the same size, cardinality, and density, but are of different orientation (90° difference). Considering the fuzzy modeling GrC structure, every granule consists of a linguistic rule, in this case: *GrA*: *if* X = *X2* then *Y* = *Y2*; GrB: *if* X = *X1* then *Y* = *Y1*. Rule GrA has more output sensitivity (narrow space) and less input sensitivity (wide space). One can direct the algorithm towards solutions of type A or B granules, depending on modeling requirements.

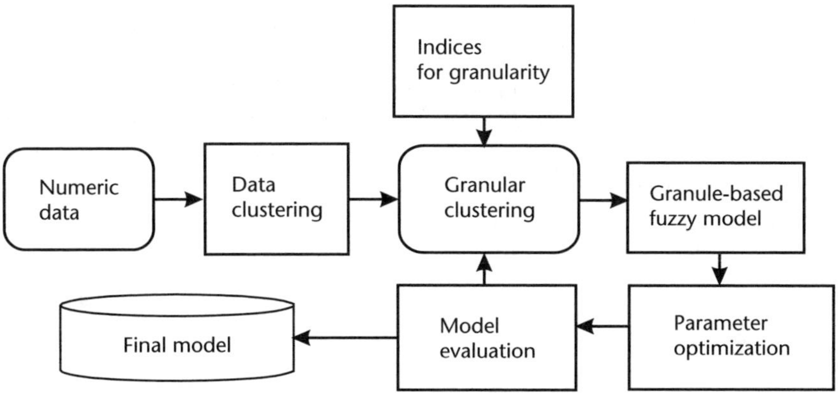

Figure 10.4 A schematic diagram depicting the structure of the granule-based fuzzy modeling methodology.

gaining momentum. It is hoped that through the different synergies described in this book between systems engineering and medicine, other equally interesting and challenging synergies will be considered in the future to build and maintain an everlasting *rapprochement* between engineering and medicine.

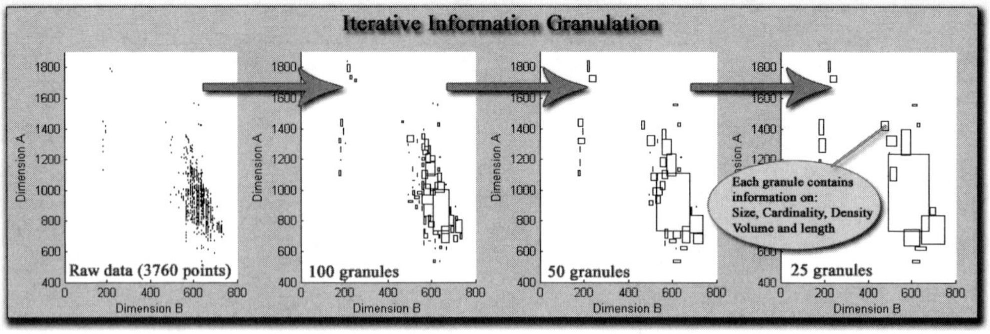

Figure 10.5 The iterative process of information gathering.

References

[1] Luczak, H., and F. Raschke, "A Model of the Structure and the Behavior of Human Heart Rate Control," *Biological Cybernetics*, Vol. 18, 1975, pp. 1–13.

[2] Luczak, H., U. Philipp, and W. Rohmert, "Decomposition of Heart Rate Variability Under the Ergonomic Aspects of Stressor Analysis," Chapter 8 in R. I. Kitney and O. Rompleman, (eds.), *The Study of Heart-Rate Variability*, New York: Oxford University Press, Oxford Medical Engineering Series, 1980.

[3] Barber, D. C., and B. H. Brown, "Applied Potential Tomography," *J. Phys. E: Sci. Instrum.*, 17, 1984, pp. 723–733.

[4] Brown, B. H., D. C. Barber, and A. D. Seagar, "Applied Potential Tomography: Possible Clinical Applications," *Clin. Phys. Physiol. Meas.*, Vol. 5, No. 6, 1985, pp. 109–121.

[5] Frerichs, I., et al., "Heterogeneous Distribution of Pulmonary Ventilation in Intensive Care Patients Detected by Functional Electrical Impedance Tomography," *J. Intensive Care Med.*, Vol. 13, 1988, pp. 168–173.

[6] Harris, N. D., et al., "Applications of Applied Potential Tomography (APT) in Respiratory Medicine," *Clin. Phys. Physiol. Meas.*, Vol. 8 (Suppl. A), 1987, pp. 155–165.

[7] Bargiela, A., and W. Pedrycz, *Granular Computing: An Introduction*, Boston, MA: Kluwer Academic Press, 2003.

About the Author

Mahdi Mahfouf is a professor of intelligent systems engineering in the Department of Automatic Control and Systems Engineering at the University of Sheffield, United Kingdom. The author and coauthor of more than 110 professional papers, five book chapters, one edited book, and one book (research monograph), Professor Mahfouf is the recipient of five paper prizes, including the 1992 prestigious Institution of Electrical Engineers (IEE) Hartree Premium Award. He received an M.Phil. in 1988 and a Ph.D. in 1991, both in control systems, from the University of Sheffield, United Kingdom. Among the many engineering research facets that fall within his interests, neural-fuzzy systems with special applications to biomedical processes remain the subject of his focus.

Index

Recent Related Artech House Titles

Advanced Methods and Tools for ECG Data Analysis, Gari D. Clifford, Francisco Azuaje, and Patrick E. McSharry

Fundamentals and Applications of Microfluidics, Second Edition, Nam-Trung Nguyen and Steven T. Wereley

An Introduction to Micromechanical Systems Engineering, Second Edition, Nadim Maluf and Kirt Williams

MEMS Mechanical Sensors, Steve P. Beeby, Graham Ensel, Michael Kraft, and Neil M. White

Microfluidics for Biotechnology, Jean Berthier and Pascal Silberzan

Nanotechnology Applications and Markets, Lawrence D. Gasman

Nanoelectronics Principles and Devices, Mircea Dragoman and Daniela Dragoman

Text Mining for Biology and Biomedicine, Sophia Ananiadou and John McNaught, editors

For further information on these and other Artech House titles, including previously considered out-of-print books now available through our In-Print-Forever® (IPF®) program, contact:

Artech House
685 Canton Street
Norwood, MA 02062
Phone: 781-769-9750
Fax: 781-769-6334
e-mail: artech@artechhouse.com

Artech House
46 Gillingham Street
London SW1V 1AH UK
Phone: +44 (0)20 7596-8750
Fax: +44 (0)20 7630 0166
e-mail: artech-uk@artechhouse.com

Find us on the World Wide Web at: www.artechhouse.com